Praise for *Building Profits in the Construction Industry*

Though contractors don't always realize it, marketing always has—and will continue to be—the best way for them to distinguish themselves from their competition. Now, for the first time, everything a building professional needs to know about marketing construction services has been pulled together in one easy-to-follow book. A "must read" for success in the twenty-first century!

DAVID W. "WOODY" WOOD
Past president, Construction Writers Association
Publisher, award-winning construction newsletter Words from Woody

As the owner of a mid-sized construction company, I continue to see first-hand how critical an effective marketing program is to one's success.

This marketing guide is developed specifically for the construction industry. It offers comprehensive and straightforward advice about successful marketing procedures for the user who is developing a new program as well as experienced marketing-oriented firms that are interested in fine-tuning their existing program.

Regardless of your focus—building, municipal/utility, heavy/industrial, or highway construction—this book's in-depth discussions will be of great value and will help you remain competitive and successful in the years to come.

JOHN F. "JACK" KELLY
Nickerson & O'Day, Inc., Bangor, Maine
Chairman, AGC Marketing Committee

Building Profits in the Construction Industry is an up-to-date, comprehensive source of information invaluable to all segments of the building business, from contractors, construction managers, and design firms to owners, owners' representatives, and developers.   JAY BOTHWELL
Senior Vice-President, Hines Interests Limited Partnership

During this exciting and transitional time, particularly as it applies to the construction industry as a whole, this book presents a broad landscape of trends, insights, and practical applications. It should be mandatory reading for everyone in the company—from the CEO to the Superintendent.

MICHAEL RAFFERTY
Vice-President, Blake Construction Co.
President, AGC of DC

Successful selling today requires building a lasting partnership with clients. *Building Profits in the Construction Industry* shows how to build and sustain the relationships that will take you above and beyond the competition.

MICHAEL J. REILLY, CPSM
Director of Marketing, Symmes Maini & McKee Associates
President, Society for Marketing Professional Services

The pace of change we see today is similar to a constant kick in the pants for anyone marketing construction services. This book will help make that change a little less painful. A must read for anyone who wants a road map for success in the twenty-first century. JOE PERINI
Vice-President, Perini Building Corporation

Today, it's everyone's role to market and sell. This book will help you with everything from implementing a marketing plan and building strategic alliances to selling virtual construction and design-build methodologies. What an effective and powerful tool for the construction industry!

LISA MOYLAN
Business Development Manager, Daniel Mann Johnson & Mendenhall

This is the first construction industry book I've seen in a long time that puts it all together—technology, marketing, communication, delivery systems, and more. It's a very useful reference for anyone in our industry.

DAVID M. BUSH
National President, 1999 Associated Builders & Contractors
President and Chief Executive Officer, Adena Corporation, Mansfield, Ohio

Building Profits in the Construction Industry

Michael T. Kubal

J. A. Jones Construction Services
Charlotte, North Carolina

Kevin Miller

Frost Miller Group Inc.
Bethesda, Maryland

Ronald D. Worth

Washington Building Congress
Washington, D.C.

McGraw-Hill

New York   San Francisco   Washington, D.C.   Auckland   Bogotá
Caracas   Lisbon   London   Madrid   Mexico City   Milan
Montreal   New Delhi   San Juan   Singapore
Sydney   Tokyo   Toronto

McGraw-Hill

A Division of The **McGraw·Hill** *Companies*

1 2 3 4 5 6 7 8 9 0 DOC/DOC 9 0 4 3 2 1 0 9

P/N 134104-8
PART OF
ISBN 0-07-134985-5

The sponsoring editor for this book was Larry S. Hager, the editing supervisor was Stephen M. Smith, and the production supervisor was Sherri Souffrance. It was set in Century Schoolbook by Joanne Morbit and Michele Pridmore of McGraw-Hill's Hightstown, N.J., Professional Book Group composition unit.

Printed and bound by R. R. Donnelley & Sons Company.

McGraw-Hill books are available at special quantity discounts to use as premiums and sales promotions, or for use in corporate training programs. For more information, please write to the Director of Special Sales, McGraw-Hill, 11 West 19th Street, New York, NY 10011. Or contact your local bookstore.

This book is printed on recycled, acid-free paper containing a minimum of 50% recycled, de-inked fiber.

To all of those professionals committed to enhancing the quality of marketing and selling in the construction industry

Contents

The New Realities of Construction Services Marketing

Introduction

Things change. And this book meets the challenge of this change in marketing construction services by providing marketers with the resources and ideas necessary to adapt to the numerous paradigms changing the construction industry.

Marketing, including marketing of construction services, has changed substantially over the past few years. Marketing is now as likely to take place electronically as it is in person.

Just a short time ago, few marketers had access to e-mail, but today it is so common that brochures and qualifications statements are sent electronically and can be received instantaneously by potential clients. With over 90 percent of companies now offering their employees access to e-mail, it has become a standard marketing resource. CD-ROMs are replacing brochures, overhead presentations are computer presentations, and oral presentations have morphed into video conferencing.

In fact, marketers can perform their services today without ever having to meet the customer face to face (virtual marketing). For example, remote construction marketers, regardless of their location, can interrelate with federal government representatives, located in Washington, D.C., electronically using the Internet. It is entirely possible that a sale can result without the customer ever having met with a contractor's representative.

Web pages and the Internet have also become standard marketing tools. Not only do customers have their own Web sites from which marketers can gain insight into their companies, but every major contractor now also has a Web site to transmit marketing information to anyone throughout the world.

Clearly, marketing is not what it used to be, and the intention of this book is to introduce marketers to the new realities of construction marketing that enhance the successful marketing practices that will never change.

This Book's Presentation

Construction marketing techniques and resources will continue to change in the immediate future. This book presents the procedures to help marketers adapt to cyberspace marketing. Before anyone can adapt to marketing in this age, an understanding of how the construction industry is changing to meet the new paradigms of the virtual age is mandatory.

Chapter 2, Virtual Construction, is a detailed analysis of the industry's changing customer profile, the way customers conduct business, and what these mean for contractors and designers. Equipped with a basic understanding of the virtual age paradigms, a construction marketer can tailor, develop, and formulate a structured marketing program that increases opportunities for success.

Having first reviewed how things have and are likely to continue evolving in the industry, the book discusses the necessary structuring of a mission statement. Any aggressive program is founded on the organization's mission statement and strategic plan. The mission statement is a road map of directions for a firm's future success. Without a plan or direction, a firm will be lost in the virtual age, and this topic is presented in detail in Chap. 3, Creating a Market-Driven Sales Culture for Growth and Change, and Chap. 5, Creating Your Marketing Plan.

A mission statement prepared and accepted by management will not automatically lead to success unless it is practiced daily by all employees. Mission statements must become the firm's bible for growth and success, referred to and adhered to, not left on a shelf to gather dust. Clearly the mission statement must become the nucleus for any successful marketing plan, but at the same time a complete marketing culture within the firm must be established.

In marketing-driven organizations, everyone is a salesperson. Aggressive contractors take the necessary steps to involve all their personnel in marketing, including superintendents who are trained to network with subcontractor representatives at the site for other sales opportunities. Their project managers become experts in continuing relationships with existing clients. Even the accounting staff recognizes the importance of servicing customers to ensure a steady flow of repeat business. In fact, the more progressive marketing-driven companies in construction establish a marketing budget for every ongoing job site to facilitate marketing opportunities as they arise. Those companies that will continue to be successful into the next decades will base the structure of their organizations on an environment of growth and constant renewal, as presented in Chap. 3.

The idea of a market-driven culture for companies in the building industry was almost unheard of just a few years ago. Today there are national associations such as the Associated General Contractors; a recent survey of its membership highlighted the critical need to instill this "culture" within each member of the team, including those in the field, project managers, estimators, secretaries, and particularly any member who interfaces with the client on a regular basis.

Progressive construction firms of all types are effectively enhancing their culture to reflect the dynamics necessary to succeed in today's competitive environment. Chapter 3 includes several "best practices" of leading construction and design firms, with their visions, goals, and aspirations for leading their companies forward for growth and profitability.

Chapter 4, Research Is the Foundation of Your Marketing Plan, is the foundation of the marketing plan. Today we are presented with a wealth of new opportunities in the preparation of a game plan that will minimize risks, provide solid direction, and lead companies to favorable results. This chapter strives to balance information overload with a targeted approach to research. *The New York Times* in a single issue has more information than a person in the 1700s was exposed to in a lifetime. Today we can download the entire Library of Congress in one day. (That's if you had a computer big enough to hold it.) Knowing where to look for specific information you are searching for, accessing this information from any location, and developing beneficial on-line dialogs with individuals who can assist your search, or having this information prepared for you daily and downloaded to your PC, will allow you to develop a marketing plan that is solid, proactive, and achieves the goals you've established.

Once the research is complete, it is time for a company to develop an executable plan that is realistic, establishes schedules, and has action plans that can be addressed by the individuals who are to be held accountable. Chapter 5 discusses implementing your plan and "making things happen." The last thing a company wants to do is expend the time, effort, and emotion toward developing a marketing plan and then have it sit on a shelf. This chapter taps into the wisdom of industry leaders who have written some of the best marketing plans in the country and it tabulates the results into one location. The appendix also provides handy reference to the enormous flood of information that has been drafted on this subject.

The research has been completed and an exceptional plan has been developed. Now what does a company do? Chapter 6 reviews at length preconstruction services and the power of networking. These two issues are growing in importance for companies that want to break away from hard-bid, traditional-bid scenarios. Companies must find ways to separate themselves from their competitors. This can be done through industry leadership, niche markets, segmentation strategies, relationship marketing, joint venturing and alliances, or any one of a host of other options that will be discussed.

Chapters 4, 7, and 11 take the reader through a detailed review of marketing in the cyberspace age. Of the top 100 general contractors listed in the yearly special issue of the *Engineering News-Record* devoted to the top 500 contractors, not one is without a Web site. No longer is marketing electronically considered something of a futuristic novelty; in this new age it is a necessity. All federal government agencies also maintain active Web pages that provide bid and proposal announcements as well as electronic transmission of plans and specifications.

The changing concepts of marketing construction services and cyberspace capabilities and resources are explored in further detail. Marketing plans are developed and then reinforced through continuous networking and the use of the latest technology to ensure a company's marketing plan never becomes stale. Considering that the average piece of computer hardware or software is outdated in less than 3 years, marketing plans today can no longer be projected for 5 years, as was common practice in the past. Plans require constant renewal, and these chapters form the foundation for revamping an organization's marketing from passive plans to proactive plans that lead to success.

Chapter 4 deals primarily with using the Internet to gather strategic information about markets, competitors, prospects and clients, new market territories, new leads, and the latest in products and services for your segment of the industry. Chapter 7 provides the latest information on software and hardware that is being used to revolutionize the marketing and sales processes. Information about personal information management software, digital cameras, presentation software and hardware, and a radically evolving communication arena that is bringing us all closer together is presented in that chapter.

Chapter 11 gives you a step-by-step guide for developing your own company Web site. It will give you practical reasons why every firm today should have a Web site, how to set goals, how to organize your site, and how to build it. There are references and pictures of several Web sites in the chapter to help get you started. Finally, the chapter focuses on what is a great mystery to many: how to get people to see your Web site once it is developed.

This book is designed to help you plan, organize, and develop many of your own marketing materials. Chapter 8, Marketing Communication Plan, will be your first building block in developing effective marketing support materials. The planning process—establishing your goals and objectives, identifying your target audience, determining your competitive advantages and key selling points, and developing a strategy—is an essential first step for any company planning to market itself. The information you gather as a result of reading this chapter will help you develop consistent marketing materials from now on.

Although marketing is, at its core, the strategic plan of an organization, sales are the necessary end result of any effective marketing plan. Chapters 5 and 8 explore the ability of well-prepared marketing and communication plans to produce the profitable sales necessary for organizational growth.

No longer can a contractor depend upon a group of estimators, local or national bid reports, and plan rooms to obtain work. Although low-bid lump-sum contracts are still available, the contractor or subcontractor who depends solely on this method of obtaining work in the twenty-first century will be in a limited market. Chapter 12 looks at changes in marketing traditional fixed-price, low-bid projects, and for those who will still depend on this sector for a large part of their business, it discusses how to successfully market lump-sum work.

Today, construction contractors should implement marketing programs that reach project opportunities well beyond these traditional fixed-price projects. Marketing directly to potential building owners, architects, and even contracting consultants is now a necessity. Chapter 12 presents in detail how to target these industry associates and potential clients.

Design-build contracting is a solution for faster delivery of the building product that is so essential for every industry today. Consumers have come to expect instantaneous delivery of services: photographic film developed in hours or minutes instead of days, eyeglass prescriptions in an hour instead of weeks. The construction industry must follow the trend of expectations and compress the time to produce the finished product. It is no longer acceptable to wait several years for a manufacturing facility to be designed and built to manufacture a product that can become extinct before the building project is complete. Chapter 14 prepares the marketer to enter this specialized but growing market segment.

The largest user of construction services in the world, the U.S. government, will likely increasingly choose contractors based on an ability and capacity to become an effective team member during the design and construction phases. The government is moving toward more generally accepted commercial practices of contractor selection, including the ability to perform rather than just being the low bidder.

Forced by downsizing to resort to outsourcing, privatization, and new contracting methods, the government has created a vast new arena of contracting opportunities. Chapter 15 covers in detail these new opportunities and discusses how to market to the federal government successfully.

The new virtual-age paradigms have facilitated the growth in strategic alliances that enable contractors to target a wider variety of opportunities by increasing their capabilities without having to add to their existing resources. Alliances permit contractors to contract larger projects, using design-build contracts, in more expanded geographical areas than they would be able to without the alliance partnership. Chapter 16 presents an introduction to strategic alliances and how they effect a marketer's responsibilities.

Finally, subcontractors are also being affected by these new industry paradigms, including the move toward design-build contracting, where contractors select team subcontractors during the design phase and long before plans are complete, rather than selecting subcontractors based only on price issues. More and more subcontractors recognize today that they must implement marketing programs to ensure their participation in opportunities such as these, establishing a relationship with not only contractors but with designers and potential clients as well. Chapter 17, Relationship Building between Subcontractors and General Contractors, addresses these new paradigms.

Basic New Concepts

Several new paradigms have changed the way construction services are marketed. These paradigms were for the most part driven by the customer's perspective and requirements. These new concepts must be adapted into any strategic marketing strategy for a construction marketing program to be successful:

- The global village
- Electronic capabilities
- Commodity services
- Bonding rather than selling
- Risk transference
- Time compression
- Partnering

The global village

The ability to communicate directly to anywhere from anywhere through electronic connectivity has created a global village, and for many domestic general contractors, it has created an opportunity for an entry into foreign markets. Today, approximately 25 percent of domestic design billings are derived from foreign projects. At the same time, foreign firms have captured over 5 percent of domestic billings in the United States, and this percentage will continue to grow, especially in the technology design fields.

This two-way paradigm means opportunity as well as threats to a company's existing business. Domestic firms not already owned by foreign corporations are eagerly teaming, partnering, and forming strategic alliances with them to compete more effectively. This global village paradigm is prompting the industry to hire foreign-speaking and foreign-educated professionals to increase their domestic in-house capabilities and facilitate competition abroad. Competition is no longer just local; it is now global as well.

The global village, in combination with electronic connectivity, enables individuals to work in real time with anyone around the world. This capability permits engineering firms to bring the best available talent to a project regardless of a person's location. It permits organizations to staff virtually—bringing talent to a project only as necessary without having to maintain a completely diverse full-time staff capable of completing all disciplines of engineering and architecture.

Furthermore, customers no longer have to settle for the best available "local" talent; they can now have their designs completed by architectural or engineering companies around the world as if the firm was "just around the block." The world truly has become a global village and its implications have penetrated well into the construction and design industries, changing forever how work is marketed, sold, and completed.

Electronic capabilities

Computer hardware and software have revolutionized the way business is conducted in the construction profession as well as in the companies of the industry's customers. Most work processes have been computerized as stand-alone improvements. In the construction and design industry the most effective improvements cannot be implemented until all team members are capable of connecting together electronically to share all project-related files.

Connectivity between all team members is necessary to make full and effective use of computer hardware processing capability and instantaneous communications through networks. Advancements in computer connectivity are only now beginning to facilitate vast improvements in managing the entire construction process—some 40 years after the invention of the computer. The electronic-age paradigms are covered in detail in Chap. 2.

Commodity services

Today's design and construction profession is frequently treated as a commodity service, despite the fact that these professionals have ultimate control over the client's total direct costs of building. The architect and engineers often exert the greatest influence on costs expended during the construction phase. The design team is often followed by the contractor or construction manager in descending order of influence on overall cost control.

Unfortunately, clients often place too much emphasis on the fees charged by these professionals when actually their costs in comparison to the project's overall costs are minuscule. Averages of a wide range of projects reveal that designing and construction management fees generally represent less than 10 percent of total building costs. The remaining 90 percent is directly attributable to hard, or direct, cost of construction. During the early stages of a project, including the selection of team professionals, clients must be cautioned to place an emphasis on value and not merely on price.

However, given the current trend toward commodity pricing, that is exactly what frequently happens. Spurred on by fierce competition and the need to reduce operating costs and increase profits, commodity pricing has spread from department stores such as Wal-Mart to medicine (HMOs), law, and construction. Commodity pricing is fine for assembly line products but not for services that require human beings to exercise critical thinking. You may not care if your hamburger comes from McDonald's or Burger King, for instance, but you surely care whether your family physician has a medical degree.

Architecture, engineering, and construction management are not off-the-shelf products that can be packaged and sold like widgets; they are highly specialized, thought-intensive services that must be specially tailored to each construction project. Unfortunately, the pressure to compete at the lowest common denominator has led many owners to regard these services as mere commodities and to base their selections on low prices from a short list of firms that meet minimum qualifications. Rarely do these owners realize the ramifications of forcing inadequate fees on design and construction professionals, who must limit their services and investment in a particular project to remain profitable and competitive.

That is why it is often incumbent upon construction marketers to "sell" the owner on the virtues of a firm's ability to control overall project costs.

The following are some points about which a marketer should make potential clients aware:

Successful cost control on a design and construction project is a result of teamwork by all involved parties. To increase value for the owner you have to ensure that all team members are working toward the same goals. They must not be preoccupied with maintaining their profitability due to pricing services as a commodity.

If fees are higher but include services to better manage overall costs and quality, the owner may benefit by spending more in the initial stages of a project to reduce costs in the field construction stage.

The employees make a firm, not vice versa. In design and construction services, the issue of costs generally boils down to paying greater fees for greater talent and additional services. The people selected for a project will ultimately make a project a success or failure, not the companies themselves. Selecting a firm that offers lower costs often means the client might be served by less-capable professionals or entry-level employees. This will ultimately result in increased overall project costs and lower quality. Again, rather than focusing on price, the owner should be looking for value: What type of experience does the firm have, and what are the credentials of the people who will be working on their project?

The firm or companies selected should guarantee that no "bait and switch" of employees will occur after the contract award. All too often, to meet commodity pricing of services, organizations will submit qualified candidate resumes and then switch to lower-level and less-skilled employees to maintain or increase their minimal commodity-priced profit margins.

Most importantly, once negotiations on price start, they turn the entire service into a commodity. It is better to instead stress value and direct the negotiations away from an emphasis on price. Clients will defer pricing issues only if they perceive getting better value than they would from competitors. Otherwise, all things being equal, they will price the service as a commodity with no differentiation between the competitors.

Professional services need not be sold as a commodity, and marketers are responsible for convincing their potential clients that experienced professionals deserve serious value considerations. After all, one seldom gets the best for less.

Bonding rather than selling

An era of business relationships that transcends supplier and customer negotiations simply for the best price or service has become increasingly common. The global village age requires that everyone actively seeking a business relationship surpass all previously established standard practices, even beyond partnering, to earn a potential customer's business. The new global economy is changing forever the way business is initially attracted and retained, with an emphasis on bonding with clients and customers.

Private owners and major corporations have now begun to align their facility objectives with their business objectives. Today, required real estate holdings either become profit centers in themselves or the holdings are implemented and managed at the lowest possible costs to prevent deterioration of corporate profitability. The construction industry must respond to these new realities and implement the programs necessary to facilitate the aligning of facilities needs with the client's business objectives.

Design teams are spending the time necessary to help a client review multiple sites for adaptability to their physical plant requirements before any conceptual designs are begun. Contractors are supplying preconstruction services well in advance of these conceptual designs, providing estimates of construction costs based on statement of need parameters.

Additionally, all team members may be asked to participate financially in the future success of their client, deferring some of their initial fees and compensation for possible profit sharing after the client becomes established. As more and more communities offer financing to lure business into their area, the contractor is likely to be required to participate in the assurances to facilitate the sale of industrial revenue bonds.

This process takes time and financial commitments from the contractor to ensure completion of the project at the start of bond interest earning. The architectural team must design within the bond sale revenues, and all team members must contain the final product costs within these parameters. Such public-private partnerships are becoming increasingly common.

Commitments for assistance not only start earlier but end later as well. Postconstruction services are now common, particularly among manufacturing clients who demand that the entire building team participate in the turnover and start-up stage of their operations to ensure a smooth transition to the profit-making stage of all projects. Clearly the trends indicate a leaning toward performance contracts in construction rather than only completion of a building to plans and specifications.

In the next decade, the real estate, design, and construction professionals will become more actively involved in the ongoing success of their clients' operations to ensure the success of their own organizations. Relationships that lead to future business opportunities begin earlier than ever before—when the client begins strategic planning for a construction requirement that is years away. Successful marketers become partners, not just vendors.

Risk transference

During the past decade, construction industry customers increased their use of risk transference to other parties, typically through legal advisement. Transferring risk was presumed to be an effective contractual means to reduce exposure to risk during a project's construction by assuring that an owner would not have to pay for this risk allocation.

In reality, though, the opposite effect occurs. Architects and contractors attempt to transfer their assumption of risk to their subcontractors and consultants. Moreover, parties attempt to apply increased compensation costs to assume risk, thereby increasing overall project costs.

Appropriately, risks associated with professional services should be retained by the organization being paid to complete these services. This is particularly true for all risks associated with construction designs. In the typical design-bid-build contracting method, the owner is required to assume all risks associated with the preparation of the drawings and specifications, guaranteeing to the contractor that the design is free of errors and omissions. Recently clients have realized they should not have to retain or attempt to transfer design risks but rather that design risk should be retained by the architectural and engineering firms that are paid a fee for these professional services.

This situation has led to the substantial increase in design-build projects in the past decade. Design-build contracting assures the clients that all professional risks associated with the design are retained by the construction professional. Using the design-build contract format the owner is relieved of having to guarantee the design documents to the contractor, who is now directly responsible for these documents. Chapter 14 covers design-build contracting in detail.

Marketers should be aware of how and when risk should be transferred and should be sufficiently knowledgeable of these issues to know when a potential client is requiring the contractor to assume too much risk to ensure a profitable project. Risk, in fact, should be allocated to the party that is in the best position to account for and to manage the risk. For example, general contractors are best able to deal with labor issues, architects with building codes, and owners with existing and unforeseen site conditions.

There are many common contract clauses that attempt to transfer risk to other team members; those most frequently used include

- *No damages for delay.*   An owner or architectural firm that causes a delay is not responsible for any costs or damages related to the delay, including those due to time delays and related extended general conditions costs.

- *Indemnification clauses.*   Recent transference attempts include the complete indemnity of clients and architects for any and all actions they might cause. The contractor is expected to assume this risk by contract obligations, regardless of whether an accident is caused in whole or in part by something with which the contractor had absolutely no involvement.

- *Environmental impacts.*   Often, existing and unforeseen site conditions arising from environmental conditions are transferred to the general contractor, such as soil contamination from underground petroleum storage tanks.

- *Pay if paid.*   A contractor might include contract language that attempts to avoid having to pay a subcontractor or supplier if the contractor does not first receive payment from the client.

- *Actual damages.*   Owners require that contractors and designers assume all possible costs associated with delays in project completion without regard to the amount of fee and profit opportunity involved.

- *Design responsibility.* Design team members attempt to pass design responsibility on to contractors and subcontractors through specification clauses. An example is the mechanical and electrical design coordination with architectural and structural plans that may be in conflict with each other through the design team's fault.

- *Existing site conditions.*   These include subsurface and unforeseen site conditions, the most frequent being rock and underground water, for which the client attempts to have the general contractor assume responsibility.

Marketers should have a clear understanding of the types of risks involved in construction and their organization's regulations pertaining to the assumption of this risk. In addition marketers should be able to detail to potential clients the professional team members best capable of dealing with the various types of risks involved in construction.

All project team members benefit by establishing a win-win contractual relationship. Clients receive a fair price for work contracted, all parties are able to complete the project without unnecessary conflicts, and overall quality and time standards are increased in the process.

Time compression

Consumers have entered an age of instantaneous gratification. Products and services are now customized and delivered in an instantaneous manner. Because manufacturers and service providers have to contend with this instantaneous paradigm in their own industries, they in turn expect the construction industry to respond in a like manner to produce their product.

Although no building or facility can be completed instantaneously, clients are now demanding that contractors and designers greatly compress overall design and construction schedules so that they can, in turn, compete effectively in their own industries. Construction marketers must be aware of these new time expectations and confidently present their company's qualifications and capabilities to meet these new requirements.

Chapter 2 reviews the new time paradigm in detail, including how the industry's leading companies approach advancing their project scheduling compression utilizing techniques such as virtual scheduling.

Partnering

Although not unique to the construction or design profession, partnering has become an industry standard for improving a project's overall quality and reducing the number of disputes and litigation that occur over the course of constructing a project. Partnering usually describes the formation of a project team that has a set of common goals in addition to each individual company's goals.

Partnering is not a legal or contractual obligation. In a way it is a return to the "old-fashioned" way of doing business and to the basics in business relationships. It restores trust in business agreements and opens once-closed avenues of communication. It creates a team environment in which to accomplish a set of goals in much the same way a sports team works together to achieve its goals. Using the partnering concept, individuals learn to respect other team members' roles in the project and recognize the inherent risks associated with their professional responsibilities.

Partnering ultimately results in more successful business relationships and less dependency upon legal assistance. It allows project team members to once again accept their individual responsibilities, with all other team members supporting those responsibilities.

Partnering has been a success and has become a standard in the construction industry because the ultimate success of any project or company is due to its employees. Partnering builds the teamwork necessary to permit each person to obtain the assistance needed to be successful—individually and collectively as a team. Partnering promotes the following improvement goals:

- Open communications
- Profitability of all team members
- Improved (compressed) schedules
- Improved safety program
- Better business relationships
- Innovations
- Reductions in or elimination of dependence upon outside legal assistance
- Improved levels of trust among the individuals and companies

Marketers should be fully informed of the partnering process and be able to explain and promote the process to all prospective customers. Companies that promote the involvement of their marketing team with clients before, during, and after a project's completion should involve the marketer in the actual partnering process itself. Marketers can monitor the process to ensure that the client's goals are highlighted in the partnering sessions and that an open line of communication with the customers is maintained.

Although this section does not provide a complete review of the partnering process, it gives a general description of it for those unfamiliar with the process. Construction partnering most often begins with a partnering session after all the major team members involved in a project are selected. The earlier in the overall design and construction process that partnering begins the more likely the partnering process will be successful. For example, clients that begin partnering in the design phase with contractor involvement can expect greater benefits from the process than partnering that begins after the design is completed.

There are several steps necessary to start the process leading to a partnering agreement for a particular project. The first step is the completion of a partnering session that includes numerous components:

- The individual team members make each other aware of their individual goals.
- The team defines the common objective(s).

- A structured program is developed to determine how to cooperate to reach the individual and common goals.

- A method of accountability, measurement, and evaluation of these goals is established.

- Open communications are established (including complete electronic connectivity) to resolve problems before they become disputes.

Partnering does not end with this initial session; it must be renewed constantly to maintain interest in the program that will eventually lead to a successful project for all team members. Team members should establish a schedule of partnering sessions throughout the project's duration to ensure that the effort is renewed and continually updated to monitor the progress of reaching goals established in the initial session.

Communication is the single most important principle of partnering agreements. Today, partnering is based on the ability to link together all team members electronically to facilitate instant communications, vastly improving the ability to meet and exceed partnering goals more than ever before. For example, project time schedules, which clients now rank as one of their highest goals, can be greatly improved by team members monitoring them and providing input about them on a continual basis instead of the past methods of monthly or weekly meetings to review schedules.

Summarizing the goals of both the individuals and the team as a whole is the outcome of the initial partnering session. These goals are therefore set forth in a partnering document much like an agreement of work ethics formerly sealed by a handshake. A sample partnering agreement is shown in Fig. 1.1. Copies of the document, usually suitable for framing, are distributed to all team members.

All marketers should be fully aware of partnering and their management's attitude toward and support of the process. This partnering paradigm will take on even more significance when the industry fully implements electronic connectivity of all team members to improve the quality of the finished product that will be produced instantaneously to meet the customer's requirements.

Marketing Principles That Will Never Change

Although marketing has changed and will continue to change in the virtual age of electronic communications, some standard principles will never change, especially those relating to customer relationships. How we contact and communicate with a customer may have changed dramatically, but the manner in which customers are treated will never change. Before moving on to the new world of marketing presented in this book, the following enduring marketing techniques will be reviewed:

- Marketing versus sales.
- Perception is everything.
- Some customers are best ignored.

Figure 1.1   Sample partnering agreement.

We, the team members for the _____ (project name), are dedicated to achieving the quality standards defined by the contract documents (including plans and specifications). The quality standards and goals of this project will be achieved through our commitment to providing a safe and clean environment with zero lost-time accidents, open communications, and specific problem resolution guidelines to eliminate all potential disputes. It is our intent to succeed in each of the following segments of working conditions:

Open Communications
- Ensure that each team member maintains honest and open communications.
- Maintain an open-door office policy.
- Attend project meetings regularly.
- Instill this philosophy in all other project participants.
- Attend all additional partnering sessions.

Conflict Resolution
- Resolve issues at the lowest level or the level at which the problem arises.
- Problems not resolvable at this level shall be reported to the next level of project management.
- Problems not resolved up to the project management level will be grouped and concisely reported to upper management.
- Contract documents shall govern the resolution of all problems not resolved by the partnering agreement.
- Litigation and other legal assistance shall be avoided at all costs.

Safety
- Complete the project with zero lost-time accidents.
- Insist on a clean work site every day.
- All team members shall participate in the safety program and attend appropriate safety meetings.
- Involve safety inspectors, including OSHA, on a proactive basis rather than a penalty basis.

Performance Measurement
- Zero lost-time accidents.
- Complete project with no outstanding disputes.
- Finish on or before scheduled completion date.
- No punch list at substantial completion.
- All warranties, as-built drawings, and maintenance booklets delivered to owner by final completion.
- Maintain weekly updates of project schedule.
- No warranty returns to project.
- Change requests, RFIs, and other information requests to be resolved within one week of initial submittal.
- All shop drawings approved per limits set by contract.
- Change requests limited to 3 percent of contract amount.

It is further intended that additional partnering workshops be held to introduce new members of the project team to the commitments above. At minimum, these follow-up workshops will be held:
- At completion of project documents
- After selection of trade contractors
- At 33 percent completion
- At 67 percent completion

Figure 1.1 (*Continued*)

It is further agreed that after the successful completion of this project, we will all meet to celebrate its conclusion.

We, the undersigned, are committed to put forth our best efforts to implement this program, with each member acting individually in the best interests of the entire team.

OWNER _____

CONTRACTOR _____

DESIGN TEAM _____

_____

_____

MAJOR SUBCONTRACTORS _____

_____

_____

CONSULTANTS/INSPECTORS _____

_____

_____

_____

- Treat clients as friends.
- Know your customer.

Marketing versus sales

There is a difference between marketing and sales even though many construction companies manage them as one operation, usually within the business development department. Although the same employee can be responsible for both marketing and sales, management and the business development staff should recognize the differences between the two functions.

The new realities of construction services contracting demand that a contractor maintain not only the standard estimating department but at minimum a sales or business development staff that has the ability to perform marketing functions to increase the number of opportunities identified for sale closure.

Although *The American Heritage Dictionary* defines marketing as the act or process of sales, in a technical sense there is a difference between marketing and sales. The best way to describe this intended difference is to use an analogy: If marketing is a war, sales are individual battles during an ongoing war.

Marketing is the plan of attack or road map that succinctly develops the program a contractor intends to follow during a certain period of time. Marketing becomes the process that develops the sales contacts that turn into contracts. Marketing is general; sales are specific. A marketing program includes planning to attract business opportunities in a new geographical area or specific type of work and is structured to ultimately result in specific sales opportunities.

Marketing is the process that attracts customers to inquire into a contractor's capabilities, with the sales effort being the contact with specific potential customers, identified by the marketing process. Marketing programs include planning processes that position a company to be recognized as a solution for its targeted client's needs. Once identified as a possible solution, the company's sales process begins selling the company's services to this identified client.

Specifically in construction, marketing would begin with a strategic plan and mission statement (see Chaps. 3 and 5) that structures a program to communicate to prospective clients (see Chaps. 9 through 11). The process then reverts to a specific sales opportunity when a contractor receives an inquiry from a potential client (refer to Chaps. 12 through 15). For a contractor, establishing a direct-marketing mailing program to satisfy a requirement of a strategic plan is a marketing process. Preparing a proposal for a client responding to this mailing is the sales process. The contact with this client between the initial response and the request for proposal is likely to be a combination of marketing processes: providing a general qualification package to supply further information and probably a personal sales call regarding the client's planned building project.

The line between marketing and sales often becomes blurred. Essentially there is nothing wrong when this occurs because effective marketing programs should evolve into sales calls. Everyone in business development should recognize that ultimately marketing and sales are both used to increase sales opportunities for the company. Many construction firms have no marketing programs, relying instead on a sales effort that unfortunately too often exists only as an estimating department.

Although it is not mandatory for a small general contractor to create a separate marketing and sales department, it should be understood that effective sales calls begin with a clearly defined marketing program that identifies a customer as appropriate for the company's long-term success goals.

Perception is everything

The way a person or company is perceived has always been a critical factor for successful marketing programs. In today's virtual society, perception is even more important. In fact, it means everything.

The Internet has created an environment where a person can be anyone he or she wants to be by communicating electronically the factual or nonfactual information that will form the desired perceptions. The Internet allows children to be perceived as adults, adults as children, and nonexistent companies as companies.

The electronic age of communications has created what is referred to as the virtual organization, a company that exists only to complete a certain task; it then dissolves and is recreated when another need arises (detailed in Chap. 2). In effect, by establishing a Web site, a person can create a company through posted documentation but have no employees or experience. This virtual company will hire temporary workers only when an order is received from a client. Otherwise, the company does not actually exist, even though the potential clients of this virtual company perceive it as a real operating entity.

Marketing is constantly used to create favorable perceptions. Today "spin doctors" have become common in politics. These people are actually marketers communicating their side of the story, slanting it in the most positive light possible to create a favorable perception by the public. Spin doctors use faxes, e-mail, and all other forms of communication to spread their side of the story as quickly as possible, because often the first information received by people forms the basis for their perception of the facts or untruths.

Perceptions can be based on fact or fiction, and individuals can perceive another person or company any way they want. The perception, once formed, is very difficult to change, regardless of whether it is right or wrong. For example, a person might make a first visit to a department store, be treated unkindly, and perceive that this entire department store chain is poorly managed based upon this one encounter. It is very unlikely that this perception will ever be changed.

This same person might express this view to others who have not visited the store and influence their attitude toward the store before they have had an opportunity to form individual opinions. This emphasizes the recognized fact that although a satisfied customer might not result in more business, a dissatisfied customer can cause a loss in business simply because people are more likely to communicate their disappointments than their approvals.

As with all other industries, perceptions in and about construction contractors are common. Unfortunately for the industry, too many clients think contractors are change-order driven. In fact, most contractors believe that change orders are not sufficiently profitable to emphasize them, and most would prefer projects be completed without any change requests.

A marketer establishes the proper perceptions about a company by communicating, using the numerous means of communications available today, and should be able to effectively communicate a positive perception to everyone.

Consider the situation of two contractors named Good and Bad. Both have sold similar projects to similar customers. Almost immediately, after start of field construction, both Bad and Good experience problems on their projects. Bad takes immediate steps to correct each problem as it occurs, on its own initiative. Good also takes immediate corrective measures on its own initiative. However, Bad does not maintain effective communications with the owner through its marketer or project manager.

Although Bad is correcting the problems, the client only recognizes that problems keep arising on the project. Good, however, has established an effective marketing program that includes defined steps to maintain continuous communications with the client. Good's marketer continues to relay to the client that although problems are occurring, Good keeps rising to the occasion and takes immediate steps to correct these problems despite "the problems that our subcontractors are causing." Good's client feels satisfied that it selected the right contractor, perceiving that Good is taking care of their needs.

Bad has also taken the same steps, but its client happens to meet with Good's client, who complains about how awful it is dealing with Bad and the continuous problems on the project. Good's client recommends trying Good next time because that company knows how to take care of problems.

Presented with the same situation, both contractors are doing what is right, but one ends up with a negative review, whereas the other gets a positive review and a referral to a new client. The difference is perception. Both companies should have been perceived equally, but Bad was not simply because it did not communicate with the client and put the proper "spin" on the situation.

This example emphasizes the mandatory requirement of maintaining continuous communications with clients to ensure that they receive the right impression. Too few contractors recognize that marketing is as important to keep existing clients as it is for attracting new clients. Aggressive and effective marketing programs include marketing training for project managers and superintendents so that they learn to communicate effectively with their clients. Most managers and superintendents can manage and communicate with subcontractors and suppliers, but few are trained to communicate with their clients.

For perceptions to be positive and projects to be considered successful after completion, the contractor must recognize that marketing is a continuous procedure that cannot be ignored. Communications with clients through established contact with the marketer, and additional visits when deemed necessary as Good did above, are critical to creating a positive perception in the marketplace, perceptions that lead to the future success of a company.

Some customers are best ignored

In every business there are customers whose business a company is better off without. Every marketer should be able to recognize when it is more appropriate to discontinue marketing to a specific client and abandon attempts to finalize a sale.

Too often a sale is completed even though the marketing or sales representative recognizes that the client is probably not appropriate for contracting opportunities. Be wary of clients that

- Emphasize price only
- Do not recognize the value of services provided
- Have a litigious history
- Create a hostile work environment
- Do not have sufficient financial resources

- Have insufficient work experience (in the case of a contractor)
- Want to transfer too much risk from the owner to the contractor

Customers that emphasize price only will never become repeat clients because every project they contract is selected on the basis of low bid and low price. Although many contractors depend on low-bid work, a marketer probably has no business wasting precious company resources and funds to market to a client who is only interested in price.

The federal government was the perfect example of a client that companies never wasted marketing budgets on because of its dependence on a low-bid selection process, while ignoring the contractors' qualifications to complete the work. Even though the federal government no longer depends on this selection method (as described in Chap. 15), many commercial customers still prefer to take advantage of an industry that believes it can accurately estimate constructing projects that have never been done before.

Clients can certainly use low-bid contracting effectively to prequalify construction companies, but those industry clients that are only interested in "getting something for nothing" are best left to competitors. Marketers should recognize that potential clients preoccupied with price issues never represent long-term profitable business relationships for a contractor. These clients do not recognize the value of services provided and treat the profession as a commodity. Be wary of the client that is not convinced of the value of professional services, as described above.

These same clients are likely to create a hostile work environment due to their continued emphasis on price-only issues during the actual construction phase. This is especially critical in changed conditions, and change requests instituted by the client that become extensive and unprofitable bargaining sessions prevent the contractor from making a reasonable profit.

It should be mandatory for every marketer to investigate and confirm the ability of a client to supply the necessary financial backing to successfully complete the project. Spending company time and resources to market to a client that has no capability to bring a planned project to fruition only results in wasted time and effort for the marketer. As early as possible, the marketer should determine the client's financial commitments for a project. It is not unreasonable at an appropriate time to inquire into the financial support for the planned project before expending estimating time, preconstruction services, and other support requested by the client.

Marketers should be equipped to decide if the company they represent has the capacity and experience to satisfactorily complete work for the client. Large corporate clients and others often do not recognize the need to promptly pay and settle changed work conditions quickly. In certain situations the contractor may not have the financial capacity to support or fund a large project indefinitely. Although a large corporate client might appear to be a good financial risk, inadequate payment policies or bureaucratic structures that impede payments are situations the marketer should bring to management's attention. Unless the customer can provide adequate guarantees of prompt payment and support during the construction, the business opportunity should be declined.

Many clients, including the federal government, have begun to attempt to transfer as much risk to the contractor as possible. Projects are being advertised as design-build that in reality are already completely designed, with the client only interested in having the design risk transferred to the contractor. These so-called design-build projects require only the completion of some minor detailing or minimal design input such as the landscaping design. This facade is then used to structure a design-build contract with the actual intention of transferring all document guarantee risk completely to the contractor while maintaining control of the project's design.

Clients will also use extensive legal support to transfer all risks to the contractor through the contract and specifications. Marketers must use all resources made available to them, including legal reviews, to ensure that the opportunity can be completed under a win-win atmosphere rather than a win-lose situation for the contractor.

All too often marketers do not consider qualifying the potential customer, which is a requirement of their job description. It is important to the success of any company to recognize a good opportunity as well as inappropriate situations. The negative impact from just one financially bad project can negate the success of several good projects. The margins in construction are too small to take unnecessary risks, and expert marketers are just as capable of turning away from inappropriate opportunities as they are in closing good sales.

Treat clients as friends

Remember always that people buy from friends or people they know and trust. Competing doesn't take place only on technical, price, and experience levels. Very often sales are decided by who the client knows best, either personally or in a business context.

Marketers should recognize that marketing begins well before a project is planned or announced to the public. Clients are more apt to deal with contractors they have forged a relationship with based on knowledge of their needs and working together to establish win-win situations. Marketers who can create such a working environment with the client will certainly be awarded the majority of that client's work, even in situations where multiple proposals must be received to satisfy corporate regulations.

Successful marketing is not just submitting the best proposal; the best marketers recognize that many projects are sold long before the proposals are submitted. Using strategic planning to identify potential clients and then building a friendly relationship with these potential clients to earn their trust and recognition before expecting to submit on a proposal is the key to successful marketing relationships.

Establishing a relationship based on trust and friendship requires marketers to abide by five rules when dealing with clients.

1. Never forget a customer or let a customer forget you. Marketing is about constant renewal of existing relationships and forging new ones. Through the process of personal contacts, mailings, calls, and electronic marketing such as e-mail, marketers must remind customers about the contractor's ability to help them achieve their goals. In the same manner, marketers should never give the impression that they have abandoned the client once the bid has been accepted. A marketer should keep in constant contact with every existing client through the entire construction process. A marketer should never leave the customer relationships to project management; this would prevent the client from having someone to talk to should problems arise on a project. The customer must also be given alternative contacts within the contractor's organization to ensure that communications are always open. This also creates the impression that the marketer is always there to support the client, ensuring a forged relationship on the next project.

2. Take care of your customers and they will take care of you. When clients believe that they are being treated fairly and professionally, they are likely to return that treatment to the contractor. Marketers should remember the sayings "what goes around comes around" and "don't burn bridges."

Situations often arise when a past client will become involved in present business opportunities. Such situations require marketers and their entire company to treat every customer with utmost respect. Clients belong to industry associations where they exchange information with others. No marketer wants a past client to repeatedly spread negative evaluations to potential future clients either when networking with their peers or when being used as a reference in qualification and technical submittals to a prospective client.

This same treatment must be provided on a personal basis with each specific contact within a client's organization. Should a customer representative become involved in a poor personal business relationship with a marketer, it is very likely that the contractor will never be able to close a sale with this client, even if the contractor's work history with the company is excellent. Marketers must be committed to taking whatever steps are necessary to ensure that they maintain an excellent relationship with the client and all the client's representatives. If the marketer receives the impression that a particular contact relationship within the client's organization is not working out, the marketer should immediately have another person within the company become the point of contact with this representative. Bad relationships do not result in sales.

3. Existing clients are the best source of new business. A satisfied client is and should be used as a reliable source of new business leads. Often a present customer will give the marketer potential sales leads just for the asking. Clients are certainly aware of their competitors and happenings in their industry, and this information can help marketers learn which companies might be planning new projects, expansions, or renovations.

In addition, clients can also provide information on appropriate contacts within another organization through the client's friendships and business associations. A referral or personal introduction can immediately open doors and form stronger relationships that might not otherwise occur.

At the very least, a satisfied existing client can be used as a reference to gain entry and business from other potential clients. Never forget the opportunities that a present customer can provide beyond work with their own organization. Effective marketers will use these relationships strategically to create other relationships and generate more business.

4. Never forget the word *thanks*. Too often marketers in their zealousness to close a sale and move on to the next potential lead forget to thank a customer for the work received. A marketer should never forget to express the company's appreciation for the contract. Make sure the client recognizes that the marketer will be available throughout the project's progress and continue to personally follow through with the client to ensure that the client is satisfied with the services provided.

Everyone respects and appreciates working with people who are sincere about their gratitude for an opportunity provided. Even if a contractor should not be successful in obtaining a particular contract, providing a written or personal thanks is greatly appreciated by the customer, who will certainly remember this when it's time for the next project. Never leave the client with the impression that the contractor did not appreciate the invitation to compete for their business.

A simple handwritten note of thanks is still preferred over an e-mailed electronic thank you. A marketer should frequently send thank you notes to those who provide leads, information, or direct business opportunities. Sincere appreciation is the most likely to result in future opportunities for a marketer.

5. Reciprocate whenever possible. Everyone is in business to make a profit. Marketers must recognize that their clients are also required to sell work to stay in business. Marketers, whenever possible, should pass on appropriate leads or business opportunities to their clients. Nothing can cement a relationship better than providing a business opportunity for a client.

Likewise, marketers should keep this in mind for those they use as network sources for sharing leads. No marketer will continue to share leads and information with people who do not reciprocate in kind. Passing on information when appropriate is an important part of marketing; it permits the establishment of a network group that is crucial to the success of any marketing program.

These five simple marketing practices can create a solid foundation of success for any marketer. The philosophy behind these principles is that to succeed, a marketer must remember those who are actually responsible for their success.

Know your customer

Just as important as establishing effective relationships with a customer by getting to know customers on an individual basis is gaining knowledge of the customer's business and future goals. Helping a client become successful is the ultimate goal of any marketing program and in turn helps the marketers achieve their companies' own goals.

In the information age, marketing has moved away from the process of simply providing information about a construction company's capacities to detailing how this construction company can provide resources to help customers achieve their goals. Realizing that a client's priority is achieving success and not deciding who to award a design or construction contract to is a key principle in becoming a successful marketer.

Only by spending the time and effort necessary to learn the customer's goals and requirements can a marketer provide the assistance to a client that results in long-term relationships. Meeting with a client to only discuss how good a construction company performs is for the most part useless information to a client, except to verify that the firm can complete a planned project.

More valuable to the client is a marketer and a design or construction company that provides insight into making its project planning more effective and profitable. Marketers should emphasize the value of services they can provide the client rather than emphasizing pricing issues.

Offering to become involved early in the planning stages of a project and providing input that increases the value of the completed project and the client's profitability is more valuable today than being low bidder. Going a step further and offering suggestions to maximize a client's investment in physical building requirements can be the determining factor that sets a marketer apart from all competitors.

The most successful construction service marketers set themselves apart by first learning everything about the client's business, goals, and planning process and then applying this knowledge to help the client make the right decisions for its construction requirements. Once a potential client recognizes that a marketer can provide assistance based on thorough knowledge of not just construction but the client's industry as well, the client will surely respond by inviting the marketer and contractor to join the planning process and eventually the construction phase. For example, a marketer might suggest that a client release a specialty building component early for manufacture to prevent a delay in the overall construction schedule and thereby bring the client's manufacturing process on-line weeks or months earlier than anticipated and increasing the client's profitability.

The marketer in this case succeeds in moving the discussion away from commodity pricing issues to one of recognition of professional services. Marketers can also use their knowledge of the customer to sell preconstruction services that can provide a competitive advantage for the construction phase.

Knowledge of the client is also critical for responding to a client's technical proposal requests. Rarely do clients introduce sufficient information in proposal requests for contractors to recognize what they actually expect in contractors' responses. By marketing early and learning the client's expectations for a planned project, a marketer can ensure that the proposal response clearly differentiates itself from the competition and leads to a greater share of success.

Marketers who are able to immerse themselves into each of their prospective customers' businesses are infinitely more effective in achieving success in their own goals than marketers who just pass information along about their companies' capabilities to build for a client. This customer-based knowledge is the reason that many larger construction organizations separate their marketing programs by client industry type and have their marketers specialize in only a certain type of industry. This allows them to become more effective both for the client and the contractor. For example, a construction firm might have some marketers targeting health care organizations, some targeting manufacturing corporations, and still others who specialize in commercial development. Specializing permits marketers to concentrate on one particular type of industry, enabling them to learn not only customer needs but general industry needs as well. This specialization enables marketers to become considerably more effective in assisting their clients to become successful, and ultimately the contractor shares in this success.

Summary

The changing aspects of marketing construction services are as much a function of the new capabilities and resources available to marketers as they are external changes and paradigms that are changing the way construction services are provided. Although marketing techniques and construction services are changing, the principles presented in this chapter should form the foundation of all marketing programs. Regardless of whether you are marketing by personal contact or through electronic mail, the way a customer is treated never will change. These basic principles must be incorporated into marketing.

Both new and experienced marketers must implement the electronic marketing resources that are becoming standard in all industries to remain competitive. Marketers must also be equipped with the knowledge of how the industry is improving its services and final product and must ...know how to bring this enhanced level of services to their clients. *Building Profits in the Construction Industry* provides in-depth coverage of both issues and arms every construction marketer with the tools necessary to compete successfully for years to come.

2

Virtual Construction

Introduction

Construction marketers must thoroughly familiarize themselves with how the business of construction is changing. These issues must be incorporated into marketing programs to assure customers, facing similar changes in their own industries, that the contractor is capable of meeting the challenges of the virtual age.

Before embarking on a study of marketing methods presented in this book, a look at how construction is changing and the trends that will likely be encountered in the immediate future is in order. Marketers must fully understand not only how the design and construction industry is changing but also how their clients' businesses are adapting to numerous new paradigms.

According to *The Virtual Corporation,* by William Davidson and Michael Malone (Harper Business, New York, 1993), a virtual product is a service or physical product that is produced instantaneously and completely customized according to the customer's need.

The objective of a virtual organization is to compress its product or service development time by shrinking the interval between need identification for a new product or service and the date at which that product or service becomes available. To accomplish this, a system for ongoing interaction with the customer must be maintained to ensure that an exact match between the customer's needs and the delivered product or service is achieved. Putting together a virtual organization requires a company to completely revise its structure to control and instantaneously access an increasingly sophisticated flow of electronic information.

Virtual technology is becoming a reality in all industries, including design and construction. One obvious example can be found in the way information is sent and received. In just the past few years communications have progressed from standard mail delivery to overnight express deliveries to facsimile transmissions to the now common e-mail.

What Construction Could Be Today

A good analogy of the impact of the virtual age is to compare the improvements in computer hardware and software to what would be equivalent improvements in construction.

Just a few decades ago, computer processing capability cost was measured in millions of dollars, processing time was measured in minutes, storage required cubic feet of area, and memory capacity was infinitesimal. Today we have computers that cost under $1000 and have processing time measured in nanoseconds. The hard drive storage space of equipment that once occupied an entire floor of an office building is now contained in square inches of space. Memory capability in this space is now measured in gigabits, unimaginable just a few years ago. Equally important, all these improvements come with quality that is unparalleled in any other industry.

For comparable improvements to have occurred in construction, an office building today would be designed and built in 1 minute, cost pennies per square foot, and have absolutely no defects or punchlists. The only problem with this comparison is that this office building would be about the size of a Monopoly game hotel and its useful life would be less than 3 years.

Customer requirements for speed and quality will continue to increase in the immediate future. Those construction firms that respond to these requirements will become increasingly successful, whereas those that do not will be left competing for smaller, marginally profitable niche markets.

The virtual age for construction has arrived, and it changes not only how business is conducted but how construction services are marketed and sold.

Virtual Construction

There are several specific ways industry clients are changing the construction industry. For example, in real estate management, building owners are compressing the time required to upgrade existing tenant space—a key factor in maintaining a competitive edge in the rental market. By establishing strategic alliances with design firms and contractors and maintaining integrated databases of them, real estate consultants are able to relay information instantaneously among themselves to ensure that an owner's building receives the necessary upgrading and retrofitting when required. This process substantially lessens retrofit costs through early identification and streamlines the entire design and construction process.

In the same manner, through strategic alliances, design teams will soon be able to identify long-lead construction components early in the design phase and monitor the component's manufacture until the contractual obligations of delivery are assumed by the contractor. Otis Elevators has already taken a major step forward in the virtual construction process. By providing a direct link to the manufacturer's information through the design team's computer-aided design (CAD) system, the designer is able to create an elevator system—customized to meet the owner's needs—well before it is actually manufactured.

The only way to make use of available virtual capabilities is by complete electronic connectivity between all team members of the construction process including the client, architect, engineer, contractor, subcontractors, and suppliers. Electronic capabilities alone do not sufficiently improve the overall processes of any industry unless all involved parties are connected to permit instantaneous communications and virtual management capabilities to improve the overall process or product. For example, an architectural firm that has CAD does not by itself improve the overall construction process. The CAD capability only improves the architectural firm's design process internally and brings no significant improvement to the overall construction process. For true virtual construction, the entire team, including the contractor and subcontractor, must be capable of using the CAD drawings through all project phases.

Virtual construction is defined as the structuring of a virtual corporation with complete electronic connectivity for each individual construction project; it involves the teaming of all individual organizations required to complete the particular project. Virtual construction capability is centered around the customer rather than being based upon the usual hierarchical organizational chart, allowing the customer to interact with the entire team directly and instantaneously (further explained in the Intensive and Early Subcontractor Involvement subsection later in this chapter).

Virtual corporations exist only as long as is necessary to complete a particular task. Upon task completion, the virtual corporation ceases to exist until another task is required and another team is assembled to complete the task. The virtual concept began with the virtual corporation. Virtual corporations exist without physical space requirements, often operating only with computer connectivity linking virtual employees together who might be located throughout the world.

Each virtual corporation is unique and consists only of the team members required for the task at hand. A virtual construction corporation is not the individual construction company; it is the entire design and construction team assembled to complete one specific project for one specific client.

The concept of the virtual corporation has actually been applied within construction for decades, as a group of individual companies comes together to design and build a singular project for a client. The entire group then disbands, seeking work on another project. Today, however, numerous paradigms are changing the way business is conducted in the design and construction profession. They require the entire building team to connect electronically and to adapt and apply the computerization capabilities now available to improve the overall quality, speed, and cost of every individual project.

Numerous industry clients, including the federal government, are already moving to virtual construction. OSHA's "negotiated rule making" is now computerized and links all negotiating parties electronically. This facilitates the immediate review of process rule changes and waivers, eliminating the need for endless paperwork, redundant and time-consuming meetings, and unnecessary costs. The Chesapeake Division of the U.S. Navy now requests that all as-built drawings be computerized and stored on CD-ROM disks, eliminating mountains of blueprint rolls. The Department of the Army has instituted the "paperless acquisition vision," which has a goal of eliminating all paper in the acquiring of goods and services (see Chap. 15).

Marketers must be aware of these virtual requirements and adapt them to their own companies to ensure their organizations can provide the virtual services required by their customers.

The information age

A lifetime of documented information for people in the sixteenth century would today be contained in less than 1 week's worth of the average daily newspaper. The rate at which employees are exposed to information continues to grow faster each year due largely to the capability of computers to process information at continually faster rates that are now measured in nanoseconds (a nanosecond is the time it takes light to travel a distance of 1 foot).

Through virtual technology, corporations can now bring together employees throughout the world, regardless of their location, via telecommunications and computer linkage. This is referred to as "just-in-time" talent: people who are connected instantaneously regardless of distance.

The technical advances of the past two decades are requiring the construction industry to adapt and efficiently apply these technical capabilities to improve the overall process of construction. Although most firms have made use of computerization to improve individual processes, such as schedule preparation, construction firms have yet to put these technical advances together to improve the overall speed, quality, and cost of the finished building product. The industry is now just beginning to adapt to operating paradigms of the virtual age.

The virtual age

Everyone is now exposed to capabilities of the virtual age, and this exposure and capability to make use of computerization will increase exponentially in the next decade. The computer is becoming fully integrated into everyday mainstream life, often unnoticed by those benefiting from these capabilities. For example, every new automobile produced today has several times more computer processing power than the first manned mission to the moon. New products arrive in the marketplace with computer processing power embedded directly into the product to make them smarter and better to use.

At the same time, these virtual age products also become outmoded or extinct due to shrinking time parameters. As manufacturing and service organizations rush to bring new products to the market, design and construction companies must respond by building the facilities necessary to produce these products and services faster and before they become extinct.

The virtual age has three distinct characteristics:

1. Ability to communicate instantaneously
2. Business operations without physical or geographical boundaries

3. Ability to provide or produce customized products instantaneously and completely

Instantaneous communications.   Computers and the necessary infrastructure are available to communicate instantaneously with each other regardless of location or distance. Although e-mail has become mainstream in its usage, video conferencing will become as commonplace as e-mail in the next decade.

Today engineering companies are making use of this technology in much the same way because manufacturing firms have established overseas facilities to take advantage of a less-expensive and more readily available labor force. Available software allows the remote employees to work together on a single document or design instantaneously, with costs for this connectivity being negligible in the overall product production costs. For example, engineering firms now often outsource their drawing and details preparation to engineers throughout the world, using cheaper but capable engineers to better serve their clients. This is in fact a virtual corporation, connecting engineers, regardless of location, to work on a common project and then disbanding when the project is completed.

The capability to connect at ever-increasing speeds is improving continuously. These improvements make some technology obsolete before it has gained widespread availability. Fiber optics, which vastly exceed the speed of wired communications such as existing phone lines, may itself become extinct due to the capabilities of wireless infrared communications. The infrared capability now available on laptop computers permits communications such as e-mail and video conferencing from anywhere without a wired connection to the laptop computer.

Boundaryless society.   There are no longer any borders in the world that prevent connectivity between team members, regardless of their locations. Instantaneous communications, including the Internet, permit employees to participate in a project from anywhere at any time. This capability permits progressive design and construction firms to bring the best available talent to work on a specific project for the benefit of the client.

For example, an engineer who has the best experience on a particular structural engineering concept for a project might be located in India. An engineering firm in the United States could bring the talents of this engineer to the project virtually to provide the client with the best available resources without any additional costs involved.

A boundaryless society permits employees to work for a firm without having to physically relocate to the organization's headquarters. In fact, some virtual organizations no longer have a physical headquarters, existing only through electronic connectivity.

These capabilities will move into mainstream design and construction management to ensure that a client's project is completed using the best available talent and at the same time improve the overall quality, schedule, and cost parameters of the completed product.

Instantaneous and customized products. Consumers are already experiencing the virtual age by purchasing instantaneous and completely customized products. We are able to order pizzas with our personal choice of toppings and have them delivered within a half-hour. Prescription eyeglasses and the development of photographs can be obtained in an hour, and even clothes are now manufactured "instantaneously" and completely customized to a customer's specific requirements. Consumers can now visit a department store, have their body scanned by a computer, and have an exact-fit pair of Levi's manufactured and delivered in just a few days. Available on the Internet are specially manufactured vitamin tablets created to the customer's specifications. In Japan, automobile manufacturers are restructuring their capabilities to permit customers to order customized vehicles that are built and delivered to their homes in 3 days.

Instantaneous production time varies widely within individual industries. Obviously a building cannot be designed and built in 3 days, but the construction industry is striving to meet their clients' needs for "instantaneous development" within this decade.

It is no longer acceptable to take 3 to 7 years from design through construction to complete a typical building project. As the industry's clients move to respond to their customer's requirements of instantaneous production, so too must the construction industry respond to its customers.

Construction is already a customized product; adapting to the virtual age requires the industry to move toward instantaneous production. Progressive industry firms have begun to respond to the instantaneous trend by offering design-build packages that are taking years off the design and construction cycle.

The virtual age has begun to affect the industry externally. In the next decade the virtual age capabilities will begin to improve the industry internally.

Virtual technology trends

Technological advances and improvements that are changing the way business is conducted in the design and construction industries include the external changes caused by clients' requirements and internal changes caused by technological applications introduced to improve the construction processes.

External changes

Corporations are placing an increased emphasis on infostructure rather than infrastructure. Physical highways will become less important to an organization's success than the information highway. Whenever possible, companies are eliminating physical assets. For example, in communities across the country, telecommuting is becoming as common as commuting to an office for work. Access to cyberspace by installing the necessary infostructure to facilitate the electronic connectivity so business can be conducted around the world from any community in the country is expanding nationally and globally.

The infostructure paradigm will also place a greater emphasis on virtual office space that will continue to replace physical office space. The number of employees working from home or other distant locations continues to increase yearly, now measured in tens of millions of people. Not only are employees telecommuting to work by connecting with their office from home, but other concepts of the virtual office are affecting the industry as well. These include the concept of "hoteling," which provides temporary physical office space at a company's location to employees on a shared basis. Rather than providing one office for one employee, corporations now can combine the use of telecommuting and hoteling, giving employees access to a physical office or conference room on an as-needed basis. Numerous other employees share this space, thus the term *hoteling*.

The federal government is also embracing the virtual age by establishing telecommuting centers in remote suburban locations to allow employees to work closer to home using virtual technology. This eliminates long commutes for thousands of federal employees, particularly near Washington, D.C., and assists the government in meeting increased EPA clean air standards less expensively than by other means. Computerization of more and more business processes is influencing the shelter needs of all companies. For example, legal firms no longer allocate valuable space to house hard-copy law book collections. Legal research is now completed via computers. In addition, most lawyers now use personal computers, limiting the need for clerical staff. The same can be said for many other professions, including design and engineering. In fact, virtual technology has effectively reduced the physical office space requirements of corporations by tens of millions of square feet for the next decade.

Virtual offices also permit locating an employee directly in a customer's office until a specific assignment is completed. The advantage of having the employee work literally next to the customer optimizes meeting the requirements of the specific project, with better and faster results. This capability has actually taken partnering to virtual levels; one virtual organization is formed until the specific project is completed. Numerous engineering and construction firms have taken advantage of this concept, working through the early planning and design stages directly with the customer to facilitate the faster completion of the building project and to increase overall quality and customer satisfaction.

An example of the use of this virtual office capability is the completion by J.A. Jones Construction Company and Benham Engineers of an Ocean Spray juice syrup facility in Nevada. To meet the schedule demands of completing the manufacturing facility in record time from concept to production, Jones and Benham placed their key project personnel in Ocean Spray's offices to facilitate the planning, design, and construction of the project. From concept to production, the team partnered directly with the client to complete the project in severely curtailed time parameters and meet the goals of designing and installing new food processes based on the owner's product requirements.

The concept of virtual offices has directly affected the construction industry by requiring less and less physical office space, particularly in city centers. No longer is it necessary for an organization to have a high-rise headquarters building in a downtown location to be "noticed." In the electronic age, it is more productive to be noticed on the Internet than on a city street.

It is evident that high-rise construction in metropolitan centers, although not extinct, has been curtailed compared to the 1980s. Companies realize that it is no longer practical to make employees commute to a central location, preferring instead to bring the work to the employees by providing suburban locations for their physical office needs. This closely parallels the telecommuting and virtual office space paradigms, creating again a direct external impact on the construction industry by changing clients' needs for office space.

A major external impact on the industry is the continuing simplification of complex issues by both hardware and software companies. Many virtual age capabilities have not yet been implemented within the design and construction industry because of the complexity of the technology or the inability to find or train employees to handle the technology. This simplicity paradigm will generate considerable change in the industry, particularly in bringing mobile computing power directly to the physical construction location. For example, superintendents are now able to access a specific detail of the building's design through an eyeglass monitor when they are in the field inspecting the construction.

Although we have seen vast improvements in technological capabilities over the past 25 years, the next 25 years will be spent learning to take advantage of and implementing these capabilities within the industry. In much the same manner that VCR manufacturers are attempting to make their product easier to program (and eliminate the flashing 12:00 familiar to many users), technology firms are managing the process to bring their products into mainstream usage. A company may have a specific unique technology, but if the average employee cannot use the technology, the company will never be able to sell or make the product a success. Costs have also made certain technology impractical to implement for daily usage, but the future will see many of these technologies become available for everyday use.

Although used selectively today, virtual reality capability will become readily available over the next decade. It will become common for architectural and engineering firms to have their clients "walk through" their unbuilt projects. Contractors and other project team members will use similar technology to review a project before and during construction to increase quality and reduce completion schedules.

Many design firms are already making effective use of design programs that use animated three-dimensional geometry from CAD files to enable clients to inspect proposed designs before construction begins. This computerized walk through lets architects apply various colors, textures, and material finishes to proposed designs to determine what is the optimal scheme for the client.

CAD drawings and specifications will become fully integrated into project management controls. Contractors will be able to use virtual reality software to visually use the project schedule rather than rely on critical path programs. CAD files will also enable contractors to determine a perfectly accurate quantity of materials to better estimate and control project costs. Smart or expert software will be incorporated into the project management controls that automatically direct the computer to check for inconsistencies in designs, including conflicts between the architectural and structural or mechanical drawings.

These improvements will all be used not only to increase the overall product quality of the industry but also to compress project schedules to meet the growing demand for consumer instantaneous satisfaction. Instantaneous satisfaction will extend beyond consumer products such as pizzas to all industries, including construction.

No longer can a computer chip manufacturer wait 3 to 7 years for a facility to produce a computer chip that has a useful life of less than 1 year. These external improvements will be used to continually compress the overall schedule of a typical project, including planning, design, and construction, to less than 1 year. This may seem unlikely, but historically it has been done repeatedly when necessary. It will become mainstream only if more aggressive construction firms establish this goal as their standard.

Internal improvements

Although much of the influence during the next decade will come externally, many advances in construction products and services will come from within the industry. Material improvements that offer better quality and/or price continue to evolve. Metals will become less important in construction because plastics are increasingly being used for structural components. Plastic composite beams are now replacing steel in some designs. Plastic piping for plumbing is increasingly being approved by building codes for use.

Improved concrete mixes are now being tested that include self-healing concrete that has the ability to close small fissures and cracks that appear in the finished product. Concrete with limited flexural strength is also being tested for commercial use. Paint with the ability to clean itself of mildew, stains, and even graffiti is now available. Such products will become more prevalent on the market as construction material manufacturers adapt the use of computer modeling and testing to improve their product lines.

In conjunction with these new products, typical construction products will become "smart" products that have the ability to adapt to changing external conditions as necessary. For example, glass used in certain walls will darken itself during the sunny portion of the day and then become lighter and allow more light to pass during other times. Exterior wall components will be able to monitor for unsuitable conditions, alerting the building engineer of conditions such as excess moisture that might be evidence of leaks. Flooring may contain sensors to adjust the climate and lighting controls in an individual room when foot traffic enters.

Subcontractors' roles

Operating in the virtual world precludes any organization from operating in a vacuum isolated from both suppliers and customers, and this is becoming increasingly true in the construction industry. To meet the demands of virtual scheduling and instantaneous gratification of the industry's clients, contractors must respond by becoming connected with all team members and, in particular, with sub- and trade contractors.

Few contractors today are in reality true general contractors self-performing all work on a project; rather they act as construction managers and subcontract the majority of work on a project. Subcontractors have become the niche experts within the industry, providing the technological knowledge of a specific building trade. This knowledge must be incorporated into the early phases of project planning and design to meet the demands of the industry's clients for faster and better delivery of the construction product.

As technical experts in their specific niche markets, subcontractors can directly assist design firms in selecting products and components that enhance the overall design by making pertinent suggestions about what can be done to reduce construction schedules and costs while improving the overall quality of the finished product. The participation of subcontractors in early planning is even more critical for design-build contracting.

Design-build contracting

Design-build contracting can trace its beginnings back to the master builders of the Egyptian pyramids. Some 4000 years ago, the master builder completed both the design and construction and self-performed all field activities. Under this contracting method there was clearly one single point of responsibility—the master builder. In the past few decades the construction industry has moved away from design-build and self-perform work, not due to problems with the method but to transfer legal liability to other parties as the legal profession became entrenched within the industry.

Architects and engineers, by using the low-bid separate construction contract, transfer the design obligations to the owner, who must indemnify the general contractor for design errors under the contract. Contractors, in turn, subcontract most if not all of the work to pass on liability to subcontractors and suppliers. In the past few decades, as this transfer of liability increased plans required separate specifications from the drawings, not to improve the quality of the finished product but to pass on liability from designer to contractor using the written word and then allow the contractor to pass this liability on to the subcontractors. Plans of decades ago were limited, with all written documentation appearing directly on the plans. The voluminous plans and specifications required today to complete even the simplest project have caused projects to become bogged down by disputes and extended overall completion schedules.

Industry clients have responded by turning back the clock to the proven methods of design-build, which places all liability on a single source and improves delivery schedules considerably. The increasing demands for faster and faster delivery of the construction product will certainly nudge the industry toward wider acceptance of the design-build method of contracting. Although the requirement to use design-build is an external preference, it may become the contracting and design method preferred by progressive firms that implement programs to respond to the need for instantaneous satisfaction.

Time compression

The increasing need to compress schedule parameters of a typical building project is forcing the construction and design industry to implement techniques to reduce the overall time needed to produce a completed project. Critical path method (CPM) scheduling by itself is no longer capable of meeting these scheduling demands. The industry will begin to implement virtual scheduling techniques.

In construction, as the time value of money grows, there is an increasing dependency upon effective computerized scheduling software. In a move to virtual scheduling, major construction organizations now tie their priority scheduling programs directly into their subcontractor's and supplier's production control systems. A contractor can then receive a continuous stream of data directly from the supplier's production line. By constantly tracking the component's manufacture, the contractor can integrate product-specific information directly into the overall project schedule to accurately reflect the component's actual delivery and installation date.

This eliminates costly trips to the supplier's production facility to physically verify production schedules. Should production problems arise, this tracking makes it much easier to develop an alternative action plan to keep the project on schedule. The increased use of virtual scheduling will have a tremendous impact on project delivery, resulting in a faster, better construction process for the entire industry.

Virtual scheduling demands that all participants on a project be electronically connected to provide instantaneous communications, improve building scheduling, and meet the requirements of the virtual age. Combining CAD files, CPMs, and all other project files into a communications database that is shared by all members of the project team will soon become standard in the industry.

For example, virtual schedules will be instantaneously updated to track the production of precast panels at the factory, piece by piece. Smart software now available will continually update the schedule, alerting team members as required to any changes and preventing slippage of the schedule. This same software will enable subcontractors that have early start dates to be automatically kept informed of precast material shipping earlier than originally scheduled. Virtual schedules will not be I-J node charts that hang on job trailer walls and are never used to improve the timely completion of the project. Virtual schedules will be totally interactive and continuously updated; they will help the team to proactively manage rather than react after it is already too late to implement the necessary steps to maintain the schedule.

Virtual Construction Industry

What will tomorrow's construction job site look like? It will probably be paperless. All drawings, schedules, requisitions, transmittals, and other "paperwork" will be on the computer and linked electronically through wireless transmission so the entire project team can have access to any information about the job, regardless of physical location.

Technological trends

The two major technological trends changing the industry are electronic data integration (EDI) and wireless communications. EDI allows the user to integrate several previously incompatible software programs to perform complex functions in unison. For example, by integrating a word processing software program with CAD drawing files and estimating software, the computer can be used to create a quantity total of every nut and bolt of the design. This will make the estimating function, subject to inaccuracy due to human error, completely accurate and completed in seconds versus days or weeks when done by hand. This technology could shift the function of estimating from the contractor to the architect in the virtual age.

Software such as Lotus Notes is another example of virtual technology already available. The program enables the creation of a communal database. It can be established on a desktop computer and linked to all project team members. The software incorporates all data created during a project including CAD files, specifications, correspondence, schedules, shop drawings, spreadsheets, and any other electronic data. Notes creates the database that permits team members to search all the records for a specific item, including the entire history of documentation related to that specific item. This brings partnering to the virtual age, allowing all team members to access and share pertinent information instantaneously. At the completion of the project, the contractor can turn over one or more CD-ROM disks that contain the entire historical documentation, including as-built drawings.

Improvements in wireless communication also accelerate access to information directly in fieldwork activities. Handheld wireless computers and eyeglass monitors will become standard tools for superintendents. At the construction job site of the future the superintendent on the tenth floor of the structure can review the placement of concrete. The superintendent accesses the CAD drawings for a related reinforcing detail and transmits a clarification memo to the engineer, who in turn transmits back to the job site the revised detail and simultaneously updates the project's records instantaneously.

This virtual communications technology requires a complete disregard for hierarchical organization charts on the project. As in the case above, the superintendent would be able to directly communicate with the structural engineer without having to submit the request to the contractor's office and then to the architect, who would then direct it to the engineer. When the question is answered, this archaic process is repeated in the reverse order. Such outdated communications channels will be abandoned in the future to permit communications to become instantaneous. Employees at all levels of an organization must share the responsibility for processing this virtual information to eliminate the red tape that slows the process to a crawl today.

These internal and external changes will create a radically different construction industry in the next few decades, consolidating five key virtual management techniques:

- Direct computerized linkage
- Linear project organization
- Intensive and early subcontractor involvement
- Compression of design and construction schedules
- Emphasis on information processing to increase profitability

Direct computerized linkage. All project participants will electronically link together to share the project's information database to improve the overall quality, cost, and schedule of individual construction projects. Each project team becomes a virtual corporation, coming together to complete a project, then disbanding to form virtual corporations for other projects. Virtual linkage is instituted in the conceptual stages of a project when subcontractors can participate to improve the design process that leads to improvements in the construction phase.

CAD files will become the nucleus of the virtual construction organization. All planning, design, purchasing, scheduling, and communications will grow from the project's CAD files. Estimates in the planning stages will be continuously and instantaneously updated as the design progresses or is changed, with quantities of materials being automatically adjusted concurrently with the design. This enables contractors to provide the client with a constantly updated project cost through the design phase to better control costs in the construction phase.

Smart software also permits self-monitoring of designs to detect any mechanical or structural mistakes or inconsistencies between the various trades. This software also instantaneously and automatically verifies building code compliance.

Contractors and subcontractors will use these CAD files to institute early release and ordering of long-lead items identified by the computer and use the program to continuously monitor the production of these items.

Files and correspondence between team members will be transmitted instantaneously, with each team member always aware of all actions that may affect progress during the construction phase. However, this communication linkage is effective only if project management is structured linearly; it eliminates the hierarchical organization common in construction today.

Linear project organization. In every virtual corporation, a linear organization is structured to utilize the closest and fastest line of direct communication between the client and the employee responsible for the work. Virtual

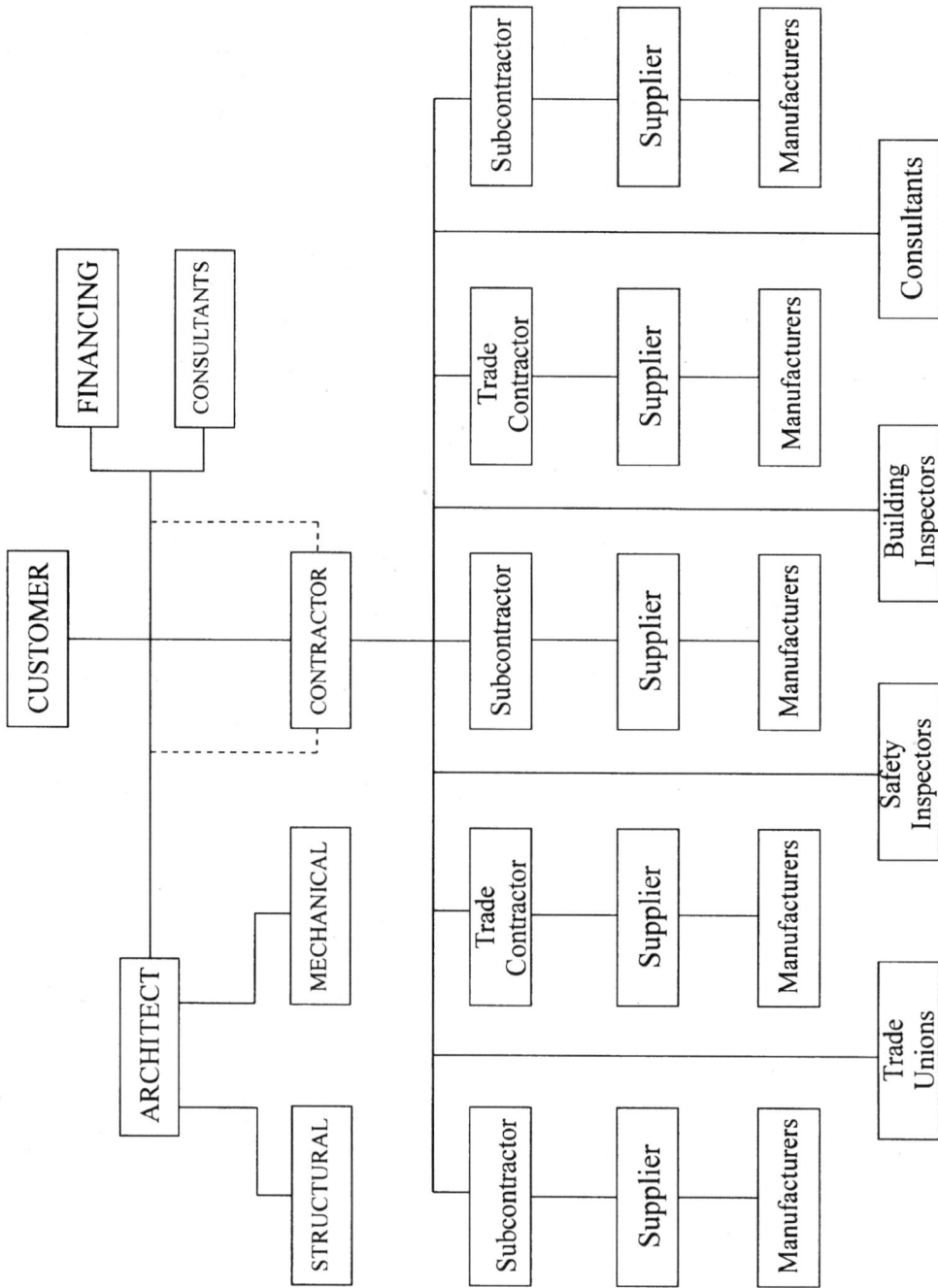

Figure 2.1 Typical organization chart implemented by general contractors.

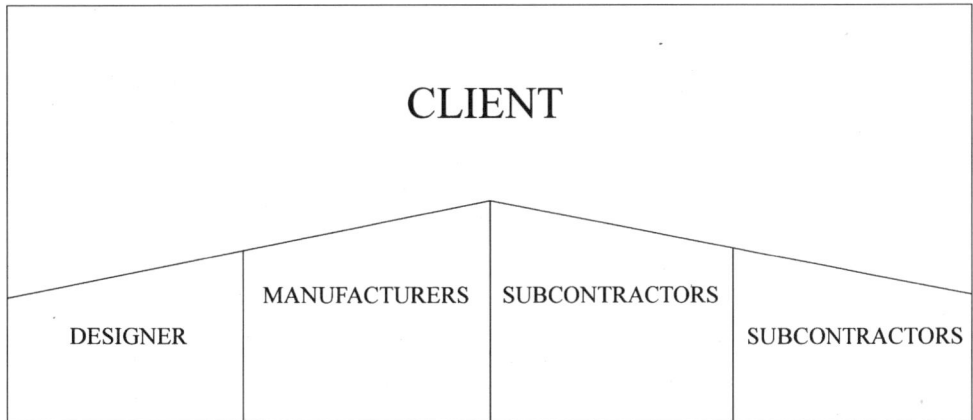

Figure 2.2 Linear organizational chart.

corporations do not have clients struggling through a hierarchical and bureaucratic organization to provide needed input and decision making.

The organization chart shown in Fig. 2.1 is typical of that implemented today in the industry. Obviously this organization is not conducive to communicating effectively or implementing virtual scheduling techniques. Contractors implementing the linear organization for virtual management must realize that subcontractors need direct links to the client, and the contractor must be directly linked to the subcontractor's suppliers. The ultimate purpose of a virtual organization is to permit instantaneous communications, which eliminates any form of hierarchical organization such as shown in Fig. 2.2.

In a linear virtual organization, the subcontractor will have considerably more input in the overall project success. The top-down management style (as depicted in Fig. 2.3), created from the need to pass liability down to lower-tier subcontractors, is not effective in a linear organization. The virtual construction organization must implement the upside-down contracting method of managing to be successful in the future.

Intensive and early subcontractor involvement. Although relatively ignored in construction, manufacturing and service sector industries have literally been turning themselves upside down and inside out in the revolution of reorganizing corporate structures to compete in the virtual age. Changes in management philosophy are occurring based on an integration with suppliers that encourages extensive outsourcing to improve overall quality and competitiveness.

At its foundation, this revolution is based on changing from intraorganization (within one company) to interorganization (among several companies) management, which promotes strategic linking with suppliers, subcontractors, and consultants. This networking affords instantaneous links with talent throughout the world, referred to as "just-in-time" talent. Intereffective corporations provide services or products that are superior in quality,

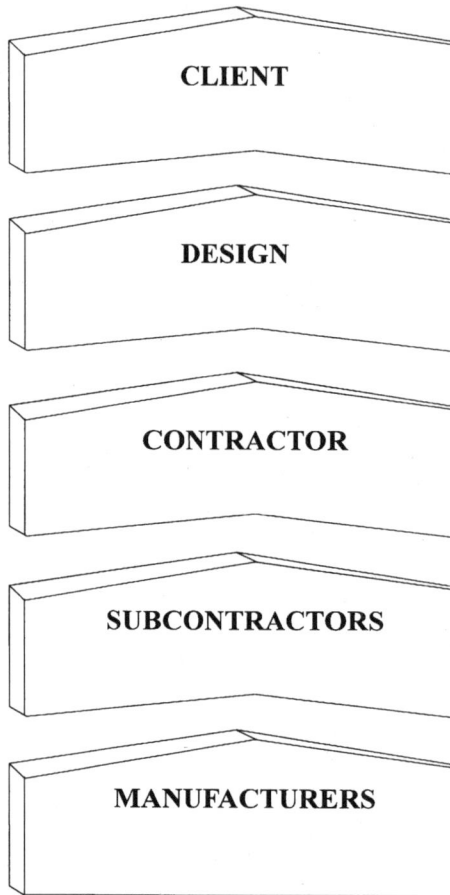

CLIENT

DESIGN

CONTRACTOR

SUBCONTRACTORS

MANUFACTURERS

Figure 2.3 Top-down management.

instantaneously produced, and less expensive because of teaming relationships with their suppliers and subcontractors.

Suppliers, now an integral part of automobile manufacturing, are included in the process from conceptual design to quality controls. Manufacturers select team members not on price issues alone but also on networking capability and performance abilities. Supplier relationships have become so crucial that automobile manufacturers' employee performance reviews now include evaluations of the employee's ability to deal effectively with suppliers.

Decades ago major automotive manufacturers produced nearly 100 percent of their automobiles' contents. Today most are directly responsible for less than one-third of actual production, but the overall quality of its product has improved. As a result, the "manufacturing plant" is transformed into an "assembly plant" with virtual capabilities, collecting talents and products globally in a just-in-time manner. This continues to revolutionize an industry that once treated suppliers and subcontractors as "low-life" creatures to be kept in the dark, an attitude unfortunately still prevalent in the construction industry.

Previously, automobile manufacturers designed components, prepared specifications, and distributed them to suppliers that would be selected based solely on price. However, bidding low only gained the supplier entry into the selection process; rounds of bid peddling and shopping took place before a contract was awarded.

Moving away from these ineffective relationships has enabled the automobile industry to improve quality, costs, profitability, and customer satisfaction. In construction, many subcontractors and suppliers control the majority of actual field construction but continue to operate in contracting relationships in a manner reminiscent of the bygone automobile industry. Contracting practices continue to emphasize selection based solely on price, with no subcontractor input into the design or construction processes in terms of scheduling, quality control, or innovation.

The interaction of these relationships will change as the industry's clients and progressive designers and contractors realize that the advantages of interorganizational processes are essential to success in the virtual age. Subcontractors and suppliers will become integrally networked with contractors and design teams, forming strategic alliances to change the direction of the industry. Benefits of upside-down contracting that place a greater dependency on subcontractors include

- Cost reductions from subcontractors' input during design and preconstruction services
- Faster response time to new innovations
- Superior quality improvements
- Time compression of project schedules
- Improved communications
- Better profitability
- Improved networking and alliances

Management changes such as the automobile industry's response to the virtual age will begin to occur in the construction industry. Management will be compelled by the need to establish networks with suppliers and subcontractors to manage construction processes effectively. Successful firms in the next decade will succeed in managing the information highway and in receiving, processing, and responding to electronic data in an instantaneous manner. Information that is driven from the bottom up will make contracting more successful through positive relationships and interdependency on suppliers and subcontractors.

For example, major suppliers, manufacturers, and subcontractors are now linking directly to architectural and engineering firms to incorporate their designs and details directly into the CAD drawings. The assistance of the suppliers and subcontractors in the design phase is considerably more effective than having the architects and engineers working in a vacuum with no technical niche expertise input from the suppliers. Establishing such virtual networks provides knowledgeable assistance when it can most directly affect the success of a project, long before problems occur in the field due to inadequate or miscommunicated information. This capability can only be successful by implementing the linear organization of project management, as shown in Fig. 2.2, upside-down contracting.

Whereas client concerns previously focused on the time value of money, they are now concentrated on the money value of time. Clients whose own industries are facing the virtual information age successfully are demanding that the construction industry begin implementing the instantaneous information highways necessary to make the most of time value. This process is being achieved in other industries by direct networking with the suppliers and subcontractors, allowing this instantaneous information processing to achieve the time value standards necessary to compete in the future.

Compression of design and construction schedules. Imperative in a world increasingly demanding instantaneous gratification is the ability to meet overall project completions that will in the future be less than half what they are today. For clients whose products become outdated in years or months and the general need for just-in-time physical buildings, virtual scheduling will become imperative.

Already, the continued rise and acceptance of design-build contracting that enables fast-track construction starts shows the need to satisfy clients' demands for compressed project schedules. Only with the implementation of electronic teaming and virtual scheduling can schedules be further compressed to meet customers' requirements.

Available software now permits direct connection between the CAD files and project scheduling programs. This connectivity enables the contractor to visually construct the project using the drawings sequenced according to the schedule and to view the results on a computer monitor. If any inconsistencies exist, they are immediately recognized (e.g., a wall being constructed before the floor slab is completed) and the schedule is corrected. A visual schedule can also be used for quality improvement programs, enabling the work crews to visually see what the final product should look like before they commence construction on a particular phase of work.

Virtual schedules will also be electronically connected to all team members, including suppliers and subcontractors. This connection allows the general contractor to follow the progress of manufactured off-site items to ensure they are meeting planned delivery dates. Virtual schedules will become common and effective tools to compress the overall project schedules, including the design phase, to meet the time standards of the industry's clients.

Emphasis on information processing to increase profitability. The virtual age has created an extremely profitable economy based upon information processing rather than manufacturing. Today the most successful and profitable firms are those that process information or provide the ability to process it. The Sabre Reservation System, which provides instantaneous information on airline reservations availability, is more profitable than any of the airline companies themselves. Microsoft, the company that produces the programs that enable computer hardware to process information, is more profitable than any of the hardware computer manufacturers.

The virtual information age may likewise shift profitability from the actual physical aspects of construction to the information processing of management systems that enable contractors to improve their capabilities to produce a finished product. Progressive contractors will continually expand the scope of their services not only to the industry clients but to competitors as well. Virtual contractors will provide information-processing services including connectivity services for project management.

Preconstruction services may become key to a project's overall success, with virtual contractors offering the connectivity and access to a virtual team, including upside-down management, during the conceptual phases of a project. Virtual contractors will form strategic alliances with clients to provide project management by processing information throughout the project. Team members who participate in a project may in fact never actually be part of the physical construction, only offering their services through cyberspace. The construction industry will be facing numerous new paradigms, and those firms that step outside their "boxes" and create new services complementary to the virtual age will become the industry leaders in the next decade.

Summary

Undoubtedly the virtual age is encroaching upon the construction industry. Its effects on how construction is completed will become obvious, and changes to how construction is sold and marketed in the twenty-first century will also occur. Not only will the digitized electronic connectivity present new and mandatory methods of marketing for construction firms, but internal management changes to adapt to virtual world techniques will change our clients and internal customers.

General contractors will market teaming relationships to subcontractors and suppliers, architects will aggressively target progressive contractors for strategic alliances to compete in design-build, and subcontractors will market developers and other industry clients directly with the availability of virtual communications capabilities.

This review of the virtual age and virtual construction management was necessary to establish the criteria for the new techniques to market and sell construction and related services in the next decade. It is important to realize how our industry will change before properly preparing a strategic marketing plan for the future. The concepts addressed in this chapter are not capabilities that are years away; they are available today and are changing the industry forever. Successful construction marketing and sales personnel must not only be able to sell their firm's virtual capabilities but must also use these virtual capacities in their marketing techniques.

3

Creating a Market-Driven Sales Culture for Growth and Change

The fundamental pattern of behavior laid down in our early days as hunting apes still shines through all our affairs, no matter how lofty they may be.

DESMOND MORRIS
The Naked Ape

Introduction

What is a company culture, why is it important, and how do you create a culture of innovation and a company climate of success? As we have stated earlier, companies today consistently seek ways to stay competitive in a constantly changing world. The heart and soul of an organization is in its culture. Over the last several years we have worked with many companies that were striving to do everything "right" but lacked the inner culture to carry out the principles that had been established. Culture, according to the dictionary, is being in a state of refinement or growth.[1] It is this state of continual innovation and development that we want to strive to create. Each individual within an organization needs to be working toward this goal either consciously or subconsciously as a result of the training and corporate philosophies that have been established. This is important because, in a study, Rosabeth Moss Kanter, Harvard University, found that "companies with reputations for progressive human-resource practices were significantly higher in long-term profitability and financial growth than their counterparts."[2]

Culture, then, is the pervasive ability of companies to look beyond their own organizational box, view the organization as a whole, and experiment with ways to influence the outcomes of other segments in the building industry, such as owners, architects, engineers, and contractors, through their involvement. What this means is that you can leverage your capabilities and strengths in these segments of the organization to enhance your success and the overall success of your company. In this chapter we want to expose you to how you can influence behavior and opinion (and ultimately the end result) without the direct use of commands. At the same time you want to build a mood of accepting change and integrating innovation and change into every activity you do, always improving each aspect of your company's daily activities, getting more efficient, effective, and productive. This is especially true for marketing because what worked yesterday may not be the right choice for today. Flexibility of decisions and utilizing all of the human resources within your organization can make a considerable contribution toward obtaining that next contract or in laying the groundwork for the perception that you are in a company of winners and are willing to look beyond the basic paradigms that have been established for the building industry.

Joel Arthur Barker wrote a book on business paradigms that is especially relevant to our industry. Paradigm, according to the dictionary, is "any framework of theories and concepts."[3] Those theories and concepts (also known as procedures, standards, and routines) do two things. First, they establish boundaries. In a sense, that's what a pattern does; it gives us the edges, the borders. The building industry has many preset borders: This is the way everyone does it. We've always done it this way. This cannot be built this way. Second, these rules and regulations then go on to tell us how to be successful by solving problems within these boundaries.

Mr. Barker states: "It is our paradigms, our rules and regulations, that keep us from successfully anticipating the future. We try to discover the future by looking for it through our old paradigms."[4] An example he uses is the Swiss watch industry, which had a virtual monopoly in the watch-making market. When the Japanese invented the digital watch, it was ignored and thought not relevant to the way you make a real watch. Within 20 years the Japanese, through the cultivation of this inexpensive, efficient method of watch-making, took over the entire industry, leaving the Swiss with less than 5 percent of the market.[5]

This thinking is especially true in the building industry, where change has never been its strong suit. The risks are high, and stubborn personalities are the norm. Inflexibility and avoidance of anything outside of the current state of the art has been ignored. The building industry has been built around a series of occupational segments: field, accounting, project management, estimating, marketing, sales, management, and office support. Each does its job well but is traditionally separate from what the others are doing. Each is expected to stand independently with little or no interaction. Add to this the conditioning that each department is a kingdom unto itself. There is a built-in bias that no one from outside of the department "really knows" what is important to the success of that department. Change is a threat to the status quo and is to be avoided at all costs. Barriers are erected to stifle any interchange. Suggestions, ideas, and criticism from outside of the department are ignored, and thus the opportunity to truly grow and develop an innovative culture is stagnated. The traditional building culture of segmentalism and territoriality inhibits solutions and innovation. Companies that have organizational departments will not go out of their perceived territory due to inherent company political subcurrents. They limit creativity by condemning everything but preexisting procedures and methods and not encouraging any new ideas from outside of this norm.

An entrepreneurial philosophy within companies is quietly stifled in these segmented companies. Many times a free-spirited individual is viewed as radical and not a team player when in reality this person is simply proposing something new that may actually improve the department or adjacent segment of the company. In the past the building industry has perfected this wall of defense, especially in regard to marketing. New ideas are considered as being not really in touch with what is happening or not viable to the workings of the organization. The sad part of all of this is that marketing is usually the only segment that is being exposed to the latest ideas and innovations occurring within the industry due to the outward focus of their activities. In marketing one hears about a wide range of new methodologies being proposed by competitors or material suppliers within the industry. Marketing is also literally going door to door, constantly searching for answers to problems, future trends, and new opportunities.

Instilling this belief is not easy, but you can start now to cultivate a culture that permits flexibility and encourages an active dialogue and interplay between the different segments of your company. A goal is accepting change as a given, integrating the dynamics occurring within our present environment into the corporate mix. Marketing and sales may not have the traditional power within an organization, but it is essential that they influence the whole company through their experiences and expertise.

Overcoming the Architecture of the Building Industry Culture and Breaking Out of the Mold

The architecture of the building industry has not changed in over 50 decades. That is all about to go through a radical transformation now, as we mentioned in the earlier chapters. The procurement practices and organizational structure of the industry are flattening out to form a new shape of mutual partnering versus the primeval structure of the past. Contractors are not better than subcontractors, architects are not superior to contractors, and real estate developers cannot function in a vacuum without the direct input and assistance of the marketplace around them. This new order is changing a tradition of inbred rules and regulations. The individuals and companies who can redirect their efforts to this new pattern will succeed in the years ahead.

Each author of this book has come from a corporate environment where the overriding premise was to keep your head down and keep working. This is actually a comfortable manner of doing work in that you just concentrate on the task at hand and give minimal thought to the environmental dynamics that are occurring. If you were an estimator, you completed your estimate and turned it in to the sales or management departments, who in turn bundled it into a nice, neat package and forwarded it on to the bid authority. Individuals within companies usually did not concern themselves with what was happening among other comparable companies. There was security in their jobs as long as they kept completing the task at hand.

Then the 1990s hit and all rules were basically discarded as companies and individuals fought to survive in a new world order for the building industry. Money supplies dwindled, downsizing became the norm, new projects became scarce or nonexistent, few jobs were available within the industry, workforce problems became much more prevalent, productivity dropped to a new low, costs kept rising, and margins became increasingly tighter. No longer did security exist within any organization. At a recent gathering within the building business it was noted that in the 1990s, 12 million individuals left a construction industry that had been demanding a minimum of 200,000 new workers each year. In addition, the workforce that was left in place after the downturn held the lowest literacy rate in construction in over 40 years, with only 27 percent meeting the skills needed in the trades. Across the country a reduction in technical educational options reduced the incoming talent needed. The national labor unions and schools have remained unable to meet the severe challenges felt by the building industry. To obtain work, you had to get out in this arena and make things happen. You tried everything because you didn't know which of the options would generate an order for your company. On top of that, a sort of musical chairs took place where many were suddenly thrust out of the comfortable nests in which they had spent the last 20 years and had to fend for themselves with a new employer, new methods of doing work, and the distinct knowledge that the security they had previously enjoyed was gone forever.

The industry that emerged from this dynamic reorganization was faced with a complete restructuring of its current culture. New external pressures were occurring that had been downplayed in the 1980s. A significant realignment was occurring within the industry where, as we mentioned in Chap. 1, a fragmentation was occurring. Design/build, construction management, and build-to-suit made different members of the team take on new roles. The concept of total quality management (TQM) was bantered about and resulted in the inclusion of many value-added activities and services in the overall building process. This required even more services on top of a compressed margin structure. Partnering was not only discussed but actually implemented on many job sites around the country.

Technology became a driving force within the building industry as a method of improving productivity, given the reductions that were occurring with staffing. These new technologies altered the landscape of design and construction. Suddenly architects were able to provide 3-D CAD simulation and visualization to express their ideas and concepts. Owners like the Cleveland Clinic purchased special software to view construction materials, colors, and placement within their campus setting before the ground was broken. Integrated project management tools and communication and information technologies were allowing office managers to view multiple activities from their desks across all disciplines within their organizations. The cross-disciplinary approach created new ideas about how business could be conducted and provided initial training to individuals who had never ventured across this unknown turf before.

Culture changes continued as sustainable materials, protection of the environment, and globalization altered the way we viewed our world locally, regionally, and internationally. Technology has brought this world to us, exposing us to new ideas, construction methods, tools, and resources. The United States still dominates with 36 percent of the global market, but it was NAFTA that created a new market for the industry.

A renaissance within the culture of the building industry is occurring after decades of slow, gradual change. The Construction Industry Institute relates a direct correlation of this change of over 13 percent cost reduction to the adaptation of better project planning using new hardware and software tools. A 2 to 5 percent cost reduction occurred because of the standardization of the modularization of design components. There was also a 20 percent improvement in communication among team members, 4 to 33 percent better use of information technology over the life cycle of projects, and a potential 10 to 30 percent increase in the improvement of worker utilization.[6]

It is an exciting time to be in the industry and experiencing this radical change in a culture that has not experienced change for decades. It was gut wrenching, but the outcome will position America and the industry for tomorrow. The industry still has a way to go. Three enormous tasks need to be done:

- Break down destructive industry paradigms within the different segments of the A/E/C community.

- Overcome the extremely poor image within the industry that has created a shortage of workers.

- Continue to foster a culture within the industry of innovation and creativity.

We'll now take a look at what other companies are doing to handle this new culture through their use of best practices and vision.

What Innovative Companies Are Doing to Prepare for the Future

Best practices: Case studies

An article in *Architectural Record*[7] entitled "Lessons from America's Best-Managed Firms" commented that "in the 1990's the firms with the best numbers are often the ones that pay their associates well, offer generous benefits packages, and give to their community." Shocking? Not really. People want a good company to work for. Most people in the industry spend more time in the office working with business associates than they do at home with their families. Is it any wonder, then, that they want this environment to be stimulating, challenging, and rewarding?

Good management makes good sense today. We are seeing a wealth of new management "systems" sweeping the industry: total quality management, best practices, entrepreneurial companies, partnering/collaborative management, chaos management, fast track, client-driven and value-added systems, benching-marking, and empowerment for all. Some of these styles will come and go, but there is a major revolution taking place that is restructuring what successful companies are doing today.

In this section we will review what several major construction and design companies are doing to prepare for the future head-on. Many are trying some of these new buzzwords within their new marketing and business plans. Some new ideas will work and others will need to be readdressed and adjusted to meet the realities of a company's individual markets and economies.

Mortenson, as an example, is working hard to integrate the delivery of design and construction services to facilitate this new market reality. It is developing each department within the company to be technology driven, expanding its services, and requiring continuous education of all members of its team. On top of this it is experiencing a proliferation of strategic alliances with current and potential partners for future opportunities as they evolve.

An article in *Contractor* discussed how contractors could market their companies in the "century of the consumer."[8] It mentioned that the 130 contractors in attendance at a marketing meeting agreed that contractors must differentiate to survive.[9] The industry has emerged from a devastating depression in the early nineties, and the best of the best have survived and are battling for control of every dollar remaining in the market. All these contractors have similar costs, margins, and quality; the winning contractors are the ones who have differentiated themselves in the details.

So how do you differentiate and concentrate on perfecting the details? The *Contractor* article told of a customer who had a problem with his septic tank on a hot Saturday in the summer. He called the septic tank contractor, which immediately sent over a plumber, wearing a wet suit, who fixed the problem before a group of guests were due to arrive.[10] Who do you think this guy called in the future for all of his plumbing work? Yes, this is beyond the call of duty, but it is this unique approach that nets the next order. Most companies are doing what is required to produce quality work today; successful companies are going considerably further, with the development of the finer details to beat the competition. In this chapter we will examine several firms that are going beyond what is expected of them to meet the demands of today's business.

Many new forces are pushing companies to become much more innovative. Customers today require flexibility and the use of an expanding arsenal of new technological tools to be effective in designing and constructing buildings. (We will address the technological factors in considerable detail in Chap. 7.)

Customers and clients today have become much more sophisticated in their approach to the building industry process in large part because of the sharp increase in product delivery times, which creates further pressure to improve delivery and construction services. Not only are buildings required to be built faster, but they need to incorporate the advances of technologies that affect the end users of these structures and also the life-cycle costs for maintenance, operations, and even potential reuse and sale of these structures. Microsoft is a good example of one such client that puts projects on a fast track on its Redmond, Washington, campus. With product delivery times for computer products in the 4- to 6-month range, owners must get their latest structures up, running, and producing as soon as possible. Twelve months is not unheard of for a 100 million dollar high-tech project. A new market has been created for a construction service called Flash Track that reduces a more traditional method like fast track.

Technological flexibility is another phenomenon that is shaking our paradigm of what an office building needs to consist of. The new Arthur Andersen regional headquarters in Reston, Virginia, is being designed to meet this future. Only a small percentage of the building will have fixed walls. The building is being designed with an underfloor duct network of fiber optics to allow critical offices and conference rooms to spring up overnight. Temporary offices are built around these modules and "teams" are allowed to function in this setting until the job is done. This can be in 1 day, month, or year.

The concept of hoteling in the office environment is furthering the trend toward more and more telecommuting and job sharing. Office workers have their offices on carts that contain a computer, desktop accessories, and files. Workers check the carts out on the days they will be in the office and set up a virtual office within the existing space wherever it is convenient. Otherwise, they work on notebook computers and with cellular phones and fax machines at home, in the car, or while in the air.

One company, Hilti (an equipment power-tools company), has its workforce in mobile vans. The local sales representatives, on an as-needed basis, can use a small local office that has a showroom, a service center, and a small bank of offices. But generally the workers have all that they need with them in the van. This is simplified even further in that the phone number for all the national sales representatives is the same, leading back to a receptionist at the Oklahoma headquarters, who pages a representative when a call is received; the representative returns each call while on the road. Usually most clients have no knowledge that this is a virtual corporation that is being managed from a national headquarters that is thousands of miles away from the local offices.

The Ceco Corporation is an example of a company that is restructuring to meet the needs of a changing and dynamic world. Owned by Heico Industries of Chicago, Ceco represents the formwork division, which is centralized in Kansas City, Missouri. At that location safety, contract management, accounting, human resources, and payroll are handled for the various satellite business units. These business units are further refined around regional hubs, located in San Antonio, Seattle, Chicago, and Cincinnati. This is further broken down into smaller strategic business units (SBUs) that have a staff of 1 to 5 for a $2,500,000-size office.

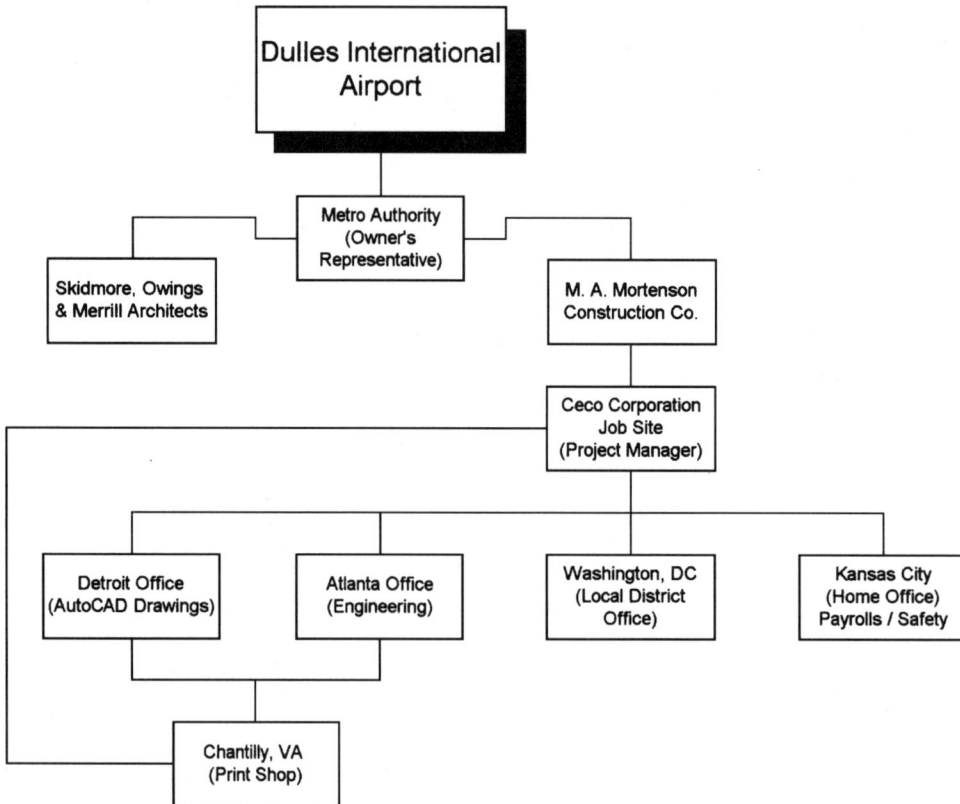

Figure 3.1 Virtual construction site with reach across the United States.

One large project managed by Ceco is the expansion of Dulles International Airport in Chantilly, Virginia. Ceco acted as a subcontractor to M. A. Mortenson using the above concept of reduced staffing, dispersed personnel, and technological advancement. The organizational structure they use is shown in Fig. 3.1.

Drawings produced by Skidmore, Owings and Merrill in its Washington, D.C., office were put on disk for, or electronically downloaded by, Mortenson, which dispersed the plans to the various suppliers and subcontractors. Ceco, for example, downloaded the drawings to an AutoCAD detailer in Detroit, Michigan. When the detailer completed the necessary work, the drawings were downloaded to an engineer, who was in Atlanta, Georgia. He reviewed the details, made modifications, and downloaded his details to a print shop in Chantilly and to the project manager at the job site at Dulles. The project manager in turn checked the drawings received and gave Mortenson a paper copy and a disk overlay for the project. As project changes evolved on a daily basis, this same procedure took place, and a finished product could usually be presented to the contractor and the architect later that same day. Pricing of the changes was usually done locally by the project manager.

Weekly payrolls were electronically sent to the central Kansas City office, prepared there, and expressed back to the job site the next day. Most communication was done electronically and documented accordingly. An electronic link to over 20 other office locations could assist the project manager in finding answers to the problems encountered; usually someone else had had the same problem in the past. This large and complex project was managed basically by a project manager and a superintendent, who worked in concert to resolve the job needs and do it as efficiently and quickly as possible. Ceco worked with a small core of executives and full-time employees, and assigned much of the group's fluctuating work volume to outside contractors, part-time help, and consultants. What a difference the technological age has made to our projects and businesses today.

What some companies are doing.   Turner Construction has consistently been in the forefront of the national construction arena due to its ability to be flexible regarding industry needs and trends. It, like Ceco, has established a series of core offices around the country with a strong client focus. Turner has established a reputation for building top-quality projects on time and within budget, but it takes this a step further by developing partnerships that can last a lifetime, according to Turner's Tom Paci. The office that develops a relationship follows the client wherever it goes. This customer alliance has allowed Turner to lead the country in sales revenue for a number of years in select market segments.

NAVFAC (Naval Facilities), under the leadership of Rear Admiral David Nash, has been transformed from a meek government agency to a twenty-first-century organization. Nash has ushered in new forms of procurement like design-build and enhanced NAVFAC's ongoing partnering programs, and is minimizing past problems with litigation. His spirit and enthusiasm are contagious, and he has instilled a team feeling within the construction divisions of the agency. What is especially of interest here is that this new spirit has spawned public/private ventures with the opportunity for civilian ownership. The culture of innovation and enthusiasm engendered by Nash is spreading to other governmental divisions as his agency interacts with them. Captain Julian Sabatini, with the Naval Construction Division in Washington, D.C., states that he is experiencing a closer partnership with the design/construction community. NAVFAC is stressing best-value procurement methods, including design-build and negotiated projects, that are attracting a qualified base of builders to work with in the future. This innovative segment of the government is leading the way for other agencies.

Rand Construction is a minority business enterprise (MBE) that has grown tenfold over the last several years under the leadership of Linda Rabbitt. Construction has traditionally been an industry of strong perceptions of how work is to be done and sheer brawn. Rabbitt has led her company by utilizing the newest techniques, such as networking. She learned early-on that clients want a construction company that can provide a new facility as fast as possible, with the highest quality and the least cost. (Not an easy feat by any

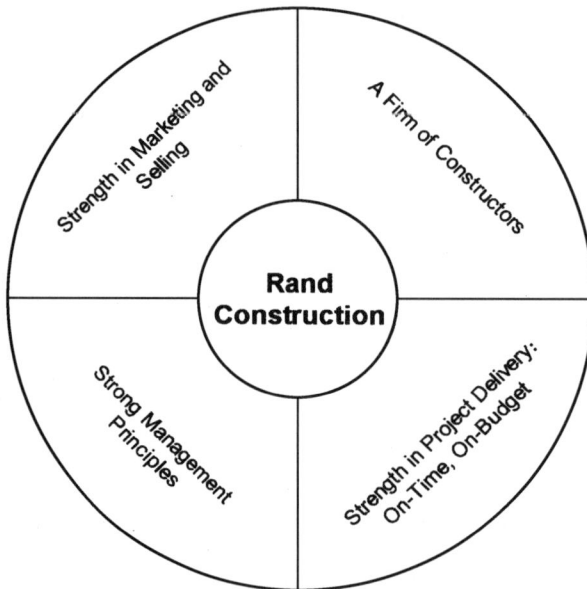

Figure 3.2 Rand Construction's model for success.

means.) Constant development of relationships and Rabbitt's challenging her team to develop innovative construction methods based on the best-practices concept has allowed Rand to grow to be one of the largest female-owned construction organizations in the country. The company attributes most of its success to flexibility and dedication to quality. This flexibility has meant the ability to do all types of projects, from very small to large, effectively, on-time, and profitably. Rand never stops searching the market for new opportunities, in order to maintain a continuous flow of work and consistently increase market share. (See Fig. 3.2.)

TDX Construction is a Manhattan construction management company that has focused on only a couple of segments of the market and doing the very best it can to service that niche. TDX has achieved success by developing a close relationship with a key segment of the construction community, the New York Dormitory Authority. Logan Hurst, a partner with the firm, states that its reputation for quality and performance has been gained slowly but steadily. TDX utilizes the latest technology in many aspects of the construction process, including estimating, project management coordination, cost control, scheduling, and project documentation. Relationships are the driving force for the company, and mandate close ties with client representatives in a tight, competitive market. Hiring the best, ongoing training, and a culture of developing new business opportunities have been an integral element of the firm's philosophy.

Symmons Industries, a manufacturer of residential and commercial building products, goes even one step further by providing clients with a guarantee that lasts for the life of the home in which each product is installed as long as the original owner lives in the house. In advertising, Kevin Symmons, president, illustrates this concept by standing in a white bathrobe with a hammer and chisel in hand in front of a stone monolith engraved with the inscription "Built to last 1,000 years." Check its Web site at http://www.symmons.com. A guarantee is difficult to provide, but if a company is truly serious about quality and service, this can create an invaluable market niche.

This guarantee shows how one company takes the industry standards and expands the depth of the details to its distinct advantage.

As you can see from these various company perspectives, the attitudes, practices, and beliefs of these companies are evolving to meet the realities of today's market economy. They can't be overlooked as being insignificant. Rick Dutmer, Director of the Quality and Productivity Improvement Group for FMI, has stated that companies that experience long-term profitability have a culture of positive attitudes. These attitudes allow employees to understand the relationship among profits, benefits, and salaries.[11] These are key aspects of any business and are even more relevant in the competitive, fanatic market we are faced with today.

You Control Your Future and the Future of Your Company

We all have the ability to influence our future through the actions we take to accomplish the goals and visions we set for our companies and ourselves. We don't always recognize this influence, but it can have a considerable impact on the outcome we desire. Let's look at a few examples of how individuals influence outcome or perception.

To lead change within your organization you must be able to differentiate the authority you have and the influence you have from the actions you take. We have all heard of the Peter Principle, which says that you rise to your level of incompetence. When you are first promoted to oversee a group of employees, let's say in the estimating department, initially you manage this group through your authority. But as an individual rises in the corporation, he or she is not able to physically oversee what each of these individuals does and must rely more and more on influence to sway their thoughts and actions. When you discover the feeling that you are taking on more and more personally and feel you are not in control of the results, you are starting to experience the variance between the authority that has been given to you and your influence over the outcome of these activities.

Cold calls are another good example of how each of us uses influence in a situation where we lack authority. We walk in to a potential client's office, introducing ourselves, our services, and the benefits that we can bring to their company. You are not always successful, but you are influencing the situation through your direct actions 100 percent. We all see individuals each day who flex their authority but wield no influence. When this action doesn't achieve the desired results, they become more stiff and authoritarian and strive to push more, again achieving very little in the process. Without your personal influence, little can be achieved.

An individual we worked with in the past consistently missed deadlines. He knew the time required to get his estimates done but never seemed to be able to meet the dates set by the office manager to have his estimate done and ready for review by the office and field staff for a bid. Many times the project would be bid from a marginally reviewed estimate because the time left to review it was nonexistent. He was consistently told that he missed the deadline. But that did little to change his actions. Another approach would have been to sit down with him and let him set the date for the staff meeting, in a time frame that is conducive to total staff review. Then he is involved in the decision-making process and is taking responsibility for his own actions. He would have an open door policy to discuss the assembling of the bid and what it would take to make sure the bid is due by the date he establishes. You would be influencing his actions by allowing him to assume responsibility for his actions and receive the credit for bringing the estimate in on the date he established, not one set arbitrarily by the office manager. From this example, we're sure you can see many examples of items that occur around you daily where you can start to affect the outcome through your personal influence.

As the earlier example implied, each of us has influence ability, but we don't necessarily use it to its full impact, just as we don't use our body or our brain to its full ability. An increase of 5 to 10 percent of your personal influence would have a considerable impact on your productivity, success rate, and satisfaction.

Your future leadership ability is only limited by your inability to take action to solve each dilemma you face. Tackling each impossible task one by one by use of your personal influence can greatly maximize your influence. Achieving what others cannot achieve is the way to differentiate yourself within your company and with your clients.

You are leading and influencing others 100 percent of the time through your words and deeds. You control what others do through your actions and how you attempt to solve the problems you consistently encounter. This is a constant in your life. You must take responsibility for the results that occur from the tasks you are responsible for because it is your influence that will mark the net result. You are in control of the future for your personal success and the success of your company.

Influence also instills a leverage factor to the actions you must take to achieve the results you are seeking. Influence allows you to lead in the change. If the company is not headed in the right direction, it is up to you to assist with the redirection that is necessary. It allows you to manage the climate within an organization. Jim Clark, CEO of the Clark Group, is known for his dedication to the very best prices from the subcontractors to ensure the lowest possible price and the greatest likelihood of success on bid day. He does this by being present at each bid, talking with each individual person involved with assembling the numbers, and then continually stressing how important success is for the entire organization and the families of every employee of the company. The effort being made is not just a bid, but a way to help fellow workers and continue to ensure the success of the company.

The last factor of influence is the leverage attainable from commitment. The dynamics of today's business mandate that all employees provide constant commitment to the outcome of each assignment and event. Each person within the building industry today wears many hats and each is just as important as the next. Because of overlapping activities, jobs, and projects, a domino effect happens when any one individual does not do his or her job properly. It is critical to ensure total commitment from everyone involved. You can ensure this through their commitment to the assignments they need to handle. Your influence on their actions to achieve positive results can make the difference to the entire company.

Commitment necessary to sink or swim

None of us like commitment. When a spouse asks if his or her mate will promise to be home on time that night, the mate might think, Oh no—I want to commit to being on time and able to be counted on, but I know that the general makeup of my industry and the unknown demands of the job can easily prevent me from honoring this seemingly simple request. This single example provides a fact of life: You can't have total commitment. There are unknowns with everything in life and as such, there will always be times and occasions when you will not be able to maintain the commitment you have made on the job, to your boss, or to clients. This is not bad; it is just a fact of life. How you handle these unknowns is the key to your influence on the result you want.

Given this fact of life, we then need to remember that all of those around us want to be committed 100 percent but can't always do what they promise. This is not a reason to not trust them or their judgment; it is just something we need to build into the plans we formulate because inevitably we will not achieve all our commitments and will need to make modifications and take alternate routes to achieve success. Commitment brings about several inherent benefits that help the team achieve the goals that have been established, from improved communication to getting the results you have set out to obtain.

Commitment is a key ingredient that your influence will need to achieve success. Base your beliefs on the fact that those around you want to have commitment to the established cause. Think about what you can do to ensure this success through your personal influence on the situation and what actions will be necessary for the results sought. And it is up to you to promote the commitment of the plan you are proposing or that is being worked on by the team. You can assume the leadership in developing that total commitment and ensure the success of the endeavor.

What can you do to ensure you are on the right track?

As we have discussed, it is a radically different world today than just 5 years ago. You must properly prepare yourself and your company to survive and thrive in the changing and evolving business environment we find ourselves in. Finding solutions to the myriad of issues and problems facing your segment of the business needs to be addressed at lower levels of the organizational chart than previously ever considered. Downsizing has eliminated the multilevels at most organizations, and all individuals must make their own decisions relating to the complex business dilemmas that constantly develop.

In that same train of thought, the entire team must now be focused on common goals that your industry, business, and company must face daily. Unity of approach can help narrow the chance of ineffective decisions being made and minimize the potential for conflicts between individuals within departments or different segments of the organizational chart. This effort will encourage more ingenuity from the team and will self-initiate and enhance motivation. The net result for your company is a direct impact on your bottom line, improved productivity in the office and the field operations, and ultimately higher levels of quality for the finished product and a satisfied end user of your product or services.

Establishing a Climate of Success for Your Team

This section discusses what you can do as an employee, manager, or owner to improve your ability to get the results you want. There is no easy fix or solution. It takes time and concerted effort on your part to get others to take actions you want them to take in your behalf. Everyone is distinctly different, therefore the methods will vary for each individual or department you are striving to influence. What you must keep in mind is the reality that you alone control the ability to make the changes necessary to effect the results you are seeking.

Mission statements

The concept of developing the right mission statement for your company is in vogue today. The statement needs to be short, catchy, and zing with where you are going. From your local government to subcontractors, everyone is struggling with the effort to define for themselves the perfect mission statement. In many ways it is viewed as a panacea of all that ails companies. It sets you apart from the pack and blows away your competition. The mission statement is simply your company's philosophy. It is what your company does and why. You may already have a mission or a philosophy in your head. But, if you write it down and share it with your employees and customers, they will help you make it happen. Stephen Covey, in *The 7 Habits of Highly Effective People,* created a national trend when he took the concept of mission statement and applied the concept to "personal mission statements."

Your employees must know where the company is going to help drive it there. If they understand your mission, their decision making becomes easier, more focused, and more organizationally cohesive. Second, the makeup of your employees begins to take shape. Some individuals do not fit with some company missions; let them find a match somewhere else. You want people who are going to support your mission.

When your customers and prospective customers know your company's mission, they know what to expect from you. They know what you do, where you stand, and why they should or should not call you.

There are several great books on the market that will help you write a mission statement in a concise manner, focusing upon issues like your position in the marketplace, defining the business you are in, identifying your target market, and crystallizing your company philosophy.

After you have written your mission statement, ask yourself one question: Could any of my competitors use this statement? If they can, you're not done. Your mission statement must include your differential advantage, what makes your company different from any competitor.

Mission statements are not always the perfect solution to providing company direction. Most are never used once developed, or have been created by upper management to be followed by the troops in the trenches. Mission statements that come from the entire organization and are inbred in the company philosophy will have a chance of succeeding. They must be relevant to your core business, implemented, and adhered to.

If you really want to galvanize your efforts, you need to think even bigger by developing a vision statement. Vision statements are less specific than a mission statement. They are a statement of what you will become as a company if you repeatedly and successfully accomplish your mission. You create your vision by asking, "What will accomplishing our mission do for our company, our industry, and our community? What impact will we have?

Think of the best company you currently deal with (supplier, customer, etc.). What does it have that makes it stand out? It is likely that the employees have a uniform sense of purpose. The company's business dealings always have win-win results, it contributes time or money back to the community, and it always acts in a fair and ethical manner. Just plug a company name in the last sentence and it is darn close to being a vision statement.

Don't be afraid to beat the path for employees to follow. They generally thrive when they know exactly where they are going.

Notes

1. *The Tormont Webster's Illustrated Encyclopedic Dictionary*, 1st ed., 1990.
2. Based on a 5-year comparison of "progressive companies" utilizing their return on equity, return on capital, 5-year average growth in earnings per share, 5-year average growth in sales, 1-year net profit margin, and debt/equity ratio. From Rosabeth Moss Kanter, *The Change Masters* (Simon & Schuster, New York, 1983), p. 394.
3. *The Tormont Webster's Illustrated Encyclopedic Dictionary*, 1st ed., 1990.
4. Joel Arthur Barker, *Discovering the Future: The Business of Paradigms* (Charthouse Learning Corp., St. Paul, Minn., 1990).

5. Ibid.
6. Construction Industry Institute, *Study on Opportunities for Improvement in the A/E/C Industry,* 1995.
7. Charles D. Linn and Clifford A. Pearson, "Lessons from America's Best-Managed Firms," *Architectural Record,* January 1997, pp. 106–117.
8. Rob Heselbarth, "Contractors Must Differentiate to Service 'Century of Consumer,'" *Contractor,* July 1998, p. 3.
9. Ibid.
10. Ibid.
11. Rick Dutmer, "Profit Is an Attitude," *FMI Management Letter,* July 1998, pp. 2–3.

4

Research Is the Foundation of Your Marketing Plan

Introduction

An article in *Engineering News-Record* stated that pressure never lets up on A/E/C officers and their firms regarding the implementation of information technology in the new millennium by leveraging computers and communications effectively.[1]

This chapter will explore the ways you can ensure a foothold for you and your company in this roller-coaster ride called business development within the building industry. If you are still not sold on the importance of this topic, consider an investigation by a private eye on a 23-year-old mother that gathered "a five page computer printout from her name alone. The private eye had found her Social Security number, date of birth, every address at which she had ever lived, the names and telephone numbers of past and present neighbors, even the number of bedrooms in a house she had inherited, her welfare history, and the work histories of her children's fathers."[2] And he was just getting warm.

The information highway is rapidly building a roadway to connect us to a wealth of information that exists in society today. As in the twentieth century when a national highway system was developed and soon linked all urban and rural areas, the information highway is doing much the same thing. Old rules for and methods of finding information to plan, enhance, and perform our business lives are taking the form of an electronic "hot-button" that can tap any question, answer any inquiry, or solve any problem we may encounter. The advent of the new highway system resulted in the reduction of our past reliance on the railroads and shipping lines, just as the information highway is reducing our dependence on fax machines, telephones, and print media.

The building industry is experiencing a radical shift in the paradigms we view as the norms for business and will be forced into a new way of doing business. Those businesses that can master this new means of doing business will thrive and grow; those that don't will slowly wither and cease to exist. What we are talking about is a total revolution, not unlike the industrial revolution that reshaped our world at the beginning of the nineteenth century. The future is coming, and it promises to be one filled with chaos and change. Workers who want to thrive in this new economy must be flexible and able to quickly adjust their workstyles. The workplace is evolving into a combination of home, office, and road, creating a "connected and linked" lifestyle for those brave enough to venture out into it.[3]

An amazing fact is that the typical individual of the 1800s was rarely exposed in his or her entire lifetime to what is published in a typical daily newspaper. Today we are experiencing an information glut, with news coming at us from all directions, and this is expected to keep increasing exponentially in the years ahead. The design and construction industry has been slow to integrate information technology into its office and field operations. But as we initially mentioned, this adaptation will be critical for the success of companies tomorrow. Companies are making the transition from using computers strictly for accounting estimation to using them for nontraditional construction services like voice recognition for daily logs by superintendents. This has been a difficult road to follow given the antiquated machinery currently in use, such as central computer servers with maintenance-intensive custom hardware and limited flexibility. Leaving this conformable base to update and modify existing systems is extremely stressful for most firms. The pressing need to make the "right decision" at all times prevails.

Small firms have traditionally had a distinct advantage over larger firms in this area. They are able to upgrade and retrofit their existing technology easier and faster than larger firms. Less overhead expense and politics gives these firms the agility to make quick changes to their in-house resources. This modification of technology will allow them to provide more quality and speed to the end product for their clients and will promote a new interconnectivity for their consultants, all the while easing the transition from in-house staff to a much more flexible situation with outsourcing to professions on a temporary basis. The net results for companies that can achieve this transformation will be that they are highly price competitive and much more productive and efficient with their overall operations. This can provide a considerable competitive advantage because it sets you apart from your competition.

The sudden explosion of electronic data is a natural extension of the rapid changes and advancements in the technology world. It has been barely a decade since the microchip transformed the way we think about, analyze, and transact business. Now, business owners can connect to all office workstations through remote control. Research initially begins with the development of local networks to improve data sharing and office communications. It isn't long until management discovers the advantages of easy access to posted project schedules, job estimates, and projections using software similar to Timberline; monitoring the status of AutoCAD drawings with Arch T and Precision Estimating; and accessing OSHA MSDS sheets, UPS delivery status, government agencies, subs, and suppliers. Many companies are maintaining dynamic/living schedules for immediate feedback of personnel and resources. On top of this are additional in-house resources: Dodge Services (both electronic plan services and project tracing), daily logs and reports, time sheets, payroll, change orders, RFIs, submittals, and job photography (digitally uploaded and stored on the company server).

The construction industry has much to gain from technology that provides overall performance improvement within the industry. The new technologies are allowing the documentation and regular monitoring of activities in a real-time mode and for all levels of job involvement. The complexity of this technology has limited the use of it to a few individuals within each company. Smaller companies, due to the nature of their jobs and the past cost of the computer technology, limited the addition of technology at the field level and provided only minimal networking within the office. Larger companies have also been cautious to add technology to everyone within the organization due to cost of training and equipment. Several new advances are changing that for the field, and in turn are providing a new arena of information for data manipulation while enhancing communication among clients, competitors, and coworkers. These items include personal digital assistants (PDAs), a radical increase in all areas of the industry of company Web sites, voice recognition technology, the improved Project Information Managers (PIMs), teleconferencing and on-line learning, product-specific Web sites that are literally a warehouse of information on companies and materials, and the new use of "intelligence agents" to sift through this mass of data.

Marketing Research: Building Your Foundation to Protect Your Assets

Marketing research in the past versus the new reality of research

What a difference a few years make. Only 5 years ago most of us, as we pursued information about our clients or a new territory, would go to our friendly library, sit down at the card catalog, and search for topics relevant to the areas we needed. Subject, title, and author were the parameters and the librarian would help with this series of quests. Usually the information was stored somewhere else, and we would have to come back and review it later. We would sift through magazines or view the information on microfilm. However, a quiet revolution has been taking place. Since the 1960s information has been gathered electronically by a variety of commercial services, the pioneers creating somewhat clumsy databases that were utilized primarily by law offices, government agencies, private industry, and university libraries.

The growth of the Internet to millions of individuals now is pushing these database services to review their past practices of selling this proprietary information in a competition with the new major free search sites like Yahoo!, Alta Vista, Infoseek, Hotbot, Excite, and Webcrawler. The money to be made today for these search engines is in methods similar to banner ads. These new search engines have been free, but companies and individuals are quickly discovering that using them can yield an enormous amount of information. A site that charges for services and can provide premium content for the user is perhaps the direction these new research tools will take.

The ease of using these new sites is staggering. It is really no different from calling 411 and asking the operator for information. You can organize these search engines under your bookmarks on your Internet provider and readily access them at your convenience. No longer do you need to drive somewhere to work on a research project you are developing. Within minutes, from your personal computer, you can acess the entire library. As we discussed earlier, with each of us doing more and more with less and less time, this can be an exceptional method for us to do more from the convenience of our desk at work, at home, or on the road.

Planning a research program for your company

Chapter 1 discussed the need for organizations to focus on business development and the bottom line. Companies need to be nimble, given the dynamic nature of the current market and that each decision and action need to be as correct as possible. Whatever your strategic plan, there are good research methods available to you on the Internet. Staff development, competitor research, market expansion, personnel background check, monitoring market share, keeping up with the latest trends and developments within your segment of the building industry—all of this and more can be organized by your company. As with any marketing activity, you need to perform a series of steps before you jump into the "Internet water":

- Identify and comprehend what information you specifically are looking for.

- Brainstorm with fellow staff, existing clients, prospective clients, and local trade associations.

- Prepare a list of potential sources you will be checking into to obtain the statistics and data that you are searching for.

- Start gathering this information and organizing it into topic and subtopic areas.

- After you have gathered a significant amount of information, meet again with your office staff, review the information uncovered to date, and try to draw some conclusions about areas needing additional support and new directions that have surfaced.

- Continue the search process to fill in the blanks.

- Meet again with your company team to assimilate the information gathered, summarize it, and focus on conclusions that have been drawn.

- Bring as much of the organization as possible into play so that different interpretations and other potential scenarios can be discussed.

- Start the development of a strategic plan based upon the copious amount of research you have gathered.

This looks like a lot of work, but the results for you and for your organization will be focused and based on fact and can provide the direction needed to make the serious decisions necessary for a strategic plan with a real potential of adding to your bottom line.

Developing a plan is critical to the success of any research you may intend to pursue. It will provide a planned, organized roadmap to use to collect and analyze information and timetables to follow to meet an expected end product. The plan will help you make better marketing decisions.

The driving goal of all research is to help the company increase anticipated revenue. It is also an excellent means of helping to focus the organization on the right things to ensure that profits are maximized. A company that is about to undertake a marketing research project of any size is well advised to consider, during the initial planning sessions, its strengths, weaknesses, opportunities, and direct market threats, and thus formulate a SWOT analysis (see Fig. 4.1). Such an analysis provides a method for the building team to systematically review the internal and external elements that can have a direct influence over a company and, ultimately, success (or lack of it). Members of most organizations are good at defining the strengths and opportunities of their companies, but are less able to specify and be honest about company weaknesses and potential industry threats. Examples of each part of a SWOT analysis follow:

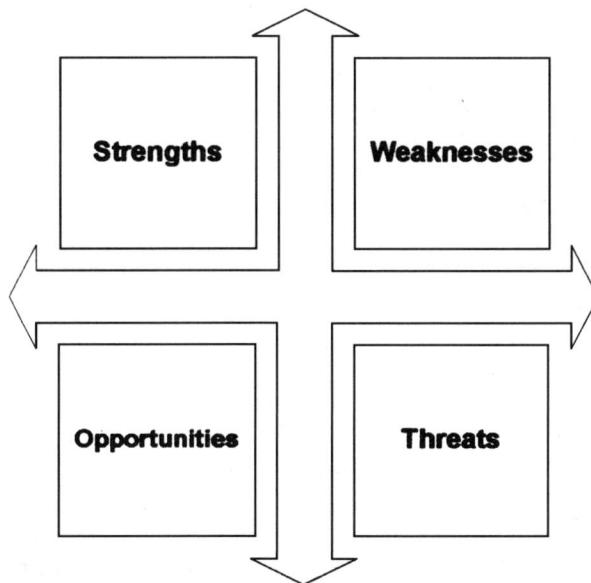

Figure 4.1 SWOT analysis.

Strengths
- Average experience of individuals in the firm is 8 years
- Expertise in stadium structures
- Repeat business with 75 percent of client base
- Able to adjust to changing market conditions

Weaknesses
- No marketing plan developed
- Limited experience in hot new market segment
- High level of new employees with no experience
- Low repeat business ratio

Opportunities
- No direct competitor in institutional market segment
- Successful completion of current work in new market area
- Strong networking ability within regional associations

Threats
- Three new firms moving into area with specific market experience
- Low unemployment rate
- New procurement methods to which you have limited exposure

Once a SWOT analysis has been accomplished, the process will become more common and simple. Also, all aspects of this initial planning process may not be applicable. In addition, the research should not be started in the middle of your busiest time of the year. Strive to plan ahead and start this process when you have the time to devote your full attention to it. Technology has simplified many aspects of this process for you so that it can be done systematically and somewhat easily as you go along.

Set priorities and proceed step by step. Try not to take on too much all at once. Split the pieces among your staff and support personnel. Establishing a priority of tasks will focus you and your team on the item to be accomplished. As you go about your regular daily assignments, meetings, preparation of estimates and proposals, and general problem solving, you will encounter pieces of the research puzzle that you can put into your files and carry on. Research may seem like a tedious process, but given the "connected" society that is rushing in your direction, your competition will know *all* about you and your organization.

Using the Web for Information Gathering

Customer + competition + markets + database

Now that you have completed the planning process with your team, reviewing existing market conditions, company position, potential environmental forces, and the specific information you want to research, another enormous amount of information is available to you. In 1997 there were an estimated 80 million people connected to the Internet in 166 countries, and this number is growing exponentially.

Paul Doherty, in his book *Cyberplaces,* suggests there are over 14 million computers linking more than 85 million people around the world.[3] With those kinds of numbers, this is an enormous and growing world of electronic media that can provide you with the solutions to problems, products and services, and direction to plot a course of success for your company. As has been mentioned in many articles, one of the compelling reasons not to ignore the Internet is that you don't want your competitors to have the only visibility and access out there. We'll review the many diverse sites available to provide you with the information you need; you will be exposed to everything from accounting statistics, legal services and settlements, and market penetration by your local competition to the projected trends in your segment of the market for years to come.

How to get there

If you don't have an Internet connection now, you can get one for as little as $19.95 per month that will provide you with the ability to search the Internet, design your own personal Web site, and have an e-mail address. The number 1 services in the country are America Online (http://www.aol.com) and GeoCities (http://geocities.com). Many companies are using these two services as an easy first venture into cyberspace.

Another alternative is to develop a personal Internet address with your Internet provider. For less than $100 per month you can have your own company registered; you can reserve your domain name now through WorldNIC (http://www.worldnic.com). You can also check the entire Internet to see if your name is still available by going to http://rs.internic.cgi-bin/whois. What is even more amazing is that you can be up and running in less than 24 hours. Needless to say, this is for a very basic site. Sites can easily run from thousands of dollars for companies wanting to make a grand statement to millions for a one-of-a-kind, complex, graphically intense site.

Internet search tools

Browsers.   There are many browsers out there and selecting one is a personal matter. The two leading browsers are Microsoft Internet Explorer and Netscape Navigator. Both can be downloaded over the Internet for free for a trial period. These two continue to battle it out for leadership, but with Microsoft now firmly embedding Internet Explorer 4.0 within its latest operating system, Windows 98, it will be even more difficult to unseat it. However, many will not be willing to leave Netscape, with its superb integration with the Internet.

A new browser feature is called "off-line browsing." It allows individuals to download to their desktop computer a Web site of their choosing and accessing it at their own pace. This can speed the process of access and is an easier way to access that Web site in the future. Changes that have occurred to the Web site since it was downloaded to the desktop will be automatically updated when you access it online again in the future.

For marketing research of information on the Web, each search engine has its own strengths and unique characteristics. You'll need to experiment with each and see which one fits your specific needs during that particular search. Another simple method of saving the sites you discover during your research is to mark them. Right-click on them with the mouse. This will allow you to save them as your favorites sites.

Directories.   Yahoo! (http://www.yahoo.com) is the largest and most well known of the directories on the Web; it was the original pioneer search index. You can search for anything using this tool, and it will provide you with an index of potential sites. One limit with this extensive site is that only Web sites that have registered with Yahoo! will be found during your search for information.

Search engines.   Search engines differ from directories in that they search the Web pages looking for words that match the words used in your marketing research. This is slowly taming the Internet beast by allowing you to find all references to the topics you search for; the drawback is that you can uncover thousands of sites that match your search specifics. You need to find methods to narrow your search so that you are not flooded with sites that have no or limited importance. Utilizing a *boolean search* methodology can direct and focus your research efforts. Keywords such as AND, OR, and NEAR can enhance your search (see Fig. 4.2). A few examples of boolean search strings follow:

Economic Development AND Cleveland: ensures that both terms are present in documents searched for

Associated General Contractors OR AGC: ensures that at least one of these two terms is present

City Hall NEAR Boston: ensures that these two terms are within 10 words of each other

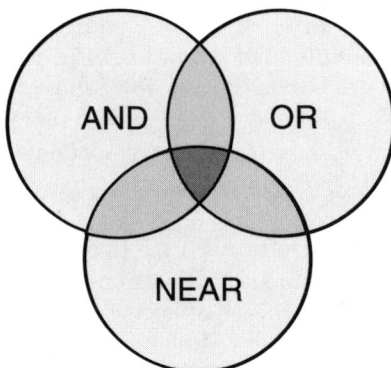

Figure 4.2 Boolean search.

Listed below are some of the industry search engines that have a reputation for getting you positive results. Each has a distinct advantage in the research process.

AccuFind—Java Script search engine. (http://nln.com)

Alta Vista—Very large, well-known search engine with over 8 billion words and filling over 30 million Web pages. Exceptionally fast. Digital's mother lode of Web sites and newsgroups with minimal advertising. (http://altavista.digital.com)

Amazing Environmental Organization Web Directory—Number 1 environmental search engine. (http://www.dickson.com)

BigBook—Large Yellow Pages for business information. (http://www.bigbook.com/)

BigFoot—White pages search engine. (http://www.bigfoot.com)

Building Industry Exchange—Directories of building-industry business, people, publications, products, and employment opportunities worldwide. (http://www.building.org)

BuildingOnline—Compiles more than 72,000 Web pages in such categories as contractors, consultants, product manufacturers, wholesalers and distributors, and associations and education. (http://buildingonline.com)

BuildNET—Comprehensive directories containing more than 150,000 pages of building-industry organizations and resources. (http://www.abuildnet.com)

Constructor Equipment Online Directories of equipment dealers, manufacturers, specifications, and trade names. (http://www.coneq.com)

Construction SuperNetwork—Produced by the Aberdeen Group's electronic media division; comprehensive site caters to the concrete and masonry industries. Search the entire site or specific directories to locate manufacturers, distributors, contractors, concrete producers, industry associations, industry and product news, upcoming events, classified ads, and archived articles from *Concrete Construction*. (http://www.SuperNetwork.com)

ConstructionNet—Search for contractors, manufacturers, distributors, and trade associations. Represents more than 850,000 companies. (http://www.constructionnet.net)

Contractors Hot Line—A nationwide database of construction equipment for sale. Includes an equipment locator and price guide, auction results, and state department of transportation bids and lettings. (http://www.contractorshotline.com)

c/net's SHAREWARE.COM—A great search engine. A shareware program to meet your needs. (http://www.shareware.com)

Deja News—Search engine for Usenet newsgroups. This is a great source to track down experts in a particular field; this site is your best opportunity to sort through the Internet's bulletin boards. You can also with the click on a button listen in to what others may be saying about your company. (http://www.dejanews.com)

Electric Library—For a nominal annual fee you can have comprehensive access to a library of information from thousands of printed sources ($9.95 a month or $59.95 a year, after a 30-day free trial). (http://www.elibrary.com/id/2525)

Excite—Like Yahoo!, allows you to search by concept or single word a vast archive of material. (http://www.excite.com)

Gamelan—Java resources search engine. (http://www.gamelan.com/index.html)

GTE SuperPages—Expansive yellow pages for business across America and over 60,000 Web sites businesses. (http://superpages.gte.net)

HotBot—The next dimension of search engine. If your initial search returns several hundred sites, you can use that first search as a basis for a second. This site can make the Internet universe more manageable. (http://www.hotbot.com)

IBM InfoMarket—Incredible search ability over the news media and private industry. (http://www.infomarket.ibm.com)

Inference Find—Groups searches by Web sites. (http://www.inference.com/ifind)

Infoseek—Highly rated business search engine. (http://www.infoseek.com)

Intropro Resources, inc.—Unique search engine for locating construction-related information and sites, services, and products. This site is entering and indexing information such as building material and product vendors using CSI's 16-unit structure and is linked with over 9,000 architectural firms. (http://www.ipr.com)

Internet@address finder—Looking for an Internet address? This is a good place to look. (http://www.iaf.net)

Newsbot—The ultimate guide to news online (business, health, entertainment, etc.), all within the last 24 hours, past week, or past month. (http://www.newsbot.com)

Northern Light—Sixty-five million pages of Web index with an extensive collection of over 2900 publications. The search is ranked by relevancy, with folders dynamically organized that contain your results by subject, type source, and language. (http://www.nlsearch.com)

Lycos—Another highly rated search engine developed by Carnegie Mellon University. Allows you to zero in on your competitor's Web sites to find out how they position themselves. (http://www.lycos.com)

Specs-Online—Directory of specified construction products for commercial, institutional, and industrial projects primarily in North America. (http://www.specs-online.com)

Search tips

Most search engines are divided into two general types: indexed or category-based. Indexed search engines, such as AltaVista and Excite, use automated Web crawlers to index the words on Web pages. You enter key words in a box on the search page, and the engine returns links to pages that contain those words. Many engines support full-text searching of the sites included in their databases. Others go even further by indexing certain fields as well (such as title, author, and abstract) just like a traditional library.

A category-based search engine such as Yahoo! has a staff of reviewers who organize Web sites into a hierarchical structure. Instead of searching for specific words on a Web page, these engines search for words in a category name or site summary. The search results begin with general categories that branch into more detailed subcategories. Category-based searches are useful for collecting general information on a topic quickly. To find information about topics that aren't so easily categorized, you're better off using an indexed search site.

The following sites are a great place to start if you would like more information on search engines, such as what they can do for you, additional methods to refine your search, and what are some of the latest search engines available.

Search Engine Watch—Explains in detail how search engines index Web sites and provides a thorough review of all search engines, reviews, and tutorials. Sign up for its free monthly update of new search engines. (http://www.searchenginewatch.com)

Internet Searching Strategies—Rice University has developed this site to help students utilize the Internet. It is easy to navigate and provides a step-by-step approach to maximizing research activities. (http://www.rice.edu/Fondren/Netguides/strategies.html)

MetaCrawler—Doesn't depend on one search engine but many, and thus has the potential of allowing you to find more information pertaining to the subject you are seeking. This site also strives to rank the information it provides. (http://www.metacrawler.com)

ProFusion—A search engine for Internet search engines developed at Kansas University. Susan Gaunch, associate professor of electrical engineering, designed a program that would help narrow the search to specific terms within a title of a Web site and to terms within a specific page without representation. (http://profusion.ittc.ukans.edu/info.html)

As we have discussed, each site has its own advantages and benefits for your research activities. Yahoo! has maintained leadership by consistently attracting the most visitors to its site for the last couple of years.

Associations

Associations are often overlooked as a source of information. Industry associations can furnish you with a variety of data to facilitate comprehensive market research. Most associations can provide calendars, back issues of publications, contacts for staff and members, and information on current code and regulatory issues, and many have a help line that allows you the opportunity to ask questions regarding problems you are experiencing.

AIA Online/AIA Web site—Directory of all member architects around the world, over 50,000. Listed by firm, individual, area, and expertise. Also provides daily search of the *Commerce Business Daily* (CBD). You can use the local AIA chapters as another source of ongoing information. (http://www.ais.org)

American Consulting Engineers Council (ACEC) (http://www.acec.org)

American Institute of Steel Construction (AISC) (http://www.aiscweb.com)

American Marketing Association (http://www.ama.org)

American Society for Concrete Contractors (ASCC) (http://www.ascc.org)

American Society of Civil Engineers (ASCE) (http://www.asce.org)

American Society of Mechanical Engineers (ASME) (http://www.asme.org)

American Wood Council (AWC) (http://www.awc.org)

Associated Builders & Contractors (ABC)—National merit shop association of contractors and subcontractors. (http://www.abc.org)

Associated General Contractors of America (AGC)—Serves as a "one-stop shop" for all your construction-related business products, publications, standard form contracts, training, and ad opportunities. Links with the majority of the membership base of over 40,000 nationally. (http://www.agc.org)

Construction Specifications Institute (CSI) (http://www.csinet.org)

Greater San Antonio Builders' Association (http://www.sabuilders.com)

Institute of Electrical and Electronics Engineers, Inc. (IEEE) (http://www.IEEE.org)

National Association of Home Builders (NAHB) (http://www.nabh.com)

National Society of Professional Engineers (NSPE) (http://www.nspe.org)

Society of Marketing Professional Services (SMPS) (http://www.smps.org)

Urban Land Institute (ULI)—Provides custom-tailored responses to bibliographies, available literature, and specific experts. Exclusive information sources with access to the library of over 5000 books, 300 periodical titles, numerous databases, and special document collection. Ready access to ULI research staff and key real estate professionals from among ULI's 16,000 members. (http://www.uli.org)

Washington Building Congress (WBC)—Includes 1000 members representing all facets of the real estate, construction, and design community within the Greater Washington, D.C., region. Access to regional professional associations, member links, and current articles on trends and new developments, along with a calendar of events. (http://www.wbcnet.org)

Construction-related sites

The Internet is gradually developing a series of sites dedicated to market-specific products and/or services. These can greatly facilitate your search. You may also want to regularly monitor your market segment's Web sites to keep up on current trends, market changes, and perhaps what your competition is up to.

Aberdeen Group—Single sources for the concrete industry. (http://www.supernetwork.net)

Architecture Engineering Construction Network—Communication medium for the building industry. (http://www.aecnet.com)

AECInfo Center—Headquartered in Canada, AEC InfoCenter is international and considered the world's largest professional business center for general contractors. Includes listings of related topics, material suppliers, and industry links and information about permit fees, professional regulations, interactive Q&A forums, and classifieds with building product vendor information organized under the Construction Specification Institute (CSI). (http://www.aecinfo.com)

Boston Design Center—Boston's link to the architectural community. (http://www.bostondesign.com)

Build America (http://www.buildamerica.com)

Build.Net—An expansive information service, which offers access to over 1200 pages of building industry resources, services, and companies. (http://www.abuildnet.com)

Building Information Exchange—Offers many products and services to the industry. (http://www.building.org)

Building Site Network (http://www.buildingsite.com)

Commercial Construction Site—Represents many facets of the industry involving engineering, construction, and architecture and includes articles and information relevant to the industry. (http://verticalserv.com/construc.html)

Construction Monthly Online—This construction industry site provides articles about the industry and related events and has many industry links. Features e-mail, chat networking, links to WWW sites and bulletin boards, downloading from news, list servers and UseNet groups including box lunches, seminars, and forums online. (http://constmonthly.com)

Construction Site—More of an overseer of information on the building industry. It develops and documents industry trends, upcoming events, and changes in rules and regulations. (http://www.constructionsite.com)

Global Construction Network (http://www.gcn.net)

PSMJ Resources—Market Web site that lists marketing and consulting publications to purchase and seminars that are available. (http://www.psmj.com)

Government databases

Government databases can provide the statistics you may be searching for on a new market territory, market size, or potential revenue sources. Usually somewhat tedious to navigate, they can offer a treasure of contacts and data.

American Society for Testing & Materials (ASTM) (http://www.astm.org)

Census Bureau—supplies a lot of statistical data regarding all locations within the United States. (http://www.census.gov)

Council of American Building Officials (http://www.cabo.org)

Federal Trade Commission—Good links to other business-oriented sites. (http://www.ftc.gov)

The Federal Marketplace—Run as a small private company; allows you to search for current government contracts up for bid, for a fee ($75 to $400 per year). (http://www.fedmarket.com)

Federal Web Locator—Need to find government information but not quite sure of an agency's name or its overall purpose? Try this first. (http://www.law.vill.edu/fedagency/fedwebloc.html)

FedWorld Information Network—One stop for all government publications. (http://www.fedworld.gov)

Occupational Safety and Health Administration (OSHA) (http://www.osha.gov)

House of Representatives (http://www.house.gov)

Library of Congress—Information on all books about the building industry. (http://lcweb.loc.gov)

National Institute of Building Sciences (NIBS) (http://www.nibs.org)

Securities and Exchange Commission—Includes access to the Edgar database, which contains the full text of reports that public companies are required to file with the SEC, including quarterly financial statements. (http://www.sec.gov)

Small Business Administration—Links to many small-business development sites, including a large quantity of shareware and access to the Service Corps of Retired Executives (SCORE). (http://www.sbaonline.sba.gov)

THOMAS (http://www.thomas.loc.gov)

U.S. Business Advisor—Gateway to all government resources if you want to do business with the government. (http://www.business.gov)

USPS ZIP+4 Code Lookup—Gives you the ZIP+4, compliments of the United States Postal Service. (http://www.usps.gov/ncsc/lookups/lookup_zip+4.html)

White House—Send the president an e-mail. (http://www.whitehouse.gov)

Resource sites

Listed below are several industry Web sites that can provide background research material for your next marketing plan or budget.

Advertising Age—Information on advertising and marketing, plus the latest in marketing news. (http://www.adage.com)

American City Business Journals—The source for 28 local business papers. These can provide a great source of information on markets you may be considering moving into. (http://www.amcity.com)

American Demographics/Marketing Tools—Browse previous versions of publications that will help you learn how to focus your marketing efforts intelligently. (http://www.marketingtools.com)

The Blue Book—Leading publications' sources linking the industry together. Allows access to more than 500,000 classified listings for manufacturers, material distributors, equipment dealers, architects, engineers, general contractors, and subcontractors in all major specialties. (http://www.bluebook.com)

Businesswire—Information, primarily in the form of press releases, about American companies both large and small. (http://www.hnt.com/bizwire)

Commerce Business Daily (Loren Data Corp.)—Lists of contracting offices in the government and of contracting consultants, with free job listings. (http://www.ld.com)

Construction Resources On-Line—Online marketing business resource center. Sign up for the free e-zine. (http://www.copywriter.com)

Dataquest Interactive—Your direct link to high-tech industry. (http://www.dataquest.com)

Dun & Bradstreet—With a database of 10 million U.S. businesses D&B has plenty of tips about a variety of business-related topics and the lowdown on just about any public company. (Approximately $20 per business background report complete with sales figures, employee information, targets markets, even special promotional events.) (http://www.dbisna.com)

EXPOguide—Enormous list of trade shows and conferences. (http://www.expoguide.com)

Future Enterprises, Inc.—A leader in information technology training and consulting. With nearly two decades under its belt the company has trained over 350,000 individuals. This is a great way for you to get your team up to speed. One of the most frustrating things about the current status of technology is how little we really know about the powerful tool that sits on our desks. Change that situation today. (http://www.fei.com)

Hoover's Online—Massive site that provides information about thousands of companies; you'll pay a fee for most of the information. (http://www.hoovers.com)

How to Price Your Products and Services—Solid, elementary reference from the small business administration on pricing strategy.
(Gopher://www.sbaonline.sba.gov/00/business-development/general-information-and-publications/obd5.txt)

IndustryLink—Excellent compilation of links to sites of interest to those in a number of specific industries. (http://www.industrylink.com)

PointCast Network—Best news-gathering source available to keep track of customized news feeds that are delivered directly to your desktop computer when you boot up. Will search the news within 35 industries and is like having your own newswire customized to keep track of your competition. Software is downloaded free of charge. (http://www.pointcast.com)

Minority Business & Professional Directory—Thorough directory of businesses owned by minorities and women. If you're an MBE or DBE, add your company for new business leads. (http://www.minbizdir.com)

PR Newswire—Excellent source of press releases for thousands of companies. (http://www.prnewswire.com)

Thomas Register—American register of manufacturers including 1.5 million individual products, services, and companies. All for free.
(http://www.thomasregister.com:8000)

Understanding Your Market—Another SBA site that walks you through the process of understanding your clients.
(Gopher://www.sbaonline.sba.gov//00/business-development/general-information-and-publications/obd3.txt)

The wonders of Web pages

The World Wide Web is a subset of the Internet that was established in 1994 at CERN, the European Laboratory for Particle Physics (http://www.cern.ch/CERN/WorldWideWeb/WWWandCERN.html) as a way to simplify the Internet and provide a means for organizations to list their information on Web sites or Web pages. When you access a Web site, you will get the home page first. It is the most important segment of the Web site because it is often the only page most people will see. With this new ability to electronically connect with companies, you can get the information you need without having to know exactly where to look. You can list a topic or click on a picture or subject area and you are automatically taken there by a "hyperlink" with other sites that can transport you to wherever you need to go.

Hyperlinks have made the complexity of the Internet more user friendly for the business and the general public. A much more in-depth discussion of the development of the Web can be found at Thomas Boutell's World Wide Web FAQ—an A to Z for the World Wide Web (http://www.boutell.com/gaq/). You can also review Yahoo!'s World Wide Information and Documentation site (http://www.yahoo.com/Computers_and_Internet/Internet/World_Wide_Web/Information_and_Documentation) and the WWW & Internet Manuals, Demos site (http://techa.unige.ch/info-www.html).

Publications. This subsection lists some of the most useful and handy publications for your "favorites" listing on your homepage. Each of these can be read online, saving you the actual cost of a subscription. You may also review back issues.

Architects' First Source (AFS)—Comprehensive building product listings with specifications and contacts. (http://www.afsonl.com)

Architectural Record Online (http://www.archrecord.com)

Building Material Retailer Magazine (http://www.bmrmag.com)

Business Week Online (http://www.businessweek.com)

Design Intelligence (http://www.di.net/index.htm)

Design Architecture (http://www.cornishproductions.com)

Engineering News-Record (ENR)—Online directory of design and construction firms. This proprietary search engine will help owners and facility managers find companies faster by location (state), project specialty (buildings, bridges, etc.), or type of company (engineer, design, etc.) Enormous back vault of articles and information on the building industry with numerous links. (http://www.enr.com)

Interactive Construction Online (http://www.inconstruction.com)

Metropolis Magazine (http://www.metropolismag.com)

ProFile—The sourcebook of U.S. architectural design firms. (http://www.cmdg.com/profile)

Companies. Listed below are some of the best company Web sites in our industry. These illustrate ideas you may want to incorporate into your own Web page.

Andersen Windows (http://www.andersen.com)

Bentley Systems (http://www.bentley.com)

Holmes & Narver (http://www.hninc.com)

Ellerbe Becket (http://www.healthon-line.com/eb.htm)

3D International (http://www.3di.com)

HNTB Corp. (http://www.hntb.com)

Law Engineering and Environmental Services (http://www.lawcom.com)

Marvin Windows (http://www.marvin.com)

National Gypsum Company (http://www.national-gypsum.com)

Steelcase (http://steelcase.com)

The McGraw-Hill Companies (Construction Information Group)

F.W. Dodge—DataLine, provides the industry with a continuous supply of high-quality construction project information every day, day after day. Incorporates the traditional Dodge reports into an electronic media that is easy and quick to use. DataLine also links to the Market Leader Construction Marketing software package (similar to a personal information manager, or PMI) so you can control your sales effort by tracking projects and contacts and create target mailing. It includes over 600,000 projects, spanning over 1200 reports and correspondents who file or update more than 5000 projects every business day. (http://www.fwdodge.com)

Sweet's Group—CSI 16-section organization of all major business materials and products. Easy to use for research and has pictures and additional information for background information. (http://www.sweets.com)

McGraw-Hill Bookstore—Every book you can want on the building industry is here. (http://ww.bookstore.mcgraw-hill.com)

Continuing Education Center (http://www.mhcec.com)

Construction market data group (http://www.cmdonl.com).   Direct competition for McGraw-Hill, this similar database of project leads is well organized and easily accessible on-line. It also provides products and services that enable architects, designers, manufacturers, contractors, and other industry professionals to make the necessary intelligent and informed decisions required to be successful today. The Group has many resources for developing cost information, project leads, estimating services, ongoing marketing intelligence, industry contacts, and building product information. Some of the services provided are.

- The Architects' First Source—A comprehensive source of information for architects, specifiers, and designers.
- Clark Reports—Project leads and market intelligence throughout the United States.
- ProFile—An enormous reference book of architectural firms throughout the United States. Available in hard copy or over the Internet.
- Daily Commerce News—Project leads for coverage throughout Canada.
- Construction Market Data (CMD)—Project leads and intelligence throughout the United States.

R.S. Means.   A division of the Construction Market Data Group. R.S. Means is the granddaddy of construction cost guides, statistical analysis, technical publications, and educational services (http://www.rsmeans.com).

Newsgroups

Newsgroups function on the Internet as a public forum of sorts. Groups of people post and read messages on a particular topic of interest and then make comments if they have something of interest to add. They can be a good source of information that is not known to the general market; the information is shared among the group for support and assistance. Newsgroups allow you to get the inside track on your competition and rivals by monitoring rumors and through good business intelligence practices.

Some basic hierarchies within the Usenet segment of the Internet are marketed with the suffix of the group's name and end with:

Alt (alternative topics)

Biz (business)

Comp (computers)

Misc (miscellaneous topics)

News (news and discussions about the Usenet itself)

Rec (recreation)

Sci (science)

Soc (social topics)

Talk (controversial topics)

These groups can concentrate on segments of the market territory or types of companies or can focus on trends that are evolving. You should consider trying one for a while; then you could join in offering help and asking your own questions. A good program to test out is Forte's Free Agent, which provides you with access to hundreds of megabytes of messages each day. You can use one of the search engines to sort the information you are looking for (http://www.forteinc.com).

E-Builder

The brainchild of Jonathan Antevy, this new tool for the construction system arose from demands from contractors while utilizing the Global Construction Network (GCN) initiated by the Fails Management Institute. The GCN serves as an information house and switchboard. In 1997 the system was renamed E-Builder under a pilot program developed by the new MP Interactive Corporation (http://www.mpinteractive.com) and is still managed by Jon Antevy, president and cofounder. This Internet-based communication tool dramatically enhances the exchange of information among construction project participants. Owners, contractors, subcontractors, and suppliers can read about industry-specific marketing information, providers, and upcoming events and training sessions. Companies are encouraged to form alliances with other companies through this new medium.

Data gathered and effective lists

An easy and efficient method of doing a marketing research project that will pay dividends immediately is the use of effective lists of all pertinent aspects of your marketing plan. Good lists can simplify the entire marketing and service components of your business by giving you the ability to put the information you have cultivated, organized, and maintained into a format that can be easily managed and read by your office team. It is amazing how often my marketers keep information in a small phone book in which numbers are scribbled and barely legible. The uses and benefits of well laid out and accessible lists for your marketing strategy are endless.

During your initial planning meeting, a discussion should be conducted about what types of lists would help the organization achieve its targeted goals and who will develop, select, and maintain these lists once they evolve. In Chap. 7 we discuss some of the new personal information managers (PIMs) that are allowing officewide access to these data. It cannot be stressed enough how important lists like those below can be to your firm and to you:

Client/customer.   This is the first list that comes to mind. Look around your office. Can you lay your hands on a copy of your client list, who the key participants are, the history of the relationship, current projects under consideration or in construction, or a comparison of initial project estimates versus final executed results? This is the first place to start.

Competitor list.   This is another critical list to develop and maintain to monitor your competitors on a day-to-day basis. Who are their key employees, what are their strengths and weaknesses, what is their current client base and project backlog?

Prospect lists.   Rank these lists by probability, size, complexity, margin potential, worker requirements, location, and interrelated companies or projects.

Subcontractor and vendor lists.   Rate each by level of service, products, key individuals, past experiences, and areas of expertise.

Print and electronic media lists.   These lists can facilitate an easy, quick distribution of press releases, newsworthy activities, and invitations to current company events.

Supplemental reference list.   This could include accountants, lawyers, associations, governmental regulators (i.e., OSHA), local politicians, and planning agencies.

As we discussed earlier, there are endless resources for your search and activities in developing a list. Once started, you can enlist the support and assistance of the entire office, especially if you are all linked by a server or intranet. Maintenance and diligent pruning will be necessary, but the result will be an immediately powerful tool everyone in the office can utilize, and it will focus attention on the key clients, competitors, and prospects so that the entire team is working toward a common direction for the company. Well-developed lists can be a valuable asset for any professional services firm as well as one of the foundations of a successful marketing program.

Low-cost surveys and audits

Low-cost surveys and audits are another method of market research. Ellen Flynn-Heapes, Flynn Heapes Consulting, is an industry leader in the development of marketing consulting and the use of surveys to obtain for her clients the information they are seeking for their strategic marketing plans. She advocates a series of methodologies that in our experience has yielded mixed results:

Personal interviewing.   One-on-one or with focus groups. (The Boston Consulting Group uses this method extensively in meeting with potential and current clients and competitors to obtain answers to marketing questions.) The company gathers the data and tabulates them by utilizing key quotes from interviews and looking for patterns, trends, and new opportunities.

Telephone surveys.   This can be an inexpensive and effective method to obtain information quickly on a topic or subject of interest.

Mailed questionnaires.   These generally have the lowest level of response by the selected groups. Tricks of attaching $1 to the survey form or following up with a phone call to request a response can enhance results. A method that usually gets better than average results is to make a one-page survey that can be faxed back to the sender. This seems to simplify the process and eliminates the middle step of adding an envelope and stamp.

The Associated Colleges of Construction Education (ACCE) utilized a survey on the *Engineering News-Record* Web site for eliciting information on a Futures Education Forum it was planning. All construction-related associations were notified and their members, at their own convenience, answered the questions listed. The answers were downloaded to the Civil Engineering Research Foundation (CERF) for correlation and analysis. This resulted in minimal effort, maximum exposure, and a broad coverage of the market.

With this wealth of information, there are some things that can't be done with a computer that need to be considered before you undertake any significant research project. The computer doesn't eliminate workers or save money. It has created a totally new way for us to access, digest, and manipulate information. Savings generally are found though new knowledge generated and innovations uncovered with the additional sources of information that are discovered. Computers can handle the tedious task of looking through numerous statistical materials, but it still takes time and hard study for individuals to analyze what they are really looking at.

A poorly defined problem or an inadequate sequence of steps to uncover the information will not bring the solutions sought. The initial planning stage is critical to the final successful resolution of the problems you are trying to resolve through the market research. We have all heard the adage "garbage in, garbage out." A concerted effort to focus your research at the beginning will greatly enhance the end product.

A significant trend for construction companies is to enhance the research productivity of their staffs. More and more companies are striving to minimize personnel and mandating that specified individuals within organizations do more of the decision-making by more innovation and by developing a research "mentality."

Database management for intelligent marketing research decisions

After the recession of the 1990s, we all learned quickly that to stay in business we had to cut costs and sell more services and products. A major trend has evolved called relationship marketing; it allows companies to base their direct-marketing efforts on extensive marketing databases. Companies that are doing well today are collecting detailed information about individual customers. Database marketing is the newest way within the A/E/C community to understand clients' needs and to target only the most likely prospects.

Companies today can, through an aggressive database marketing methodology, develop in-house the information needed to reduce sales and marketing staff while increasing revenue. It is not inexpensive or easy to develop. It takes time, diligence, and repetition to maintain the database and ensure the information is entered regularly and correctly as it is gathered. This is basically taking the list concept to the next level. In a world of massive information overload we can all realize the difficulty in gathering this massive amount of information and then trying to find a way to use it. Information is located all over the company and all over the world. But databases can place your firm ahead of the curve and always in the path of additional orders.

Intelligence gathering and research are critical elements in effective marketing as a result of the trends toward national accounts and international marketing and the transitions from client needs to client wants and from price to nonprice competition today. All firms have in place some form of marketing information system that connects the external environment with their executives. It is even more important to provide that link at all levels of the organization. Market research and the information that comes from it is usually not made available to everyone within an organization; it comes too late in the process to be useful, or when it is supplied by executives, it is not trusted by the support team.

A marketing information system within companies today needs to be divided into four segments:

1. The first is an internal reports system that provides current data on account receivables, payable, sales, costs, executions, margins, safety results, and cash flow. Construction and design firms have in place elaborate job-cost accounting systems that, due to the advanced development of this segment of the office environment, are being enhanced and slowly distributed to all members of the team. Buy-in and performance-based bonus structures have allowed this information segment to reach a new maturity.

2. A second research gathering marketing intelligence system is one that supplies company executives with everyday information about what is going on in the external marketing environment. As we have discussed, this can range from reading the newspaper to accessing Dodge DataLine for a projection of current projects with a breakdown on the who, where, how, and what. A well-trained marketing and sales team can obtain data from one of the online resources, hire a marketing consultant, or utilize their staff for special intelligence-gathering activities within the market; this can greatly improve the information that is coming into the company's executives.

3. A third system is marketing research, which involves collecting information that has been targeted for research by company executives and the marketing team because of a perceived threat to or opportunity within the company. This process takes five steps: defining the problem and thoroughly narrowing the research objectives, developing a marketing research plan, gathering information in both the internal and external markets, stepping back and analyzing the material and data gathered, and presenting the findings and results to management initially and to the entire staff ultimately.

4. The final system of market research, which we will not address in this book, is the analytical marketing system, which is usually conducted by outside consultants with a staff of statisticians who run the data through advanced statistical procedures and models based on similar studies and researches for comparison, analysis, and dissection.

Notes

1. "Technology and Leadership Mix," *Engineering News-Record,* April 21, 1997, p. 19.
2. "On-Line High-Tech Sleuths Find Private Facts," *The Wall Street Journal,* September 15, 1997, p. 1.
3. Paul Doherty, *Cyberplaces: The Internet Guide for Architects, Engineers and Contractors* (R.S. Means Company, 1997), p. 29.

5

Creating Your Marketing Plan

Far better it is to dare mighty things, to win glorious triumphs even though checkered by failure, than to rank with those poor spirits who neither enjoy nor suffer much because they live in the gray twilight that knows neither victory nor defeat.

THEODORE ROOSEVELT

Introduction

In this chapter we will review many of the basics that constitute what consistently successful companies are doing and how you can follow their successful formulas and make more money. When we initially started in business there was little about marketing. The perception of marketing in the early 1970s for the building industry was promotion driven. We did advertising, brochures, and made cold calls. This was the leading edge and it satisfied most employers. Sales were the consuming activity. Estimating, pricing, bidding, estimating, pricing, bidding, estimating, pricing, bidding—that was the common perception of the road to a thriving business. The concept that marketing could apply to "professional" occupations was radical, but there was a creeping subculture that was starting to discuss marketing tactics for lawyers, doctors, architects, engineers, and even contractors. Until this time, these occupations were considered *above* this seemingly unprofessional approach to getting new business.

The 1980s were a turning point for marketing because it was starting to be acknowledged as a viable means of positioning yourself and was made legitimate by Peter Drucker when he wrote (originally in the 1950s but it became the mantra in the 1980s) that the sole purpose of business is to create a customer, and that business has only two basic functions: *marketing and innovation.*[1]

This concept took root and blossomed in the 1980s and 1990s because several management books were written expounding upon the power and vital necessity of marketing to the financial health of an organization. Books like *In Search of Excellence, Creating Excellence,* and *Megatrends* each stress the concept of What business are you really in? The analogy that was touted was the example of the railroad industry that had failed to recognize its own obsolescence. The industry was not really in the railroad business, we heard. Had it reconceptualized what its real purpose and market was, it may have realized that it was in the people-moving business and changed the future of the world as we know it. It was also in 1981 that Jack Miller, of Jack Miller Seminars for the Construction Industry, wrote about the importance of the marketing plan: "A company without a plan is like a ship without a rudder. It moves aimlessly in the marketplace reacting to pipe dreams instead of real opportunities."[2] His belief was that once this plan was in place you could then start the mystical marketing process of making things happen.[3] These thoughts were well circulated among companies across the country, and influenced the development of many company planning guides.

Another industry leader in the 1980s was Hank Parkinson, who wrote about successful company executives who understood the importance of marketing. Parkinson found that the executives were involved with marketing 50 to 70 percent of their business lives, and rewarded those around them who did the same.[4]

Our own experiences follow this concept. Several technically savvy companies have lost out to the firm that had superior marketing skills, one that knew what the market and its clients wanted. Being able to deliver these services on time and within budget, it obtained the last look at the next project mostly because of the superior relationship it cultivated with customers. Successful companies are not focused on being a "bid machine" but are striving to focus their time and attention on a few quality client companies and are doggedly pursuing them with high-quality service. And, as Woody Allen is often quoted as saying, "Success is 90 percent showing up!" Implementing your plan is doing something and keeping at it.

A Definition of Marketing

Before we go any further, we should look at how marketing is defined by others. Many people have their own definition of what they consider marketing to be or not be when compared to sales. We would like to share with you some definitions that are rather applicable and pertinent to our discussion.

According to Gerre Jones, "Marketing defines, outlines, and sets the stage for sales. Marketing is to create or increase demand, to which sales—and profits—are the hoped-for results."[5] Theodore Levitt stressed that selling tries to get clients to know what you have, whereas marketing strives to develop what the customer wants at a price he or she will pay.[6] Ben Gerwick stated this another way when he noted that marketing is an essential element of the survival and growth of design and construction organizations. He defines marketing as developing strong relationships that enable your company to sell its services in the most favorable places and with the best terms and time frames.[7] The American Marketing Association's Committee on Definitions says that marketing is "the performance of business activities that direct the flow of goods and services from producer to consumer or user."[8] Finally, Herman Holtz said, "Anyone can sell cold drinks to thirsty people. Marketing is the art of finding or inventing ways to make people thirsty."[9]

These early definitions created the idea that marketing was a direct step to customer satisfaction and it provided customers with the goods, products, or services they wanted; the result was a handsome profit. How many times have you been in a meeting when the discussion turns to pushing a product or service that your company wants to sell, not what the market wants to buy? This is a common, universal dilemma. Marketing today has the mix shown in Fig. 5.1.

Figure 5.1 Marketing mix.

If marketing is based on two basic beliefs—that your company policies, operations, and planning should be oriented toward the customer and profitable sales volume should be the ultimate end result of the firm—marketing is finding methods to satisfy those client expectations and receiving a reasonable profit from that activity. The definition of marketing has evolved today into a total company philosophy that integrates and coordinates all marketing activities within all other company operations for the fundamental purpose of maximizing the bottom line of the corporation. This is not an option; it is a mandate for any company that wants to be successful and thrive today. Many company presidents believe that marketing is a simple process of hiring a marketing director and watching the contracts and money roll in. It rarely works that way.

Marketing is a much more extensive framework of resources and tools. Most new marketing directors take a year to get established; it takes at least that long for the marketing process they represent to become entrenched. Also, the new marketing manager needs more tools and resources than a company credit card to be truly effective. The company must be committed to the marketing concept. Everyone in the organization must be a part of this process and support its efforts. Regular office meetings to discuss the goals to achieve and what has happened to date needs to be scheduled. Additional marketing support material needs to be developed as an integral part of this process; direct mail pieces, the traditional company brochure, special marketing events, public relations activities, and industry involvement are all parts of this marketing process.

This process, if implemented properly, can increase market share by 30 to 50 percent. But before you jump right into marketing, you must plan. This is one step that almost everyone resents and hesitates to take, but ignoring it can stop a firm dead in its tracks. The rewards and benefits of this step are numerous and extremely beneficial. The first step in this process is the establishment of a mission statement. Just what do you want to accomplish?

Achieving What You Want through Mission Statements

So where do you start? Scott Butcher stressed the mission statement clearly when he said, "The mission statement is your destination, the marketing plan, in conjunction with the strategic plan, is your road map to get there."[10] This is echoed by Richard Sides, who believes the corporate mission statement can be the most important segment of your marketing plan because it clarifies your company's primary priority—to attract and keep customers.[11] Important? You bet. This is the guiding light that directs your efforts. It also illustrates that everyone in the organization, not just the marketing department, is part of the solution. This helps ensure that a team approach to marketing and attention to satisfying the customer is on everyone's mind.

Marketing Mix

The concept of the *marketing mix* has become the base from which marketing is developed within any company, the thought being that the mix of activities that meshes the customer's wants and needs, integrated with other external, environmental factors, will create a successful marketing program. It is this diagnosis that determines where your business is and why. The careful evaluation of where your company is in regard to the balance of the industry and in the eyes of your clients and potential clients is critical. Step back and take the time to examine this facet of your plan systematically and slowly. Bypassing this important area can negate the other key areas of your plan and blind you to the perceptions that exist about you and your position within the industry.

As you build your program and examine your marketing mix, keep this priority in the back of your mind. Your odds of success are much higher with this approach. It is simple, straightforward, and by far the most economical sequence for your marketing and sales effort. An existing service to an existing client who is satisfied and happy with your past relationship will always provide the most financial reward.

Situation analysis

A situation analysis is one of the first steps you must take before starting your journey to assemble your marketing plan. The situation analysis is no more than looking at the current competitive environment, getting a macro view of where the company is going to grow, and finding out what services, perceptions, resources, and clients your competition is utilizing. This analysis is compared to your own similar functions to see what disparity exists. It is really just a history. Where have you and your competition been in the context of the marketplace? Review your position relative to the competition. Are you a market leader, a challenger, a follower, or a niche player? The analysis should include a discussion of prices, differentiation, barriers that exist to entry to your segment of the business, and the unique advantages you may have. What trends have come and gone? What segments of the market are hot and why, and which ones are not?

Analysis can be as complex and extensive as you may need, but generally the more compact and precise, the better for the understanding of what has gone on over the past year. Some elaborate situation analyses that have been completed by consultants just sit on the shelf in the client firm. You need to make this report in a condensed format so that the entire team will read it. We will review in this chapter how best to analyze past and current marketing material assembled by your company.

Who, where, how, and why are the key questions you are after with this analysis. A thorough discussion of all segments of your business can be conducted with your office staff, clients, industry groups, and associations. The results of these discussions along with a review and summary of the data uncovered in your initial research efforts will paint a clear picture of what has transpired in the past year. An in-depth analysis of the strengths and weaknesses of the company is paramount. This should be done as an ongoing step of doing business. Regular up-to-date information on each competitor can be maintained within your company database by all of your office team. This should include pictures of your competitors' work and what your competitors' customers are saying about them. One of the biggest mistakes a firm can make, according to C.F. Culler Associates of Atlanta, is "overlooking the importance of this section, using general descriptions, and omitting market research that supports demand."[12]

Some additional key questions and issues to be addressed are

What type of business are we in now and where do we want to be in 5 to 10 years?

What is our current sales volume?

What are our current profit goals and are we making them consistently?

What are our company's strengths?

What are our company's weaknesses?

What changes would we like to add to our scope of services and/or products?

What type of potential clients are we looking for?

What is our preferred way of doing business?

What are our company goals now, and how have they been obtained in the past?

What assets and resources do we have with people, equipment, and capabilities?

What new trends do we foresee within our internal and external market?

Are we willing to expand geographically?

These questions can be organized in the form of a SWOT analysis that the entire organization can build around (see Fig. 4.1). This analysis reviews systematically all major internal and external environmental conditions that could affect an organization.

This inventory of your market should be conducted off-site in a retreat setting, where the entire staff can brainstorm and interject their own ideas and beliefs. The team should also update this inventory at least annually.

One company that has been very successful with its technique of perfecting the situational analysis to an art form is the Disney Company. Richard J. Maturi writes that Disney gathers a team together and uses "storyboarding to develop the history through a use of movable cards to organize and reorganize new and old ideas that flow from brainstorming sessions."[13] This is a simple method of allowing an entire organization to participate in marketing analysis, but still ensuring that the analysis is relevant to each segment of the company.

Market segmentation will follow the development of a company SWOT analysis. This is a placement of customers within the grids based on the strengths defined. It enables marketers to focus their time and plans on a targeted market group. This segmentation should illustrate a diverse group of

TABLE 5.1 Market Segmentation Grid

Client type	Less than $1 million	$1–5 million	$5–10 million	Larger than $10 million
Institutional				
Commercial				
Design-build				
Lump sum				
Negotiated				
Interior				
Full construction				
Joint ventures				

client types that can be further analyzed based on profitability. An example of how this information may be laid out is shown in Table 5.1.

This table can be expanded to include the total market share and what your percentage of each is and where you believe it could be. The ultimate purpose of this process is to find homogeneous groupings of potential prospects whose requirements are all essentially the same so that you can start to narrow your marketing approach.

A review of market life cycle is always an important step when making any decision at this point of the planning process. Products and services are thought of having a life that begins with their introduction, continues through the growth and maturity stages, and ends in decline. The potential life of any product or service is important to keep in the back of your mind as your analysis takes form. Initially there is minimal competition and the volume and profits of the introduction are a curve that starts an upward incline. As the service or product goes into the growth phase, the curve moves upward. Maturity and finally decline of any product or service sees a gradual and then a more severe decline of the services. Needless to say, as you analyze your position with a situational analysis, you want to attempt to position yourself as close to the introduction of any new service or product's life cycle as possible. This ensures a moderate guarantee of a significant volume of business opportunities and the opportunity to maximize your profit potential.

Product, price, promotion, and profits—the four P's

Marketing can be broken into four distinct and critical segments. It is very important to achieve the proper balance among these for maximum results.

- *Product.* The product or service that a company has designed to satisfy a client's needs. A conscious strategy needs to be in place to keep introducing new services to prolong the life of maturing products or services.

- *Price.*   The dollar value that has been finally exchanged for the product.
- *Promotion.*   The methods the company has utilized to inform potential buyers about the service or product. Make sure you convey a consistent message to your customer at every point of contact.
- *Place.*   How a company delivers (distributes) the service or product to the consumer.

Strategies + Action Plans = Marketing Plan

The marketing plan is your road map to a successful way to chart a 5-year plan of action for your organization. Without one you can get off course and waste considerable time, money, and resources. Most companies can't afford to do this in today's business arena. There are five elements necessary for a successful marketing strategy, according to Frank Stasiowiski:

1. Marketplace analysis.
2. Researching the client.
3. Understanding how your firm is perceived by clients.
4. Investigating the competition.
5. Conducting an internal assessment.

Stasiowiski goes on to state that these five elements are common to all *Fortune* 500 companies.[14] The strategic market planning process provides the total framework of a contractor or design firm. The firm has a clear direction toward the vision and mission of the firm, the steps necessary to reach the goals, objectives established, and a clear road map for the entire team within the organization.

Several publications listed in the Bibliography are devoted totally to the development of marketing plans. Planning is critical and can help define your goals for the coming year; the structure, timetable, resources needed; and an action plan of who does what. Without a plan, the marketing effort usually wanders aimlessly and much less productively than it would otherwise. Develop your own plan with all the strategic planning tools; it will drive your company toward the opportunities that await, and you will avoid the pitfalls and risks you've uncovered. Where you choose to focus your efforts and your management philosophy will dictate how you should be guiding and training your organization. Keep in mind the following key aspects of this risk versus opportunity aspect of your plan. To reduce risks

- Look out for threats
- Reduce self-performed work (outsource more nonessential activities)
- Schedule what needs to be accomplished during the year
- Administer contracts diligently

- Aggressively react to internal and external threats

To enhance opportunities,

- Look internally for opportunities within your organization
- Increase self-performance of activities you are exceptionally good at
- Plan for work to get done by an established time schedule
- Manage your plan around completed work goals (results), building your track record and reputation

The marketing plan should be considered your proactive approach to business. You are striving to think outside the box, anticipating what risks and opportunities you may be able to capitalize on. The well-worn analogy that you are marketing not the $1/_4$-in drill bit, but in reality the $1/_4$-in hole is the underlying theme that must permeate what you are developing in your game plan. The plan must incorporate all there is to know about your firm, all your disadvantages and advantages, and your competitors. Pull this information from external market research and internal resources such as audits, interviews, and call reports.

Table 5.2 is a basic outline of a marketing plan and its different components. Every plan will have at least a summary, a SWOT, specific goals and objectives, strategies, and an action plan. We will discuss these more in depth later.

Having a summary with conclusions is self-evident, and yet many firms fail to summarize their full report in a readable summary that can be utilized to focus the effort throughout the year. The summary should reinforce the mission statement of the company and provide the background material of its purpose. Of special note is that investors and top management generally read no further than the executive summary, making this even more critical to guarantee the financial support needed to implement the full plan.

The environmental (situational) analysis was discussed in large part in Chap. 4 where we examined the various components of research. But as we mentioned earlier, this is one particular segment of the plan that needs sufficient time and attention to fully grasp the depth of the industry, its players, and the trends that will be reshaping its future. Time will also be needed to review the information you've obtained from publications, newspapers, online sources, associations, government agencies, and private industry groups.

Goals and objectives provide the basis for what the company hopes to achieve from the marketing plan. This is really the meat and potatoes of the plan, which can be broken into various subplans for the company for the next 6 to 24 months. Management is tuned to the sales plan you intend to influence through this marketing agenda, but there are several areas that will be affected by your advertising, marketing research, marketing collateral, training, technology, and PR plans. It is best that the goals and objectives set be as specific and measurable as possible. Examples are

1. Positioning your company as a market leader or specialty niche player

TABLE 5.2 Marketing Plan Components

Component	Description
1. Table of contents	Listing of plan components.
2. Executive summary	Management summary of entire plan.
3. Company strategic mission	Vision, goals, and strategies of company.
4. SWOT analysis	Research and data analysis of internal and external market conditions garnered from media, private industry, government, associations, etc. Market forecasts/market share/trend assessment. Company resources (people, equipment, funding).
5. Marketing mix	Development of four Ps: product, price, promotion, place. Life-cycle service/product strategies, previous sales history and future projections, clients, prospects. Pricing issues. Promotional activities (marketing communications). Place-distribution challenges within market and target markets.
6. Strategic plan	Putting the above elements together for a cohesive plan, reviewing current resources for supplementing plan, developing a budget and tentative time frame.
7. Action plans	Goals and task assignment to achieve plan. Who, what, when, where, and how, with costs and schedule to be monitored.
8. Controls and reviews	Regular reviews to update and adjust marketing plan.

2. Scheduling development of the marketing plan

3. Creating a realistic budget

Strategies and tactics are the procedures used by the company's marketing department to achieve the goals of the plan. What specific agenda is the company establishing to achieve this plan?

Recommended actions (the action plan) are the heart of the plan. This is where the rubber hits the road and provides the specific actions, the who, how, when of the company.

Monitoring results

Now you are done with your plan and individuals have been given assignments and have taken responsibility for the actions to be accomplished. But, even more important now is the implementation of a series of checks and balances to monitor the plan and to have the flexibility to make a change in course when road blocks are encountered, and they will be. Contingency plans can be installed for what-if scenarios based on resource and people problems that may occur. Schedules can be adjusted and regular feedback and discussion can correct activities that have pointed your effort in a direction contrary to the direction in which you want to go.

Now It's Time to Do It

A good motivational book written by John-Roger and Peter McWilliams, *Do It!...Let's Get Off Our Buts* (Prelude Press, Los Angeles, 1991), is a good place to start before you embark on your marketing plan. There will be so many obstacles once you start down this road that it is helpful to read about them, develop confidence that you can overcome them, and then go out and develop a plan that will take your company where you want it to go. Chapter 4 and this chapter have covered the basic ingredients in a marketing plan. Now let's create one.

Typical marketing plan

Mission statement. We won't go into mission statements here because we have discussed them at length, but just remember they do not limit your freedom of choice in deciding where you are and where you plan to go. In many ways, they free you from having to consistently analyze the strategic decisions your company needs to face. Knowing what your business is, who your clients are, and what each of your customers values gives you and your team a clear direction to go in to obtain the corporate goals and objectives that radiate from your mission statement. Keep the mission statement in a long-term perspective to provide a purpose for your organization to pursue. As we discussed earlier, get your team away from the office, usually early in the morning for three to four hours, and brainstorm. A worksheet you might utilize to focus your firm on the most important goals consists of the following items:

Existing clients

Future clients

Market areas we currently work in and want to expand into

Products and/or services we provide

What are our financial objectives?

What skill do we particularly excel at?

What are the company's core values and beliefs?

What do we want to achieve for the employees of the company?

What image do we want to have in the market 10 years from now?

Mission statement

Market size. Chapter 4 discussed the unlimited access available to statistical data on the market area. Forecasting the total size of your potential market, while also breaking it into specific segments (i.e., retail, commercial, government, private) should be relatively obtainable. It will take some time and will require considerable analysis to make sure it is correct to meet your needs. This strategic research will be used throughout your marketing plan for making decisions about resource allocations in the pursuit of new business development. Don't depend on just what you uncover on the Internet or from the print media. Go out and talk with leaders within the market you want to conquer. Talk with busy contractors, developers, owners, and economic development agencies to find out their thoughts and vision. You can also supplement this process with brief telephone interviews and perhaps even some mail surveys. (Samples are on the CD-ROM that accompanies this book.) During your planning meeting you will be able to quote their pictures of the future, not your own, which will build buy-in from your planning team. The following market-size questions may help you develop information about your market.

What is the size of the market in square feet and dollars?

What is the size of the market segments in square feet and dollars?

How established (old) is this market?

Is the market receptive to price fluctuations?

Is the market receptive to outside fluctuations?

Is this market cyclical? If so, can you determine the patterns and their reciprocal sequences?

How difficult is it to wrest marketshare away from the industry leaders? Are there opportunities to penetrate the market given this control?

What unique skills, attitudes, and technologies are involved in the market?

Strengths and weaknesses.   We reviewed earlier in this chapter the benefits of a SWOT analysis for your company and your market. This is a critical stage of the planning process when developing a situational analysis of the strengths, goals, opportunities, weaknesses, and assets of your firm.

This same review of competitors, using a SWOT analysis, will allow you to gather important information on each of them. One subcontractor, Ceco Concrete Corporation, has developed a computer competitor analysis sheet called WINS (Winners Information Needs System) that reviews the results of each project it bids, documenting the bid results, the winner, the winner's bid, the winner's margin (against the contractor's own estimate), the winner's backlog, whether preconstruction services were provided, the number and names of all competitors, the contractor's backlog at the time of bid, the customer relationship, and the anticipated start and end dates of the project.

This document also includes all the key project players, including the owner/developer, architect, engineer, and construction manager and an address, contact, and phone number for each. The project is described in depth with the building type, size based on gross square feet, the floors, what the structural system was, an arbitrary rating of the complexity of the project, and the delivery system. Added to this is an analysis of the design data: typical bay sizes, super live and dead loads, typical story height, seismic zone, lateral system, and other technological features the structure may contain. The company is then in a position to monitor the preferences, backlog, relationships, and intricacy of its competitor base and as such to control pricing and market planning. Your analysis may not go to this extreme, but remember, your competitor has the same tools available you have, so it is best to try to obtain a good fix on your strengths and weaknesses and those of your competitor as soon as you can.

Where are we today? Where do we want to be tomorrow?   This aspect of your plan is where you will eventually end up. Many companies are so busy doing work now, executing work, putting out fires, and looking for new business opportunities that they forget where they want to be 10 years from now. Looking 10 years down the road with key employees is a valuable activity. During one of your early morning planning sessions, spend 3 hours writing a description of where you want the company to be in 10 years. Be as graphic as possible. What markets are you participating in, how many employees do you have, what market segments are you participating in, how much money are you making, how are you perceived in the market, and how do your employees feel about the organization? This document then becomes the road map to the future for your company. The next segments of the marketing plan provide the objectives and actions to be taken to get the company to this final destination.

Who are your prospects?   Another important step of any marketing plan is determining who really needs your services. Start with developing a list of who has been buying your services or product over the last 5 years. Break this list of names down into segments of the market in which they participate. Examples are commercial owners, government agencies, owner representatives, and the like. Then start to list the volume of work you have obtained from each. This should be further broken into base bid, extra and change orders, bid margins versus final executed margin, safety record, and a client maintenance factor. (Some clients demand an excessive amount of time with little or no payback in the form of new business. There is the tendency to develop a comfort level with these accounts because we talk with them regularly, provide pricing, and assist with information requests. The interesting aspect is that some of these customers can occupy an inordinate amount of your time without adding significantly to your balance sheet and as such need to be weaned.) The final step at this stage is to pinpoint your project locations.

The next step is to start the development of a list of potential customers. Who would you like to do business with? Have you thought of new segments of the market, new territories, and clients with a good reputation for professionalism, fast payment of requisitions, and low maintenance during the project execution? During your initial research, were there industry names that kept reappearing as having quality service and on-time delivery capabilities with which you could align yourself?

Summarize your existing clients and the locations where you have performed work. You need to examine actual results and strive to provide an overview that realistically provides guidance of where you have had profitable results and where you have done a lot of work but with marginal profits. One of the most difficult aspects of this industry is eliminating work opportunities that exist but do not provide profit. There is some justification to maintain this backlog of work if it helps eliminate or absorb some of your fixed costs and overhead. However, many firms have been swayed into believing a big backlog is the panacea for all of their troubles, when in many regards it only adds to the distress. Keep in the back of your mind the 80/20 rule: Eighty percent of your profits usually come from 20 percent of your customers. Managing the other 80 percent is critical to your overall success.

During this activity it is also important to analyze customer needs as best you can. Basic to the review is how each customer rated each of the following aspects of your business: price, quality, service, location or convenience to its office, financial background, training within your company, how your style or image meshed with its needs, exclusivity, product/service line, availability, warranty for services provided, ease to work with (evidenced by minimal change orders and back charges), knowledge of market, reliability, and on-time delivery of project as contracted. Satisfying each specific client need is the difference between a contract and no contract. Price is important, but there are many variables to the final decision.

Image.   Now comes the portion of the mission statement that should set the standard for all of your actions. What image do you want to give the market place, your clients, potential clients, and the construction community in general? As we've mentioned earlier, marketing is in large part perceptions. What perception do you want people to remember you by? Besides the basics of being neat, clean, well groomed, prompt, courteous, friendly, cordial, and sincere, you are a problem solver. We tend to forget that as a problem solver we act as a consultant to our clients and provide them with our ability, experience, and the resources we have at our disposal.

All firms today stress quality, within budget, on-time performance. Now, with that out of the way, what else do you have to offer? This is the time to consider any niche opportunities you may have, special market segmentation aspects your company has, and special strengths. The Davis Construction Company, Rockville, Maryland, uses the image of the owner's contractor. This firm goes out of its way to provide an environment that is a positive experience for the building owner. This effort permeates all facets of the organization from the secretary to the field forces. Who do you think owners and their friends turn to first when they have another project to negotiate?

Qualifying your prospects.   There are hundreds of prospective customers in your markets. It is your job to determine the few who will provide you with the best opportunities. Deciding what is important and worthy of your company's time and attention will pay real dividends fast. The following questions must be included on the prequalification form (sample forms are included on the CD-ROM that accompanies this book) you use for a prospective client and project:

Is this project real?   You can burn a lot of energy, time, and emotion on projects that are little more than pie-in-the-sky.

Does this project fit into your company mission and purpose?   This new project or client may seem exotic and have many good attributes, but it does not meet the marketing mix you have established with your team members and as such should be shelved for now.

Where is the project located?   If this potential project falls outside of your current territory, it may require an unusual amount of resources to oversee its execution and may impose new risks.

Who is the competition?   A review of who will be your anticipated competitors is important. Do your competitors have a close relationship with the buying authority on this project?

Will it be hard-bid or negotiated?   Will you have 15 competitors or 3? Will it be first cost or best final?

Who will be the design team?   Some designers have a reputation for flair but lack the ability to assemble executable working documents. Will you be assuming additional risks?

How will it be financed?   Will the financing be through bond issues, banks, or insurance companies? If you are to be a part of this project, you must know what the financial commitment and risks will be.

How big is the project?   Sometimes due to its shear size (or lack of it), a project should immediately be disqualified.

Action plans that generate strategies for follow-up.   Now you have a list of potential clients and locations that have a profitable track record. The most important stage of your marketing plan is the establishment of action plans to get something to happen. Planning is great, but as in all of physics, matter needs a push to start to move. Start with an overall strategy for translating your marketing plan into an action plan for the company. Break it into manageable pieces with competitive and promotional strategies, including timetables for each and an individual who is responsible for the marketing plan. See Table 5.3.

This is the step within most marketing plans that will make or break the entire plan. Regular review of the status of the marketing strategies, with critical discussion of successes and failures, will assist with the redirection of your efforts. Accountability of individuals who are given the responsibility is critical. These individuals will want to break down the action plan even further into

TABLE 5.3   Marketing Plan Strategies

Function	Activity	Responsibility	Due date	Done
SWOT analysis	Research	rdw	12/01/99	
Marketing mix development	Four P's analysis			
Strategic plan development	Assemble cohesive elements			
Action plans (to accomplish strategic plan)	Goal 1			
	Goal 2			
	Goal 3			
Controls	Review of plan activities			

manageable pieces. For example, if one strategy is to develop five new clients in the coming year, a subset of this plan will break down this objective to

Meet with x, y, z potential clients in January

- Send letter of introduction with annual report, project profiles, etc.
- Follow-up call to be made 1 week later
- Follow-up visit within 2 weeks
- Obtain at least one new project from initial effort

Meet with a, b, c potential clients in February

- Continue as January

No plan works flawlessly. They all take some modification due to unforeseen market conditions. A plan of action with a set timetable and individuals accountable will get you more quickly to your financial goals.

Persuasion, closing, and getting the order. Satisfying a client need and persuading that client that your proposal should be accepted is an art form in itself. Why should the client buy from you? Making a conscious effort to prepare a list of benefits that you bring to the table needs to be instilled in your entire team. Are you selected on quality, speed, scheduling, financial strength, a friendly and competent staff, no nickel and dime during and after the construction process, value-engineering talent, or an ability to consistently bring the project in under budget? These are just a few of the aspects of persuasion that need to be fine-tuned and developed. There is an enormous amount of time and effort expended to get to this stage; you want to ensure that you will close and get the order.

Encouraging your clients to buy now is of paramount importance. Try first to uncover the customer needs. Constantly probe the client with who, what, when, and where questions and open-ended requests for additional information.

Balance this effort with close-ended probing: How much do you want to pay? When do you want to start the project? We have two ways we can proceed....Which way would you prefer? At the same time you can add subtle items within your discussion that are integral to your proposal and that can trigger a need to buy: providing a discount if the contract is signed by a certain date; adding escalation terms to your proposal so your price will be increased or may not be applicable by a certain date; creating a verbal image of what the client will receive when it gives an order to your company, expanding on how much better or easier life would be; or subtly remind the client that profit and time will be lost by not acting now and placing an order with your company.

There is nothing illegal or unethical with these approaches. They are simply steps in moving a project from dead center to a line of action that your client has got to make and the sooner, in many regards, the better. Your ability to get a final decision on projects can work toward the benefit of all parties involved.

Monitor and make changes. A marketing plan is just that—a plan. It is a dynamic object rather than one that is completed and put on a shelf. It must be regularly reviewed, analyzed, changed, and modified. This is simply a road map for the organization to follow. As with any road map, you will encounter detours along the way, but your map should provide alternative options and flexibility for change. A 3- to 5-year plan can be an exceptional tool toward ensuring that all departments within the organization are on the same page regarding marketing activity. Each element of the company needs to provide feedback, allowing marketers to add their comments about financial budgeting modifications, human resource allocations, outside vendor assignments, and missed opportunities to the document. Each time the marketing plan is reviewed or discussed, these obstacles can be discussed and an alternative solution proposed.

The investment in the marketing activities cannot be overstressed. This is an outlay of financial and human resources that needs consistent assessment in the never-ending cycle of market planning and actual implementation. Controlling and accurately assessing the activities generated by the marketing plan are difficult and complex, but to be truly successful and so that they do not go astray, they must be measured and regularly monitored. This effort will not go unrewarded; it will generate increased financial returns and general overall goodwill and enhanced motivation to all parties involved. Monitoring methods can be both qualitative and quantitative. A review of the strategic action plans with anticipated results compared to the overall operating statement of the company needs to be made and discussed. The marketing plan should be aligned with the financial objectives of the organization (sales, growth, return on investment, etc.) and, as such, should be tied in with the compensation for the marketing staff.

With marketing an essential component of any construction or design firm, the marketing plan and how you implement it is critical to a company's overall success. Also, to develop the commitment of the plan, you must communicate this plan with passion. Companies who neglect this component eventually decline and wither away. The establishment of specific company goals and an aggressive action plan for the company to follow is one sure way to point the firm in a positive direction with a definite course to follow. As we've discussed, this takes time and considerable effort by all parties within the organization. It also takes firm leadership to provide the futuristic vision of where the company wants to go in the foreseeable future. Leadership must set aside sufficient time, devote its attention to the strategies specified, obtain assistance with the facts and details as necessary, demand accurate record keeping throughout the year, and involve all of the company's personnel and resources for the proper implementation of a serious marketing plan, making it a comprehensive, thorough, top priority within the company. As Peter Drucker has said, a plan is nothing "unless it degenerates into work."[15]

Notes

1. Peter Drucker, *The Practice of Management* (Harper & Row, New York, 1954), pp. 37–41.
2. Jack Miller, "Develop an Effective Marketing Plan for Your Construction Company," *Concrete Construction,* November 1981, p. 889.
3. Ibid.
4. Hank Parkinson, "Consensus: The Key to a Successful Marketing Plan," *Construction Specifier,* vol. 40, January 1987, p. 54.
5. Gerre Jones, "Defining Marketing," *Professional Marketing Report,* vol. 13, no. 8, May 1989, p. 1.
6. Clayton M. Christensen, "Making Strategy: Learning by Doing," *Harvard Business Review,* November/December 1997, p. 154.
7. Ben Gerwick, *Construction Engineering and Marketing for Major Project Services* (John Wiley & Sons, New York, 1983), p. 158.
8. Committee on Definitions, *Marketing Definitions: A Glossary of Marketing Terms* (American Marketing Association, Chicago, 1960), p. 15.
9. Herman Holtz, *The Business Plan Guide* (John Wiley & Sons, New York, 1994), p. 86.
10. Scott D. Butcher, "Planning for Marketing: How's Your Map? A Simplified Approach to Developing Winning Marketing Plans," *SMPS Marketer,* October 1997, pp. 4–7.
11. Richard Sides, "Marketing with Your Mission Statement," *Construction Marketing Today,* April 1996, p. 14.
12. C.F. Culler Associates, www.ebusinessinc.com, May 25, 1998.
13. Richard J. Maturi, "Disney's Legacy Lives," *Industry Week,* July 19, 1993.
14. Frank Stasiowiski, *Chesapeake Marketer,* Society of Marketing Professional Services, Baltimore, Fall/Winter 1995.
15. Peter Drucker, *Management* (Harper & Row, New York, 1954), p. 128.

6

Preconstruction Services and Promotion

Introduction

Construction within the 1990s has been totally different from anything we've experienced before. Profit levels are tight, new work opportunities are sporadic, and security is nonexistent. How can an old, overworked concept for the 1980s cocktail generation—preconstruction services—have relevance in our business world today? But these services (and more particularly, networking) are critical for you and your company. If you plan to find work opportunities that are off the beaten path, with higher than ordinary profit margins, and to develop a means of security for you and your company, what we will talk about in this chapter is critical for you. The primary activity in the implementation of preconstruction is problem solving. Preconstruction is the act of getting involved with a potential client early to show how you would approach solving a complex problem from your perspective. This takes skill, keen ability, and insight.

Preconstruction represents one of the most frustrating aspects of business today. You can spend enormous time, resources, and real money developing copious estimates for a client only to have the client take your information and peddle it to the next company.

Preconstruction has become a big part of the way we do business today. You are expected to help your potential client or support the assembled team in preparing drawings, schedules, personnel booklets, presentations, estimates, and proposals with limited guarantees of success. How you manage this new ingredient in the bid process can provide you with up to an additional 2 to 3 percent in your margin at the end of the year.

Building Value through Promotion Methods

Preconstruction has evolved as one of the key professional tools for promotional activities within the building industry. If you are already a top-quality, first-class firm, why do you need to promote? Won't the industry come to you? It should, but the industry is flush with firms who are just as capable and proficient as you are. Very few firms enjoy the luxury of having a lock on their segment of the market. Even if you do, most owners are required by their financial institutions to have at least three bidders on each project. Preconstruction can, however, give your company a considerable advantage by being considered on the team from day one. It can also usually get you at least a "last look," which will give you the opportunity to decide whether you want to take the job or not.

This proactive approach to preconstruction services is a major component in the promotional tools successful companies use today. Promotion sounds like an unethical or less than professional aspect of business, but in reality it is very important and is required to position your company so that it can land the order.

The dictionary defines *promote* as "to contribute to the progress, development, or growth of" "To advance to a higher position, grade, or honor" "To push forward; further; encourage; advance; as to promote a business venture."[1]

The perception is of someone who is a "promoter," like a carnival sideshow promoter or a used car salesperson. This could not be farther from where this concept is moving the industry today. A key element within marketing is the ability to promote your product or service so that it is perceived as the one that is best suited for that new job that you have just heard about; it is making sure that the decision makers within your customer's companies know about you, your capabilities, and experience and that they have a trust and belief that yours is the right company for its new project.

Indirectly we are all promoters, whether it is of our companies or ourselves. We are consistently striving to make someone or something aware of our unique capability to do the job ahead of any others. Someone has to promote you, your services, your products, and your company, or your competitor will gain the edge and be recognized for its unique skill. There is an old adage: Those that do—gets; those that don't—don't. It is well worth remembering this when considering preconstruction.

Promotional Methods

We will review promotional tools in Chap. 7. In this chapter we will review the promotional activities that incorporate preconstruction services and networking. It is these two aspects of the business that will have the biggest impact on your promotional efforts and will positively influence your bottom line. The ideas we will discuss are but the tip of the iceberg of available ways for you to reach your target audience. Use these as a guide and then expand them to meet your specific needs. Remember, a fundamental rule of promotion is the concept of "selling the sizzle not the steak." Everyone has steak, but you want to stand out from your competition by illustrating the unique benefits that you bring to the table.

Preliminary cost estimates

One of the surest methods to make your company a well-respected and sought-after member of the construction team is to have the ability to properly prepare a detailed cost estimate for your portion of the project. Your company will develop a positive reputation by submitting estimated costs that are on target consistently.

Projects are all analyzed initially for their financial worthiness. Your early cost advice and value engineering can be critical to allowing the project to start on a secure financial basis. Many preliminary decisions can influence whether a project moves ahead or is stopped in its tracks. Being known as a company that can help substantially at an early stage will make you a sought-after partner on the team. You should actively seek involvement and interaction with all parties involved with the assembling of the cost estimate to develop a relationship, add valuable assistance, and have an influence on the final outcome. This can be done at various stages of the estimate process. We see projects moving ahead when the design documents are only 30 percent complete. You can't get in too early, given the current schedule constraints.

Whatever the stage of the estimate, 30, 70, or 95 percent, your estimating advice should be consistently available and updated as the project is moving toward completion. A complete budget cost estimate allows a company to manage its own costs. An estimate that predicts the cost far in excess of actual bids causes mismanagement of owner's funds and can also cause an owner to suspend or postpone indefinitely a project that is perceived as being too expensive. Utilizing professional cost estimating skills—whether it is in-house or by using a specialized consultant—can be a prudent step on behalf of the project team.

A concern of all companies is that the preconstruction services that you have provided will be given to other members of the design team and that you will have expended your time and resources with no outcome or preference for you or your company. This is always a potential problem. Diligent management of this activity, monitoring how you are being received and treated by your customer, is important. One firm prepares the estimates and drawings, and has photographs of the job site, but never lets any of them leave its hands. The firm shares the information at group meetings but refuses to turn it over to the project team until given a letter of intent or cash remuneration for the effort incurred. This is a personal item, based on the trust and confidence you have in the project team. If you don't believe you can have the trust and confidence of the team at this stage, you may want to take your ideas to another team where the scope is more in your favor. You are not making these preconstruction cost estimates to benefit your competition or a project team that will strive to take you for everything it can get. It is easy to say that you should attempt to position your company on the right team, but it is always challenging for you to use your intuition and business sense. Here are some things you can do:

- *Go prospecting.* Seek new opportunities to demonstrate your abilities with cost estimates. When you hear about a job that comes in considerably over budget or about a project that is not moving ahead because the costs are so excessive, it should be music to your ears. This is your chance to approach the firm and let it know how you would have done it. Make a pitch to work with the company to find a solution to its problem. Most of the time the frustration level is high and clients are open to any possibility that can make the project a reality.

- *Never be late.*   When a date has been established to present your budget proposal, always make sure you are on time. The second most damaging thing you can do with this item besides missing the estimate is not having it ready for review when promised. This sets the entire team back and provides a perception of a person who cannot get the job done on time, which is a difficult perception to erase.

- *Go the extra mile.*   Always strive to set yourself apart from the competition by providing your clients with more than they ask for in services and end product. Many companies regularly prepare beyond the basic budget bid; they present a schedule with critical dates marked, their proposal bound in a three-ring binder with their name and the project on the cover, and a list of options that may be available to save the client more time or money.

- *Be flexible and constantly change through improvement.*   A consistent effort to improve the progress you are making with your preconstruction services by adding quality enhancements, standardizing segments to speed up your delivery time, including new features to portray your image of professionalism, and being on the leading edge of technology and future trends will provide you with clients who consistently pursue you because you make them look good.

- *Develop partnerships.*   Outsource as much of this process as you can afford to. This allows you to take on more and to develop a relationship with another team of professionals who will help you achieve your goals while laying the groundwork for additional work for themselves. There are many excellent firms near your company who can help with the presentation package, proposal writing, photography (if necessary), schedule, or a piece of the estimate that is in their area of expertise.

- *Keep learning about the competition.*   Try to get a look at what the competition has done for your client in the past and what it is proposing. A follow-up interview asking how you did and what the customer would want done differently next time will give you feedback to improve and will allow the client to vent displeasure if something did not go right.

- *Maintain visibility in your specialty.*   Let everyone know of your sincere desire to help with the preconstruction stage of projects for clients and potential clients. This can be done with a letter, on-line, through articles written for professional business publications or associations, or by word of mouth referrals.

- *Join the information superhighway.*   As discussed in Chap. 3, there are many exceptional Web sites on which you can list your availability to help with preconstruction and perhaps even list what return on investment you have provided past clients with your efforts.

- *Be available.* We will discuss next how the very nature of our business mandates that we constantly be available to potential clients for preconstruction opportunities. When an estimate or help is needed, the need is usually immediate. Being easy to reach through e-mail, voice mail, or a pager is critical. You want to be the first person called because the customer knows it can always count on you for help.

Most individuals within the building industry look at budget cost estimates as the only tool of preconstruction/promotion, but in reality there are several others. We will now review some of the other actions you can take to position your company in the best light.

The 24-hour-a-day availability

Several companies are starting to realize the opportunities that technology is making that allow them to position their firms as dedicated to customer service. Nowhere is this more evident than in the concept of a 24-hour-a-day service line. It doesn't necessarily mean that you will have your people working 24 hours a day, but the perception is that you can receive and perhaps respond to requests within at least a 24-hour period. A combination of hot lines, pager systems, fax machines, and e-mail has opened a new field of service. There is now limitless access to information 24 hours a day from anywhere in the world. Thus you and your customers can work on a computer anywhere a communication line can be found.

The industry has changed radically regarding the time people work. Many individuals work 10 to 12 hours a day. Usually the only time they have for receiving input from outside of their own company is either very early in the morning or very late at night. Also, many within the industry maintain a home office with links back to the daytime office to continue working after they have had a chance to have dinner with their families and a little time with the kids afterward. They are working long and hard hours and expect others within their project team to be doing likewise. Make it easy for them to access you to request information on your product or an updated schedule for your portion of the project, which can be sent electronically. Cellular phones are still considered intrusive when calls are made late into the night, but asking for assistance with an e-mail request for information can be effective and efficient.

Less technically savvy but just as effective is the installation of an 800 "hotline" where your client can make requests and you can control the time, place, and delivery of your response. Providing your client with a pager is another way of providing the perception of control. Most companies shy away from this, thinking that the customers or even the industry will constantly page them for their services. Our own experiences with this on a national basis have been just the opposite. The industry generally respects your time and will get in touch with you only on worthy issues and at a time that is not obtrusive.

Becoming an industry expert

Turning cold calls into warm calls.   We all hate cold calls, but this one action can make the significant difference you are looking for in your business. We are nervous about making these calls because they can bring rejection, which is the last thing we desire during our frantic work life. The problem with ignoring this specific activity is the loss of potential new business opportunities and perhaps even new friends. Often on cold calls you will meet congenial, friendly people who are willing to learn more about your company. They are usually in much the same situation you are in. They need new work. They need to keep their projects within tight budgetary constraints, meeting an aggressive schedule, managing ongoing changes to the documents, and coordinating a mirage of relationships in the field, office, and with their clients.

However, the old cold calls we used to make, dropping in without an appointment or no knowledge of the companies you are calling upon, rarely work today. To turn a cold call into a warm one you need to work much harder deciding which companies really could benefit from your services or products. You must have a marketing strategy to provide an ongoing program of services to those prospects, to help them understand you, your company, the benefits you bring to the table, and how they can depend on you to perform for them. So, how do you go about this? We are fortunate today to have so many technological resources available to assist with the management of this task. The computer was made for this. Picking one of the resources and developing a regular follow-up procedure for the firms you have targeted to concentrate your efforts on developing as new accounts will improve your marketing program. The rule of thumb is to correspond with these individuals quarterly. It can be done initially by letter to tell them who you are, what you do, and that you will be striving to be a partner with them in the future, followed up with a call, a post card, a brochure, and inevitably a personal visit. (But by this time they know who you are and are a little intrigued with the prospect of meeting a person who seems to sincerely want their business.) After this initial meeting (at which you hope to obtain e-mail addresses and business cards) you should follow up with a letter or card of appreciation and respond to any requests that may have been made.

Desktop publishing software allows you to produce high-quality pieces of research to keep your perspective clients appraised of new trends and knowledge about your segment of the industry. This material can be sent by mail, fax, or e-mail in the form of a mini-newsletter. You can set yourself apart from the market by becoming a "problem solver" for others; include case histories of some of your success stories.

These warm calls will be opportunities for you to learn more about your potential clients. In these meetings you can discuss their needs and what problems they have that you may be able to satisfy. Engage them in "value conversations" that ask probing questions to uncover what keeps them awake at night. Expand this technique to others within the organization with whom you come into contact, be it in the field, in the accounting department, or at the front desk. Once a company knows you as a problem solver and not an order taker, you will experience a change in attitude and a new client.

Niche building.   Niche marketing can be your answer to carving a successful business out of the general building industry and can mean redirecting part or all of your business toward specific areas of expertise. It follows the logic that most clients would prefer to hire companies that specialize in their area of need. Clients today cannot afford to pay a firm to "go to school" on their project. They seek out companies that have exceptional knowledge of, or a proven track record in, the type of work they need. Niche marketing also allows you to focus your attention away from just trying to maintain volume and toward increasing profits. Most clients are willing to pay a slightly higher fee to a vendor that has related experience because they know it will save them the time and aggravation of dealing with work that is performed poorly.

How do you discover your niche? Henry Ford had the model T for a while. You can find your model T:

- Survey your clients and potential clients to find out what needs aren't being served adequately.

- Stay in tune by attending industry functions, reading industry publications and general business news, and following projected future trends.

- Be creative. Decide for yourself where your market's needs will be in the future and begin to develop an expertise in those areas.

Once you have established a niche, how do you take advantage of it? The best advice is held in one simple, memorable statement: "Become famous in your field."

The only way for you to become known in a niche market is to become recognized as the leader in your area of expertise. It's easy to do if you are patient and persistent. The first step in any marketing endeavor is to prepare a detailed marketing plan that outlines your goals and your objectives for accomplishing them, with specific timetables and tasks. Here are a few suggestions that are discussed further in this chapter:

- *Publish.*   Develop a half-dozen outlines for stories that would be of interest to prospective clients—stories that give practical advice and not ones that talk about how great your firm is—and begin submitting them to publications your clients read. Also, send press releases to those same publications about new projects, new hires, topped-out projects...whatever. The more people see your name in print, the busier and more successful they think you are.

- *Teach.*   There is no better way to establish credibility in your field than teaching—particularly when you can teach prospective clients. Look into opportunities to make presentations at seminars given by associations to which your clients belong. They are usually quite interested in finding new speakers.

■ *Be visible* at functions sponsored by organizations to which your clients belong.

■ *Promote.*   Highly targeted promotional efforts, like direct mail, advertising, and media publicity, are far more effective and cost efficient than promoting the broad, diversified services of a firm to a general audience.

To survive in the years ahead as the niches you develop become increasingly crowded, you will need to develop and refine two skills:

■ Effective use of a database to ensure that you are communicating to all possible prospects

■ Flexibility to ensure that you can always stay at the head of the curve

Remember that today's niches may become tomorrow's overcrowded markets. If you do a good job of becoming famous in your field, you should be able to stay at the top of your niche even during tough times. Nevertheless, it would be wise to employ the same creativity you use to discover today's niche to find other profitable niche markets to explore in the future.

Awards programs. These programs, which have proliferated around the country, are an ideal way for you to gain recognition and respect within the market. Regardless of what award you may earn, it is important to invite your clients to the recognition dinner as a further method for networking and identifying your efforts. The awards you receive should be displayed in your office to highlight your achievements. Prospective clients and existing clients alike will see your awards and recognize the respect your company has received from the industry. Don't stop here; this is an ideal time to publish press releases or articles describing the attributes that your firm utilized to achieve this award. Obtain reprints of anything published and distribute them to all existing and potential clients; presenting a framed copy of these awards to your client is always a worthy endeavor. No one will get rid of a framed award and it will immediately be added to your client's office walls. Include a picture of the individuals responsible on both sides of the table to illustrate the teamwork that exists between your organization and your clients (even a group shot at the awards dinner can be of value).

Seminars and workshops: Their value to you.   A sure way to educate and influence a group of people is to hold seminars or workshops to explain the benefits of your product or services. These sessions can be conducted at trade shows, in conjunction with professional associations, industry groups, universities, or at a potential customer's office during lunch. The "box-lunch" concept provides a tremendous opportunity for you to take your message in-house to your targeted audience. Over a 1-hour lunch, clients can conveniently listen to your presentation of what you can do to help them do their job better, faster, or easier. On top of this, continuing education credits for professionals allow you the luxury of assisting clients meet their educational goals for the year.

Technology has allowed your workshops to be Power-Point extravaganzas with links today to Web sites, job-site cameras for "real-time" viewing of project locations, or teleconferencing capabilities so that several members of your organization can participate from various national and international locations. Added to this is the chance to leave behind additional information about your company, your employees, your unique benefits, articles written about you, and testimonials.

These public speaking opportunities, whether for a seminar or in front of industry groups, can be highly beneficial. A notice, including copies of any media coverage, can be sent to your respective target audience alerting them to your upcoming speaking engagement. You can also extend personal invitations to those you want in that audience.

This same format will work within your own organization to position you as a leader and provide you with a forum for others to follow. Helping to mentor and lead training programs for new recruits can position you as a concerned, knowledgeable leader. This effort will also help develop a close relationship with your team, and make your reputation.

Volunteer to help direct the in-house planning programs. Use these programs to express your ideas and the positive direction in which you want to help move the company. Sally Handley, LZA Technology, states

> We believe an educated client is our best customer...so we offer seminars describing roofing, curtain wall and mechanical/electrical systems, their inherent problems, and our unique approach to solving those problems. We believe that this (promotional tool) distinguishes us in our very competitive market, simultaneously communicating the specialized services we provide.

Writing to success.   There are many publications such as daily newspapers, weekly business journals, monthly magazines, and association newsletters that are looking for knowledgeable building industry experts who can write and comment on the industry. You, better than anyone else, have a good understanding of your specific segment of the industry, what the latest trends and indicators are. There thus exists an opportunity to share that information with both print and electronic media. This can take the shape of press releases announcing changes or new work your company has acquired or alerting the industry about new market audits or surveys that you have commissioned.

Another option is to regularly write articles and editorials. This develops credibility for you and your company and helps build the industry. You will also gain recognition and respect from your peers and potential clients, placing you a step ahead of your competition. Today there is a substantial quantity of local and national printed media—association trade journals, business weekly journals, and daily newspapers—that are consistently looking for industry experts to share varied topics with the business community.

Strong communications help position your company in the marketplace and let the world know what niche you are best at, what recent successes your company has experienced, what your clients are saying about you, and how you were able to bring that last project in under schedule and within budget. Excellent communications skills can place you at the head of the industry overnight, because so few in the building industry have these skills or even pursue them.

Whenever something written by a member of your office appears in the media, distribute copies of it to all of your existing and potential clients. Share with them your recent column about saving money or enhancing productivity. This illustrates your ingenuity, innovation, and support of the entire industry. It can help enhance your reputation of leadership and authority, and shows that yours is the kind of firm people want to go to when there is a problem. As we all know, this is one business that has plenty of problems to be solved. Who better for them to turn to than the industry expert?

Demonstrations and field trips. Demonstrations and field trips within the industry have recently gained in popularity. They represent another way to bring your target audience to you. There are many opportunities for you to allow field trips to your job site such as through the help of your subcontractors and vendors, with student groups, or associations events. These events need to be organized, planned, and rehearsed. It takes considerable time to prepare an environment where all of the pieces seem to flow together. Don't hesitate to invite the local media; it will enhance the experience for all and may provide some excellent coverage.

Kling-Linquist, a national design firm, has large conference rooms it allows industry groups to utilize on a regular basis, bringing a broad mix of the market into its offices for exposure and networking. The firm's name is on all the flyers, the industry groups are always very appreciative for this courtesy, and the exposure for the company is extensive.

Trade shows, exhibits, and displays. Trade shows have sprung up everywhere. They need to be cautiously managed due to the extensive costs and staff requirements needed to make them fully successful. They can also be a tremendous source of visibility for your company and can provide you with several leads about new projects and contacts. Some firms have actually built their entire marketing effort around the aggressive working of trade shows where their client base is known to be in attendance.

Teleconferences and virtual reality. Teleconferences conducted from around the globe are changing the way we view the market for our companies and how we manage our efforts. Teleconferencing is allowing us to address multiple audiences in several cities, answer multiple questions, and do it all from our offices. At the same time, we can have our clients actually take a virtual walk through a proposed project, making changes and modifications as they go.

Using your Web page as the window to the world.   When creating your Web page, keep in mind that the decision maker dictates the mode and amount of promotion that is required. You need to

Identify the decision maker.

Decide what the decision maker needs to know to decide to use your product or services.

Analyze how the competition promotes to the decision maker.

Decide what the "real" market needs are.

People sell (promote) concepts, ideas, and materials—brochures and ads simply supplement or fortify a person's efforts.

You must understand the following:

- The market size is well worth your time and effort. You can increase your market share.
- What the competition consists of and how they have been successful (and unsuccessful).
- What the decision makers need to know about your services versus those of *your competitor* to make a decision.
- What tools and resources are needed to provide what the decision maker needs to know.
- How to implement a plan to be successful in promoting your products or services to this decision maker.

The following is a list of targets to keep in mind:

Identify and quantify the market(s)

Identity the needs of the market

Quantify and qualify the competition

Identity the decision maker

Specify and implement the marketing program

Quantify and qualify results of the marketing program

Establishing Networking Goals

Networking is a numbers game. The more people you meet the more contacts, ideas, and opportunities come your way. Sounds easy, but you are probably thinking that maybe that works for insurance or car sales, but it won't work in construction or design.

The building industry, however, is one of the businesses where networking doesn't just help—it can make you a shining star, but not if you just go to industry functions and see people in the plan rooms. What most of us practice is reactive networking, with minimal prethought or planning. So what's the secret? We will review the eight key ingredients for making networking successfully work for you and gaining the results you want in life.

Proactive management through prospecting

Make a list of the 12 to 15 important customers who could be considerably important to you if they gave you more business. Who are these decision makers: financial organizations, boards of directors, presidents of companies, highway commission, city council, owners, contractors, property owners, architects, engineers, county board? Research these companies or organizations thoroughly, through Dun & Bradstreet reports, company reports and newsletters, interviews with other companies that interact with them, and even Internet archives. Find out what they like to do, who makes their decisions, and what their long-range goals appear to be. Are they active with certain professional societies or civic organizations, or do they regularly attend industry functions? Target the individuals who are the decision makers for your product or service and ask questions about what they like and don't like and about their personal backgrounds. Now you are ready to start networking, armed with statistics and information that will make the experience much more positive and rewarding for both of you. But don't expect instant results, because networking is an ongoing labor of love. Generally it is believed that it can take five or six meetings before a person starts to believe in and trust you.

Lead gathering: Prospect and qualify. You need to come in contact with many individuals before you meet the actual client who will provide you with new business or leads. This is the number's game segment of the networking process. As you feed contacts in, you need to judiciously preview each to see how you can help them and in turn how they can add to your networking arsenal. Not all contacts will provide you with orders for new business, but each individual that you can come into contact with can become part of your sphere of influence within your network. A contact may know someone who can open that closed door, make an introduction, or provide a healthy lead. Our own rule is that it generally takes 25 contacts to result in five qualified leads, which eventually will develop into two orders for your company. This is a considerable amount of work for two orders, but the benefits of volume contacts through networking can accelerate this process for you and enhance your bottom line.

Working the market. Attending industry functions and belonging to professional societies is good, but without action on your part, they will not materialize into meaningful results. Before you attend any function, you should consider what you want to accomplish, who you would like to meet, and what result you hope to glean from this initial contact. Briefly considering the questions you will ask to start the dialogue makes it easier to focus on the individual's responses without worrying about what to say or do. You may even want to use this function as an opportunity to announce some new achievement or direction your company is taking. This can be done subtly, with enthusiasm each time someone says the magic words How is it going?

Professional societies and local industry groups can provide a wealth of beneficial experiences for you to tap into. You need, however, to spend the time to become involved with committees and leadership roles if you plan to maximize the contacts you will make. What better way to show what you know, or how you can help, than by working with a person whose company you are cultivating, to develop trust and a relationship. Don't use this as a forum to toot your own horn but as a real chance to learn more about the individuals within your targeted companies, how they approach business, and what you can do to solve the problems they are facing.

Be known as an individual who gives more than you take. The old saying "always do more than you are asked to " is very true with this critical piece of networking. We all have many chances to make a difference. We can do the minimum necessary or we can go out of our way to do our best and make the other person walk away from the initial contact saying, "Wow! What a team player" or "They seem sincerely concerned about me and my problems." Sales motivational professional Zig Ziglar said it best when he said, "You get whatever you want if you help enough people get whatever they want."[2] Not a bad motto to live by.

Review your networking goals: Make at least two calls a day, which equals 500 per year. Networking is an ongoing life-style endeavor. We need to be constantly examining ways to expand our networks both personally and professionally. The first basic step is to stay in touch with your past networking contacts. Phone, notes, articles, or even a newsletter can accomplish this. Regularly updating your Rolodex and computer database with promotions, moves, and family changes is vital. Next you need to be looking for new avenues of networking opportunities. Try to go to ground breakings, prejob meetings, press gatherings, and industry socials, and keep those cold calls rolling. Try to maintain at least two new calls a day to people not included in your database. This small step alone will add 500 new contacts to your Rolodex each year, and maybe some dollars to your bottom line.

Last but not least is the personal need to expand your own network for career stability and exposure. Eighty percent of new employment opportunities come from unlisted sources. What better way to keep yourself plugged into the market and your own professionalism than by maintaining a network that can help you enhance your own probability of improving your career or providing a base to ensure some additional measure of security for you and your family? In the building industry, networking is not a fly-by-night option; it is a *must* if you intend to have any control over your future. Don't allow someone else to decide it for you.

Building your reputation as a leader

Add value to bring value in. We live in a value-conscious society today. Quality and value are not an option; they are a given. Each of your networking opportunities needs to focus on bringing value and real worth to the relationship with a long-term perspective. Not only will this enhance the quality of the relationships you are growing, but the net results will ultimately pay far greater dividends to you and your company in the future.

Make customers and clients your friends. Networking is an exceptional way to expand your circle of friends, support, resources, contacts, and grapevine. The friend aspect of networking is by far the most rewarding and fun. Each new contact has the potential of developing into a new friendship that could last a lifetime. We all need friends, people who will share our ideas, thoughts, and problems; people who will help us sort through the mass of information that is daily forced upon us from the print media, radio, TV, and the computer. Our workdays are getting longer, and the daily stress levels are high; what better way to spend those days than working with people who are our friends? And nine times out of ten, these people will be the ones your next contract comes from, without a hassle regarding the contract language, terms, schedule, or price. A few of these, dispersed with the normal orders from companies and individuals who do not know you or your company, will make your career worthwhile and profitable.

Positive attitudes

Many people, from Zig Ziglar to Dale Carnegie, have written about the merits of, and significant benefits possible with, a positive mental attitude. Why then do so many people refuse to believe the physiological advantage you can achieve for yourself, your company, and the tasks you plan to achieve? A positive mental attitude (PMA) is simply allowing your subconscious to consider what you want the final outcome of your work efforts to be. By concentrating on a successful outcome, addressing the problems, and working out potential solutions and scenarios to overcome these difficulties, you will eventually achieve success. It is said that Thomas Edison suffered over 1000 unsuccessful attempts to develop the light bulb before he was finally successful. Maintaining a positive attitude throughout these failures is the inner strength needed to overcome the exterior environmental conditions and individuals you will encounter along this journey to success.

Follow Up! Follow Up! Follow Up!

Meeting hundreds of new people through networking is great, but unless you follow up, usually within 10 days after the initial meeting, it is relatively meaningless. A personal note, a company brochure, an article regarding something that was discussed, a brief phone call are all good ways to ensure the networking effort has a chance to solidify. As we discussed earlier, it normally takes 5 or 6 meetings to develop trust and the beginnings of a relationship. A follow-up can accelerate this process if it is done with speed, professionalism, and political correctness. Strive to obtain a business card from all those you come in contact with. This will give you the correct spelling of their names and their addresses; on the back you can make a note regarding the contact and what you need to remember from this encounter. You can feed these cards into a scanner, like the Visioneer PaperPort, that will load the information automatically into a predesigned database for organization and filing. As you grow your network, you can use the database to segment the various pieces of information into leads, information gathering, peer network, associations, companies, priority of future work opportunities, and market niches.

Jack Miller, marketing guru and successful business owner, consistently preaches the merits of getting yourself in front of your clients and potential clients at least quarterly. This can be done with a phone call, newsletter (like the enormously popular *Words from Woody*; call Dave Woods for a copy at 1-800-HEY-WOODY), post cards, cold calls, or notes. These regular communication contacts may trigger an opportunity for you to get in on the ground floor with a new project or hear about some other beneficial tidbit that could grow your business and bottom line.

Monthly Reports

Monthly reports are often overlooked. They are necessary so that those you report to, or your peers, will understand the magnitude of the effort you are making in this area. The monthly report allows you to share with management or your team the promotional and networking activities you have conducted on a monthly basis, summarized annually. First, this is a good tool for review and discussion by those you respect. They should offer additional ideas and suggestions and point out areas that you may have overlooked that need to be addressed. Second, this provides a good gauge of the success you are having in different portions of your promotional activities. Third, this can be the basis of a compensation increase that may be due for your achievements, additional tools needed, and starting point for discussion. No one knows what you have done over the last month better than you do. A monthly report will make this process manageable and should draw on problem areas during the year rather than only once a year. Some companies have broken the monthly reports down to four key topics: accomplishments, problems, opportunities, and outlook (APOOs).

Networking is a considerable amount of work. Do you really need to take on more work with the limited time you currently have? Networking will only be important to you if you want to enhance your personal career, find solutions to the multitude of daily problems we all face, and build new relationships and friendships within our industry and your own community. Yes, networking is worth it.

Notes

1. *The Tormont Webster's Illustrated Encyclopedic Dictionary,* 1st ed., 1990.
2. Zig Ziglar, *Reaching the Top* (Galahad Books, New York, 1997), p. 18.

7

Marketing and Sales Technology

Introduction

If you look back at the construction industry over the past 25 years, you will observe that not much has changed in the way a building is built. Architects' vision comes from their minds. Drawings are still developed as blueprints. Engineers still lay out a job site with a transom. Earth is removed with backhoes. Structures are erected with cranes. And craftspeople still painstakingly ply their trades.

A revolution has taken place, however—in the office. And the revolutionaries are the marketers.

The world of the marketer in the construction industry has changed dramatically. Contact names are moving from Rolodex files to electronic databases. Research is seldom conducted at the library. Marketing materials are often not created on paper. Photographs are rarely stored as prints. Communication is no longer solely the domain of the telephone. And presentations are no longer made using charts and overheads.

Technology has touched almost every aspect of the marketer's job, except the most important part—the ability to understand a prospect's needs and explain how the marketer's company is best qualified to satisfy the prospect. Otherwise, it's a whole new ballgame.

Five areas of technology—the Internet, contact management software, digital cameras, communication devices, and presentation software—are responsible for most of the changes that have affected marketing professionals. And with few exceptions, the change has been welcome.

Now, more than ever, marketers can learn a tremendous amount about industry trends, prospects, and competitors; organize vast amounts of data on prospects and clients; chronicle all kinds of job site activity; and give killer presentations, using affordable, easy-to-use hardware and software.

This chapter is designed to help you integrate these technologies into your marketing program. It may teach you things about them that you did not know, things that will improve their efficiencies in ways that will help you right away.

Because technology is changing so fast, this chapter won't critique specific brands, with the exception of products that dominate the market. Instead, it will tell you how the hardware and software that you may use works, and will guide you to where you can learn more about them. When you are ready to institute these technologies, you should do research first to find the software and hardware that best fits your needs.

Internet

The Internet has revolutionized the way marketing professionals collect information and promote their services.

E-mail has enabled contractors to communicate more efficiently with prospects, clients, fellow employees, and project team members. The Internet has become a vast network of data about markets, competitors, and prospects, giving contractors an edge over competitors who aren't connected. Moreover, professional Web sites let contractors present information about their company to prospects and potential employees alike.

Using the Internet for research is covered in great detail in Chap. 4, and using the Web to market your services is the subject of Chap. 11.

Digital Cameras

Digital cameras are gaining popularity quickly in the construction industry. Like other advancements in technology, the quality of digital photographs has improved geometrically to the point where now they are nearly comparable to the quality of photographs shot on film.

How digital cameras work

Digital cameras work by capturing still images using electronic image sensors instead of film. Light reflected from the subject passes through a lens and strikes an image sensor. These sensors have two sections; a photo-sensing region and a transfer region. The image sensor and associated circuitry convert light rays into digital electronic signals that form an image of the subject. Digital images are typically stored within the camera in a removable memory device such as a CompactFlash card.

A big benefit of the digital camera is the liquid-crystal display (LCD), which lets the photographer review pictures during the shoot and directly after they've been taken, prior to storage. This lets users see their pictures instantly instead of waiting days, weeks, or even months for a roll of film to be finished and then developed.

After a picture is taken, a digital camera user typically connects the camera to a PC. Special drivers need to be installed on the PC first to allow it to read the stored images from the digital camera or the removable storage device.

Improvements in resolution

Until recently, digital cameras were mostly used to present photographs electronically, either by e-mail, on Web sites, or in PowerPoint presentations. Earlier cameras—known as VGA cameras—produced 640- by 480-pixel images, which worked out to a little over 300,000 pixels per image.

The next-generation cameras were XGA. With a resolution of 1024 × 768 (nearly 800,000 pixels), their quality was better but they still didn't compare with prints from film.

In 1998, the first megapixel cameras were introduced. These cameras capture images that contain over 1 million pixels. Many of the more popular models have pixel arrays of between 1.2 million and 1.5 million pixels. Used with a printer such as the HP PhotoSmart Printer, a megapixel camera produces snapshot-size prints that are nearly as good as photo lab prints. They can be used for some business applications like newsletters and data sheets printed on color laser printers rather than traditional presses (for more information about reproduction processes, see Chap. 9). Additionally, megapixel cameras produce exceptionally sharp photos when viewed on the Web, e-mail, or electronic presentations.

The real breakthrough came in 1999 with the introduction of digital cameras with 2 million-pixel arrays, making them nearly indistinguishable from film.

Although film is roughly equivalent to a 20 million-pixel sensor, after the lens error, it becomes roughly equivalent to an 8 million-pixel sensor, and after focus error, it becomes equivalent to a 2.1 million-pixel sensor, according to Gene Wang, chairman and CEO of Photo Access, in the September 1998 issue of *Web Techniques* magazine. A picture taken with a 2.1 million-pixel sensor is very difficult to distinguish from film, assuming that a high-quality printer is used for output.

Digital images have fewer margins for error than film, however, according to Wang. A digital image sensor will lose the highlights with only one stop of overexposure. Shadow area will be lost very quickly with underexposure. Unlike film, there is no way to correct these in processing. A digital camera, however, has the ability to display the captured image on its LCD screen, thus giving immediate feedback with respect to proper exposure. If the image is not exposed properly, the camera operator can make appropriate adjustments and take another picture.

Managing the problems with digital cameras

The downside of the higher resolution is image storage. Most models have onboard removable storage. There are a number of storage options being incorporated into camera designs, including SmartMedia, CompactFlash memory cards, and Miniature Card memory. Most of the memory cards that ship with the cameras have a very low capacity, between 2 and 8 MB, which is inadequate for most users. At the highest resolution, without compression, 4-MB cards can only hold two or three megapixel digital images.

Other internal image storage solutions have been incorporated into digital camera design, however, according to Ron Eggers, a senior editor with

NewsWatch Feature Service, in the June 1998 issue of *PC Graphics & Video* magazine. Products like Iomega's Clik! Drive will increase storage capacities without the need for expensive CompactFlash, SmartMedia, PCMCIA, or similar storage devices. A Clik! disk can store up to 40 MB of data. Look for 1 GB of removable memory within 5 years.

A problem with megapixel cameras is power management. These cameras run through batteries quickly. They require virtually 100 percent of peak battery power to operate effectively, so even a small drain on the battery can make the camera inoperable. Alkaline batteries have no real life expectancy in digital cameras. For serious shooting, rechargeable nickel metal hydride (NiMH) batteries are the best option, according to Eggers.

Unlike film cameras, which are stand-alone devices, digital cameras generally need to be connected to a PC to upload images for processing.

The recent arrival of Windows CE changed all of this. Windows CE is a modular operating system for embedded systems such as palm-size PCs in automobiles. Windows CE makes the camera a stand-alone device. You will soon see digital cameras with the familiar Windows interface on the LCD display.

Digital cameras on the job site

Digital cameras have numerous applications for the construction marketer. Although they remain an expensive up-front investment (the best-quality megapixel cameras still cost close to $1000 with enhanced memory), they can save considerably on the cost of photo development and image scanning.

One fascinating option found on some megapixel cameras is a swivel lens that can be rotated a full 280°, allowing you to take panoramic photos of job sites. No longer do you have to take five pictures at different angles and tape them together.

Job-Site Video Cameras

Job-site video cameras are also being used to monitor job progress. The cameras are linked to a dedicated server run by the company that is using the cameras, and images from them can be viewed on the company's Web site. Some of the higher-end cameras are mounted on a swivel. Several people can simultaneously manipulate the image viewed on a camera by moving their mouse from side to side or up and down. These job-site video cameras serve as interesting marketing tools and provide the contractor or owner remote access to important job-site functions like material delivery and numbers of people working on specific trades.

Presentations

If you haven't already thrown away (or stored) your overhead and slide projectors and your old easel, maybe now is the time. Your ability to level the playing field with larger competitors has been greatly enhanced by the presentation tools available to contractors today.

It wasn't too long ago that your informal presentations were done on poster boards and your fancy ones were done using computer-generated 35-mm slides. But now, you can powerfully illustrate and professionally deliver your ideas for an informal meeting, a presentation to an audience, or delivering your message over the Internet, using off-the-shelf desktop software.

PowerPoint

The clear leader of the pack in software, of course, is Microsoft PowerPoint. With its substantial R&D support and compatibility with so many complimentary software programs, a significant investment in any other presentation software program would be ill advised.

PowerPoint is a *presentation graphics program,* software with which you can create a slide show. PowerPoint helps you generate and organize ideas, and then provides tools you can use to create charts, graphs, bulleted lists, text, and multimedia video, animation, and sound clips. It also assists you in creating slide show supplements, such as handouts, speaker's notes, and transparencies.

PowerPoint provides you with over 30 templates for your content, or a step-by-step wizard to generate your own format. You can create, with desktop graphics software, a custom background incorporating your company's logo and other visual elements you use in marketing materials. Also, PowerPoint has a Web site (www.microsoft.com/powerpoint/) called PowerPoint Central where you can learn more about the software and download clip art, textures, sounds, and animations.

You can tailor one presentation to multiple audiences with Custom Shows—it creates different versions of your presentation in one master file, so you show only the slides you need for each audience. With Pack And Go, which assembles and compresses all the files you need onto floppy disks, you can easily travel with your presentations.

The PowerPoint Animation Player and Publisher, a free Internet browser extension, allows you to publish and view PowerPoint animations and presentations in your Web pages. You can take advantage of the enhanced animation, hyperlinks, special effects, and built-in sound functions in PowerPoint to build animated Web pages that stand out from static HTML Web pages. So when you want to make a professional presentation to a prospect from a remote location, the prospect can log onto your Web site and view a complete PowerPoint show.

PowerPoint Viewer enables PowerPoint users to share their presentations with people who do not have PowerPoint installed on their computers.

Two resources for learning about PowerPoint and using it are the PowerPoint area of Microsoft's Web site and a book called *Microsoft PowerPoint at a Glance,* published by Microsoft Press and sold in the business book section of most bookstores.

With practice you should be able to quickly put together professional presentations. And if you want to accelerate the process, most major colleges and technical schools now offer presentation design classes, as do private organizations. You can usually find them listed on the Internet under presentation training.

Projectors

Most marketers today prepare their presentations on their desktop PC and save it to their laptop or notebook computer to transport to presentations. Notebook computers give mobile salespeople all the comforts of a desktop computer: processing power, high-resolution graphics displays, and multimedia features. These systems are a great way to take product demonstrations or colorful presentations into customers' offices, provided you have some means of displaying the computer's images. Unless all your clients have equipped their conference rooms with projectors, you could be placed in the position of asking a group of potential customers to gather around the small screen of a portable PC.

That's why a portable LCD projector is a necessary companion for a notebook computer that is regularly used for making presentations on the road. Although the quality of a projected image cannot quite match the best fixed-projection systems, portable projectors usually are able to display clear screen images in moderately large rooms—even with some lights on to allow conversation and eye contact.

Projectors are easy to use. Better ones have one-cable connection to the computer from which you are running it. Software to run the projector must be installed in your computer. Once it is turned on and connected, it's simply a matter of making the presentation.

There are several factors to consider when purchasing a projector:

1. *Brightness.*   The lamp brightness is measured in the ANSI standard unit lumens. Projectors of 140 to 250 lumens with halogen lamps handle a 10-foot screen in a very moderately lit room (enough light to easily read a newspaper), 300 lumens of halogen or 250 of metal halide lamp projectors can handle 6-foot screens in bright rooms or possibly be acceptable on up to 20-foot diagonal screens in darkened rooms, and 350 to 500 lumens of metal halide driven projectors are comfortable on 25- (350 lumens) to 30-foot screens in darkened rooms. They can handle virtually any lighting up to 10-foot screens.

2. *Image clarity.*   Most projectors provide 800×600 SVGA capability, consistent with most laptop computers, and some provide true 1024×768 resolution. Older—and typically less expensive—PC display systems that only support VGA resolution impose a strict functional limitation because they blank out when users attempt to display images configured for greater than 640×480.

3. *Portability.*   If you go to a client's office with your notebook and a briefcase, you want your projector to be as lightweight as possible. Most of the best projectors are under 20 pounds, and some are small enough to fit *in* your briefcase. If you travel, you'll either need a super-lightweight one (under 10 lb) or

a rolling case (wheels and extension handle) that is small enough to be carry-on luggage.

Communications

Moving computers from desktops into the field has been the ultimate dream. This dream is being realized on many fronts. We all know you make or break a job in the field, not in the office.

Communication between office and field will never be the same again. You may be thinking that this has no direct application to marketing or sales. But the truth is, it has *everything* to do with marketing and sales. A satisfied client giving you repeat business is your ultimate goal. Who is better to work with than a client who is thoroughly impressed with your capability, happy with the successful project you have completed, and willing to tell others of your ability and technical expertise?

Virtual reality models and CD-ROMs

Virtual reality models and CD-ROMs are changing the way we market and sell our ideas. Today it is possible to provide clients with a virtual walkthrough of your ideas for designing and building their next project. Sitting down in your client's office and providing a step-by-step discussion of its concept (especially for the design-build industry) can give you a dynamic advantage in today's competitive market where, traditionally, price alone has dictated the final outcome. A virtual walkthrough can provide an interactive, real-time component that lets you make changes to the virtual building as the client expresses an opinion on what is shown. After the journey is complete, you have a vision based on client input, client need, and client desire.

The new Pentium and Pentium II computer processors have accelerated the affordability of this new technology that, just a few years ago, cost hundreds of thousands of dollars to possess. Now it can be obtained for a few thousand dollars and can be used on a notebook computer. John Barden, deputy program manager with Parsons Brinckerhoff's 4D Group, created a CD-ROM, which cost $20,000, for a public presentation of a key section of Salt Lake City's $312 million light rail project. The presentation, which calmed public concerns, expressed in laypersons' terms the complex process of choosing a construction management and engineering firm.[1]

CD-ROM brochures

Digital cameras, software like Macromedia Director, and the proliferation of CD-ROM recording hardware are making the development of CD-ROM brochures for marketing a reality. These marketing tools make a sensational presentation that can produce a 3D virtual reality view of your people, your projects, and testimonies from content clients. These CDs can be mass-produced and distributed to potential clients. The ability to instantly hook up to your Web site can be incorporated into the disk. These disks also make a great leave-behind after a marketer's presentation, and are good for running at a trade show kiosk or as a self-extracting file that is e-mailed to potential or current clients.

Voice recognition technology

Voice recognition technology offers considerable opportunities for the industry. Superintendents have traditionally fought the advent of computers and their accessories on their turf. A large part of this insecurity has been the basic need to type; most field individuals resent the tedious writing of reports and memos and resent typing on a computer even more. The development of voice recognition software that can take the spoken word of individuals and translate them into actual computer script is changing this previous roadblock. In its current stage, superintendents can request daily log forms verbally and then, as they dictate the information into the computer, it is put into the necessary compartments of the forms. This information is then transmitted back to the main server where it can be reviewed, categorized, responded to, and filed at the company office.

Other aspects that will enhance job-site productivity are the many daily applications of documenting the multicommunication tasks that are necessary for a successful project. Project managers can immediately apply this technology to institutionalize project meetings by recording minutes, noting attendance, and listing the actions that were decided to be taken, and distributing copies appropriately over the project intranet in a matter of minutes. They can then use the time saved to solve problems, manage by walking around (MBWA), and assist members of the project team to meet the goals and objectives that have been set.

Communications on telephones and cellular phones could be linked and all discussions recorded. At the end of each day the managers and superintendents would have complete documentation of discussions with the client, office, vendors, and support personnel. Each party would immediately receive a copy and have an opportunity to clarify segments that were unclear. This would be a loose verbal record of who does what, when, where, and how. This information could then be transmitted to the next day's schedule in a PIM, which would set off alarms to indicate actions to be taken or follow up required by others.

Some of these software products that do the above are:

- *ViaVoice Gold.* IBM's entry-level, large vocabulary, continuous speech, high-accuracy dictation software for Windows 95. This software has the ability to enhance your productivity in ways you haven't even thought of. It includes a lightweight microphone that folds flat for carrying and is built for right- or left-hand users. This software will turn what you say to your computer into text with over 90 percent accuracy. (http://www.ibm.com/viavoice)

- *Dragon System's Naturally Speaking Preferred.* A consumer-level basic system that has won the hearts of many who have tried this early entry into the voice recognition software. It usually takes 6 hours to program your computer to your words, but once complete you're on your way to dictating most of your communication tasks. This system is considered easy to use and does not require MMX for maximum performance. (http://www.dragonsys.com)

Personal digital assistants (PDAs) and personal communication systems (PCSs)

It is hard to believe we are on our second generation of PDAs and PCSs because these tools have only been out since 1990. They allow you to have a computer small enough to carry with you, relay information, and document your activities during the day, and are comfortable to use. Some of the tools we'll discuss can do all of that and check your e-mail, take dictation and transcribe letters, send pager information to peers and management, and keep you on schedule and on top of all the people, events, and critical path items you are faced with. The distinct advantage of the PDA is that it uses wireless transmission to send and receive enormous amounts of project data to and from the job site. It can monitor material deliveries, check daily job scheduling and access reference material immediately, and it can be carried in a shirt pocket or on a tool belt.

Being more productive with the limited resources available can be accomplished easily and efficiently with a PDA. Safety reports, OSHA forms, and MSDS documents can be stored on them, and safety talks and videos can be viewed. Add to this the ability to maintain accurate records of inventory at all times and from all locations, organized by field, office, year, and supplier. PDAs can assist in manufacturing with just-in-time (JIT) deliveries, which can improve cash flow, minimize limited job site storage conditions, and allow the direct placement of material in the area needed, with handling done only once.

One of the most advantageous benefits of a PDA is the chance it gives you to communicate regularly with clients, the office, vendors, and employees. Integrate it with voice recognition technology and you have a tool that can meet a variety of activities and that weighs less than a pound.

This same device can be your strategic planning tool to schedule appointments, maintain job leads and contacts, and document and track opportunities, allowing you to bring considerable organization to your frantic life. Personal and professional areas of your life can be controlled and monitored with it. When you purchase a PDA, choose by the applications you want to use daily, not the memory or hardware specifications.

PCSs have traditionally used cellular and digital wireless telephone technology and are quickly blending into PDAs. Growth in this market is being fueled by the rapid expansion of the telecommunications services that continue to push more services toward the PDAs. Two-way paging and the ability to send short alphanumeric messages are readily affordable. These devices are incorporating e-mail capability, voice mail, mailboxes, and fax capabilities. Add to this the shrinking effect that is occurring as they become almost as small as the Dick Tracy futuristic watch. Wireless technology is replacing all current traditional wire networks.

Wireless technology is changing mobile computing also. Ricochet has introduced technology that allows computer users to access anyone in the world from anywhere in the world without attaching to phone lines or other cables. New machines continue to get more powerful, thinner, and lighter, while incorporating 3-D video, real-time conferencing, and DVD video drives that allow full motion video for unbelievable presentations.

With corporations rapidly moving off the local area networks (LANs) and wide area networks (WANs) and onto the Internet, we will see LANs use the same protocols as the Internet (TCP/IP), turning companies into intranets. This will greatly assist the mobile worker today and tomorrow.

Examples of the above products are:

PalmPilot III.   3Com's popular electronic organizer that has become the number 1 PDA on the market today. It features seven built-in applications: an address book, an agenda, expense tracker, a spreadsheet program, a sketch pad to scribble your ideas, an expense report, and a LCD screen display that can be vertical or horizontal. There are also scores of third-party products that support this product. It fits neatly in a pocket and weighs only 5.7 ounces. It has a flip-up lid to protect its LCD screen and a proprietary infrared port that lets Palm III owners beam contact information and other data to one another or directly into desktop systems. It runs for 8 to 12 weeks on two AAA batteries. The only disadvantage is that it does not have a voice recorder. (http://www.ti.com)

El Nino.   Philips' new entry into the market comes with memory capability of 24 MB, a beefed-up operating system (Windows CE 2.0) and software, a fast modem (28.8 kbps), and a large, sharp screen. It has handwriting-recognition features for jotting down notes, and it has voice recognition capabilities to let you call up names. (http://www.mobile.philips.com)

Casio Cassiopeia E-10.   Also based on the Windows CE 2.0 software. It is very similar to the PalmPilot and the Velo 500 and has similar appointment-, task-, memo-, and contact-keeping capabilities and an infra-red port. The display is larger and it has a flexible data entry system that recognizes your words, so you don't have to write everything down. (http://www.casio.com)

Rex-3 PC Companion.   This PDA is not much larger than a 1.4-ounce credit card, It fits into your computer's PC-card slot and stores hundreds of appointments, phone numbers, task lists, and information. It will run for 6 months on a pair of watch batteries. The screen is barely 2 inches wide, but it displays about 30 characters across nine lines and is quite readable. It works with all the current PIMs from ACT 4.0, Gold Mine 4.0, Sidekick, Lotus Organizer, and Microsoft Schedule+. You can easily carry it in your pocket. The one drawback is that the only way to input information is to download it from your PC. (http://www.franklin.com)

Other hand-held PCs on the market are IBM WorkPad, HP 620LX, NEC Mobile Pro, and the Sharp Mobilon HC-4500, to name a few.

Beepwear. Made by Motorola and Timex. Telling time is just the start of this amazing device that receives text and numeric pages up to 105 characters long. It also can receive entertainment, sports, and business news through its affiliation with SkyTel, the nationwide paging service. (http://www.beepwear.com)

Seiko MessageWatch. Considered the smallest pager in the world. This special watch can receive stock-market closings, sports scores, and other snips of business information. It also keeps perfect time because it automatically updates itself to atomic time. (www.messagewatch.com)

IBM Cordless Modem. A 900-MHz device that works with any analog phone outlet, and you can get online as far as 200 feet away from the receiver. It can transfer data up to 56 kbpx and runs for $2^1/_2$ hours without a charge. (http://ibm.com/pc/us/accesories)

Project information managers

PIMs are catching on with owners who want to have the ability to manage their jobs while utilizing "real-time conferencing" to allow several individuals involved with the project to access the same data and add input that is instantly available to the entire team. The Maryland Suburban Sanitary Commission Corporation has instituted this procedure on its new projects, integrating digital cameras, schedules, Auto-CAD documents, and RFIs to be maintained online. This technology is removing the traditional delineation between design, construction, and the owner. The designer, constructor, supplier, and owner can simultaneously review documents, solve problems, and work out the details being sketched out on the screen. This brings to mind the picture of where we are headed: to a "virtual" job site being managed and built to a large extent by companies and individuals that are off site and around the country.

PIMs are capable of incorporating all segments of the job site: inventory, daily logs, construction documents, planned versus as-built schedules, daily weather conditions, contractor-architect-engineer-supplier observations and information, current safety and building codes that are applicable to the job, daily-weekly status reports, job progress photos, shop drawing logs, complete specifications including substitutions, RFIs, payment status, change orders and extras, and communication documentation records. Firms like Timberline Software Corp., Beaverton, Oregon, have developed new tools to integrate all job costs through Web sites. This interconnectivity can improve the overall productivities for many facets of the jobs.

Leading PIMs are:

GoldMine 4.0. Maximizes the use of the Internet so users can obtain information from Web sites and update databases. Has a sophisticated e-mail manager and sales-opportunity-management system. Top-notch scheduling and contact software. Great for sales teams but complicated software to learn. (http://goldminesw.com)

Symantec Act! 4.0. This is considered the top choice to organize your contacts, manage your time, and increase your bottom line. It supports multiple Internet e-mail accounts and integrates directly with Eudora Pro. It has its own set of Internet links that have all sorts of useful information. It synchronizes well with the PalmPilot and all other key hand-held PCs. It allows easy documentation of ongoing sales and marketing activity and can utilize word processors such as WordPerfect and Word. (http://www.imation.com)

Ascend 98 for Windows 95. The merger of Franklin Quest with the Covey Institute has reenergized this organizational and motivational PIM. It is very strong in task management and is a very personal form of PIM for individuals who want to practice its management philosophy of balanced work and personal lives. (http://www.franklinquest.com)

Microsoft Outlook 97. Works well with e-mail, but skimps on scheduling and contact features. It assists with organizing information created within Microsoft Office applications and integrates them into a scheduler and task manager mix. (http://www.microsoft.com)

Other PIMs on the market include

- AnyTime Deluxe (http://www.individualsoftware.com)
- Claris Organizer 2.0 (http://claris.com)
- Day-Timer 2.1 Deluxe (http://daytimer.com)
- Ecco Pro 4.0 (http://netmanage.com)
- InfoSelect 3.0 (http://www.miclog.com)
- Janna Contact Professional (http://janna.com)
- Maximizer 3.0 (http://ww.maximizer.com)
- Organizer 98 (http://lotus.com)
- Starfish Internet Sidekick (http://starfishsoftware.com)

Other powerful marketing software tools

There are also several software products that can change and simplify many of the most difficult tasks you have to do:

Crystal Ball. A powerful decision-making software tool that can help prioritize your outstanding leads or decide which goals and objectives you need to concentrate on for real success. You just create a model of your situation by using one of the various sample icons that represent variables to your decisions. Then you assign a numeric value to each. You will have the instant statistical analysis necessary to forge ahead. (http://www.definitivesoftware.com)

Videoconferencing. For less than $200 you can attach this device to your notebook and communicate (with someone with a similar set up) or capture digital photos with the new Kodak DVC323. The new universal serial bus (USB) ports on recent computers makes this easy to install. It has good quality and low cost—not a bad combination. [http://kodak.com/go/dvc323 and http://intel.com/comm-net/proshare (desktop videoconferencing software)]

Avantos Manager Pro 3.1. Allows you to plan and track your goals and the progress you and your team have made. People management is this program's greatest skill; you delegate key business objectives, strategies, and tactics and then monitor their status. Easily allows you to break goals into manageable parts. (http://avantos.com)

Microsoft Project 98. The number 1 project management software. It is easy to use and new users can pick it up quickly. There are over 100 third-party suppliers of add-on related products. Great for planning your strategies this year and assigning a time schedule and budget. (http://microsoft.com)

These new technologies are creating a revolution in how we get new business, view the construction process, and manage our ever-limited resources. The tools we have discussed are not limited to big company use; they will also assist small companies and have a greater return on investment for them due to the very nature of their impact: They aid in flexibility and produce more efficient staffs. However, technology is not enough. We have strived to illustrate in this book that it takes vision, the ability to change and be flexible, and an awareness of the critical importance of relationships to be truly successful.

Note

1. "The Next Dimension—Virtual Reality Is Transforming CAD Models into Moving Worlds," *ENR,* June 17, 1996, pp. 24–26.

8

Marketing Communication Plan

Introduction

The terms *marketing* and *planning* go hand in hand. Just like anything else in life, if you want to get somewhere you must have a plan—a road map—that gets you from point A to point B. You can have a good time sailing just by catching a strong breeze and riding along the bay, but you won't get anywhere in particular. If you don't have a sense of where you want to go and how to get there, a *plan,* you could end up totally lost.

In previous chapters, we discussed the importance of having business and marketing plans to achieve your goals. To develop successful marketing materials—advertisements, articles in newspapers, direct mail, trade show exhibits, or a Web site—you must develop a plan and carry it out.

This essential part of your marketing program cannot be overlooked. A *marketing communication plan* is a document that forces you to identify what it is about your company that will appeal to a prospective client. It enables you to get to know your target audience in great detail. It identifies which media outlets are best suited to reach your prospects and which best fit your budget. Then, it helps you find the best strategy for developing marketing materials that will drive your competitive advantages home in a way that will encourage your prospective clients to take action.

The marketing communication plan will help you establish measurable goals. You will be able to define—in concrete dollars and cents terms—whether or not your marketing communication program is working.

Moreover, the marketing communication plan helps you establish consistency in all of your marketing materials. Once you develop a strategy of what messages you must communicate and to whom, you can make sure that all of your marketing materials have a consistency that will allow them to work in concert to generate the results you want.

Developing a marketing communication plan is not a daunting task that can only be left to hired communication specialists. In fact, for anyone who is logical and organized, developing a marketing communication plan can be a very rewarding experience. It helps you and your staff really focus on your services and how your clients benefit from each. It enables you to get to know who is best suited to benefit from your products or services and why.

This chapter will be a step-by-step guide for those of you who want to develop your own marketing communication plan. For those of you who want to hire a marketing communication consultant, it will give you a far better sense of whether all the critical issues are being covered satisfactorily for your company.

Most of all, this chapter will be fun. You will see that the bottom line of creating effective marketing materials is simply finding the best ways to talk to other people about what is important to them. Regardless of whether you are selling general contracting services or paper towels, people buy for the same reasons: to satisfy their needs. Your mission is to identify those needs and how your products or services can satisfy them. Then, all you have to do is tell people about them.

This is also probably the most important chapter among the next four about marketing communication materials. If you do not learn how to develop a good marketing communication plan, all your marketing materials will likely miss their mark and your valuable marketing dollars will be wasted.

What Is Marketing Communication?

This may be the first time you have heard the two words *marketing* and *communication* used together in this way. Simply put, *marketing communication* is communication efforts designed to support your marketing goals. *Marketing,* of course, involves many noncommunication efforts, like pricing, distribution, planning, and sales. *Communication* often has nothing to do with marketing, like when you are talking on the phone socially, reading a book, or chatting with a neighbor.

Marketing communication puts the two common activities together. *Marketing communication materials* include anything you produce that communicates your marketing goals and objectives, such as your company brochures, newsletters, logos and stationery, direct mail, Web site, and so on. *Other marketing communication efforts,* such as trying to get publicity, would be your attempt to use communication media, like newspapers, magazines, and newsletters, to help accomplish your marketing goals.

Understanding the term *marketing communication* will also help you identify outside resources that will best serve your needs. There are several types of consultants involved with marketing, including sales trainers and marketing consultants. Likewise, there are several types involved with marketing communication. There are advertising agencies, whose primary function is to write, design, and place advertisements. There are public relations firms who write and place stories in the media, stage events for publicity, and lobby. There are also graphic designers, copywriters, Web development firms, and other specialists.

Companies that call themselves marketing communication firms typically provide a full range of services. They often integrate several marketing communication functions—advertising, publicity, and Web development—to provide a complete package of marketing support.

Establishing Your Goals and Objectives

Before you decide whether or not you are going to print a brochure or build a Web site, you have to establish your goals and objectives, because they will help guide you toward deciding which marketing materials work best for your unique situation.

Goals, in this case, are marketing goals. You should have determined them when you developed your business and marketing plans. They define where you want to be in a given time period. Following are examples of marketing goals:

- Entering a new geographic area and generating a backlog of $5 million within the next 24 months

- Introducing a new product or service to the marketplace next year

- Increasing market share by 5 percent over last year

- Generating a 50/50 mix of bid and negotiated work within 2 years

- Becoming known as an expert in your field within 15 months

From these goals you should establish clear-cut communication objectives. *Communication objectives are the responses you desire from your target audience.* You should develop as many communication objectives as possible for each of your marketing goals. For instance, if entering a new geographic area and generating a backlog of $5 million within the next 24 months is your goal, the following communication objectives could be established for you:

- Creating awareness of your company

- Creating awareness of your product or services

- Establishing the need for your product or services

- Communicating the benefits of your product or services

- Establishing a liking or preference for your product or services

- Generating inquiries

You will notice that each of these communication objectives works together to take the prospective buyer through the *decision-making process:* awareness, need, liking, purchase. It is essential that you take a prospect through each step of the process. As you will learn in Chap. 9, the different marketing materials and programs take prospects through different phases of the decision-making process. It is important that you mount a coordinated effort, often combining several materials or efforts to accomplish all of your communication objectives.

In addition to helping you decide what you need to do to accomplish your goals, communication objectives help you *measure* your results. How well you accomplish each objective will tell you whether the money you spent on your marketing materials generated the results you expected. Your objectives will help you isolate which parts of your marketing communication plan are working and which are not.

Measuring results is not as easy as determining whether or not your *goals* have been met. If your goal was to enter a new market and generate $5 million in backlog within 24 months, and you failed to do so, it is not necessarily the fault of your marketing communication plan. Variables not controlled by the marketing communication plan, such as price, quality, your salespeople's ability, and economic conditions, may all contribute to your results.

What you can measure is whether you have met your *communication objectives.* Have you established awareness? Have you established liking or preference? Let's use the above example to illustrate how establishing objectives can allow you to measure your results.

To measure how well you create awareness of your company or your competitive advantages, you might conduct a random survey of prospective customers before the marketing communication campaign and measure the percentage who are aware of your company. After your campaign, you should conduct another random survey of a similar number of prospective customers to measure the percentage who are aware of your company and its services or products. Surveys, too, could measure whether your campaign has established *liking* or *preference* for your product or service. And, of course, if your objective is to generate responses, you can easily determine the number of responses in a given time period before and after your campaign.

Identifying Your Competitive Advantages

Every company that is at least modestly successful must have *competitive advantages:* characteristics of its products or services that make it better than or different from the competition. If you think you are just like another company, providing similar services, think again. You wouldn't be in business if there wasn't something about your company that made it different from the competition. Perhaps it is quality, or maybe it's your prices. Maybe you have a technological edge, or maybe you are well connected.

Regardless, it is imperative that you *identify* your competitive advantages, because once you do, you will communicate them consistently in all your marketing endeavors. Start by gathering a group of key people in your company, those who have either worked there a long while, interact often with clients, or have a good understanding of how you sell your products or services. Meet in a large conference room and set aside at least 2 hours. Do this at a time when you will have minimal interruptions.

Get an easel with a pad of paper or use a marker board and create two columns. At the top of the left column write Competitive Advantages. Then brainstorm. Make a long list of all the characteristics that you feel are your

1. We have been in business over 30 years.
2. Our people have significant experience in interior construction and renovation.
3. We are innovators.
4. We are good listeners.
5. We have a flat organizational chart.
6. We have a lot of people with ownership in our company.
7. We are extremely service oriented.
8. We are committed to quality.
9. We specialize in working in occupied spaces.
10. We are nice people to work with.
11. We are community minded.

Figure 8.1 Competitive advantages of a hypothetical interior construction company.

strengths. They do not have to be unique to your company. They do not necessarily have to be things you do better than *every* other company, just areas in which you feel your company particularly excels. This list could be as long as you feel is necessary to identify all the characteristics, either individually or collectively, that set you apart.

Figure 8.1 shows a list of competitive advantages that a hypothetical interior construction company may identify.

Zeroing in on benefits

Once you have thoughtfully identified your competitive advantages, you must translate them into their *benefits* to your clients. This is essential. People buy the benefits they derive from a product or service, not the features of the product or service (see Fig. 8.2). People don't use aspirin, for instance, because of its features: white, chalky, bitter tasting, dry, and round. People use aspirin because of the *benefit* they receive from using it, pain relief. Similarly, people don't buy an air conditioner because it is a large, clunky, metal box with fans, pipes, and coils. They buy cool air.

Your services are no different. Yet all too often, companies, particularly in the construction industry, promote the features of their products or services rather than the benefits derived from them. How often have you seen marketing materials that promote things like the experience of our senior staff, our experience building similar projects, our preconstruction services, or our quality craftsmanship? These may indeed be their competitive advantages, but they are communicated in terms of features instead of benefits, and people buy the benefits they will receive from these competitive advantages.

The best way to identify the benefits of each of your competitive advantages is to submit it to the "So what?" test. For each advantage—experience, quality, and so forth—ask the question, "So what?" When you can no longer answer the So what? question, you have identified the real benefit. Here's how it works:

If your competitive advantage is We have experience building similar projects (a feature), ask, "So what?"

Here's a great way to hone your skills in focusing on benefits over features. Remember that features describe a product or service, and benefits describe the reward people get from using your product or service.

After you have gathered your key staff and identified your company's competitive advantages, use this fun exercise to get used to focusing on benefits:

1. Give each person a new, nicely sharpened pencil.
2. Tell them that you "discovered" this great new gadget and you think it has great sales potential—but you need their help. Act as if you have never seen this neat creation before and have some fun.
3. Make a chart with two columns. At the top left, write Features. On the top right, write Benefits.
4. Ask each person to study your neat, new gizmo and describe a feature. You'll hear descriptions like, long, wooden, cylindrical, sharp, and so on. If they say things like, "It erases," remind them that erases isn't a feature. Perhaps they meant, It has a rubber tip. Write the responses on the chart.
5. After you have generated a list of features, ask them to call out some benefits of the long, wooden stick. You'll probably hear things like "You can write with it." "It's easy to hold." "It erases mistakes." And so on. Write these responses on the chart, too.
6. Then ask the group, "what would be easier to sell?" A long, wooden, cylindrical sharp thing, or something that you can write with, that is easy to hold, and erases mistakes? Obviously, you could sell the latter easier.

While this may seem like a simplistic exercise, it is a memorable way of teaching the value of selling benefits over features. The same rules that apply to selling pencils apply to selling construction services.

Figure 8.2 Wooden stick or writing aid?

You might answer, "Well, that means we are familiar with what takes place during construction."

"So what?"

"Uh, it means we have faced similar challenges and know how to solve them."

"So what?"

"Well, we should have less down time and fewer change orders."

"So what?"

"We'll deliver the project faster at a better price."

"So what?"

"The client will save money and generate income on their completed project sooner."

Ah ha! That's the real benefit. Your experience on similar projects (a feature) allows your client to save money and generate income sooner (see Fig. 8.3).

You may think people can figure out the benefit themselves. But in reality, people are busy. What seems important to you may not be important to them. You have to tell them why your competitive advantages will make their life better.

These two words—*So what*—will forever help you be a better seller. Whether or not you ever write a word of ad copy or text on a Web page, understanding how to identify the benefits of the products or services you sell will enable you to prepare better presentations, get to the sale quicker, and understand your prospects' needs faster.

Anytime you need to convert competitive advantages that are stated as features into benefits, apply the So what? test. If your competitive advantage is quality, ask, "So what?"

You may answer, "Well, if we concentrate on quality, the owner gets a better building."

Ask again, "So what?"

You will ponder the question, perhaps get a little annoyed, and then answer, "Well, a better building requires less maintenance."

"So what?" you demand of yourself.

Now, starting to see the light, you answer, "Well, less maintenance means less expense for the owner."

"So what?"

"Less expense means more money."

"So what?" You ask again.

But this time, there is no answer. Congratulations, you have arrived at the real *benefit*. You help your clients make more money by delivering better-quality buildings that require less maintenance. Now that is a competitive advantage an owner can warm up to.

Print the words *So what?* large and in bold type on a sheet of paper and tape it to the wall in your office. It will always remind you to zero in on benefits, whenever you sell.

Figure 8.3   So what?

Most purchases, food, clothes, a car, or concrete pavement, are responses to the benefits people receive from the product or service offered. If you can easily show how your product or service can satisfy a prospect's needs, you will be far closer to the sale.

Now, apply the So what? test to each one of the competitive advantages you wrote down on your chart earlier. Your competitive advantages were listed on the left side. Under the words *competitive advantages,* write the word, *features,* with a thick, red marker. This will remind you that your competitive advantages are features, not the characteristics you should be using to sell your products or services.

At the top of the right column, write *benefits*. For each feature you listed, apply the So what? test. Keep asking, "So what?" until you can't come up with a different answer. Then, you will have identified a benefit that prospective customers can really relate to. You may find that the same benefit applies to many of the competitive advantages you have listed.

As you are doing this, you will find a cultural change taking place among the people involved in the brainstorming exercise. They will begin thinking in terms of benefits instead of features and finding ways to communicate the benefits of what they do for their customers. As they go out and promote your company, either as marketers, salespeople, or project managers, they will begin communicating this way on a daily basis to customers as well, who will then be better able to relate to the fine services your company provides.

Figure 8.4 shows how the So what? test would reveal the benefits of the competitive advantages listed for the hypothetical interior contractor in Fig. 8.1.

Competitive advantages	Benefits
1. We have been in business for over 30 years.	1. We've conquered almost every construction challenge, which saves our clients time and money.
2. Our people have significant experience in interior construction and renovation.	2. We won't make "rookie" mistakes, which delivers our jobs faster and helps our clients get money sooner.
3. We are innovators.	3. We find ways to improve the construction process, which saves time and money for our clients.
4. We are good listeners.	4. We won't make careless mistakes, which brings the project in on time and within budget.
5. We have a flat organizational chart.	5. Our clients work with decision makers, which minimizes mistakes and expedites jobs, saving them time and money.
6. We have a lot of people with ownership in the company.	6. Our people have a more personal stake in the success of our projects, which increases client service and improves quality.
7. We are extremely service oriented.	7. We always let clients know what is going on with their project, minimizing their stress.
8. We are committed to quality.	8. A quality project requires fewer change orders and repairs, improving the investment value of our clients' properties.
9. We specialize in working in occupied spaces.	9. We know the challenges, keep work areas clean, and minimize errors, thus maximizing the chance of on-time, within-budget delivery.
10. We are nice people to work with.	10. We make a usually stressful experience more pleasant for our clients.
11. We are community minded.	11. By helping people and businesses locally, we are helping the local economy and creating business opportunities for our clients. Also, by caring for others, they see we care for them as well.

Figure 8.4 Competitive advantages and benefits of a hypothetical interior construction company.

You will probably notice a distinct trend emerging when you start identifying benefits. The first is that there are relatively few benefits that people react to. The second is that you need to describe how the client receives the benefit. For the competitive advantage, We are good listeners, we identified the benefit as, We won't make careless mistakes, which ensures that projects will be completed on time and within budget. We explained how the advantage creates a benefit, which helps the prospect understand your argument.

The key selling point

After you have identified the benefits of all of your competitive advantages, you must now identify the *key selling point,* the one that is most important above all others.

Although you may feel that many of your competitive advantages are key, it is important to identify the one that is most important, because when you or your consultants are creating advertisements, direct mailers, brochures, and so on, they need to be able to prioritize your advantages. An effective advertisement, for instance, as you will learn later, must communicate one selling point clearly and then provide additional selling points if space permits.

If you do nothing else when preparing marketing materials, you must be sure to communicate your key selling point. By identifying it now, you will always know which one competitive advantage must be communicated in all your marketing materials.

After you have identified the key selling point, identify a few *other selling points* that you feel definitely need to be included in your marketing communication efforts. These will probably be the three or four most important benefits after the key selling point that you identified earlier.

Always write your selling points out in sentence form. This not only allows you to write complete thoughts, it will virtually write your body copy for advertisements and other marketing communication materials in advance.

Defining Your Target Audience

Once you have established your goals and objectives, and identified the benefits of your competitive advantages, you need to determine your *target audience,* which is *the group of people most likely to buy your products or services.* Note the words, *people most likely to buy.* Your target audience is not *every* potential buyer but rather, those who would benefit *most* from your particular products or services and who would respond best to your competitive advantages.

Recognizing that you probably have more than one target audience, it is good to prioritize the groups in terms of primary target audience, secondary target audience, and so on. This will help you allocate your resources, concentrating most of your time and money on reaching the primary target audience.

Demographics and psychographics

A target audience is defined by its demographic and psychographic characteristics.

Demographic characteristics are those that can be measured or quantified. They include characteristics like business type, geographic location, number of employees, annual revenue, and even specific job titles within an organization. Demographic characteristics are relatively easy to identify.

Psychographic characteristics group people into homogeneous segments based on their psychological makeup and lifestyle characteristics. They might include such things as interests, hobbies, beliefs, and so on. Psychographic characteristics, although they are harder to define, are often more important than demographic characteristics. As mentioned earlier in this chapter, purchase decisions, even those by businesspeople, are based on how a product can satisfy the buyer's needs. Understanding prospective buyers' psychographic traits will help you better understand their needs and how your product or service can satisfy them.

When at first defining your target audience, start with broad definitions. If you are an interior contractor, perhaps your primary target audience will be space users. Your secondary target audience may be interior architects because they often have some influence in the selection of an interior contractor. Further, you may feel commercial brokers, property managers, developers, and even base-building general contractors might comprise your tertiary (third) target market.

Demographic characteristics.   Once you have decided what type of people fall within your target markets, you should further narrow your definition to best suit your particular competitive advantages. Start with their *demographic* characteristics.

In the case of the interior contractor mentioned above, perhaps you are best at building office interiors that are between 1000 and 20,000 square feet. Furthermore, you may have determined that you want to pursue negotiated projects because you don't have the resources to prepare quantities of lump-sum bids, and you have narrowed your geographic scope to a 50-mile radius of your office to provide the best level of service.

Based on these specifics, you can narrow the demographic characteristics of your target audience. If you build office spaces between 1000 and 20,000 square feet, you can determine the size of the companies you wish to target. You might do this by calling a friend who is a broker and asking how many square feet of office space a typical employee occupies. The answer might be that typically, each employee uses about 200 square feet, when factoring in common areas. Doing a little math, you would then know that your target audience consists of companies having between 5 and 100 employees.

Further narrowing your target, you might assume that a facility or office manager within each of these companies would make decisions about hiring an interior contractor. Thinking that there are more office managers than facility managers at the smaller firms, you might decide to target them. You reason that even if there is not an office manager at some companies, inquiries sent to office managers will be forwarded to someone with those responsibilities.

Because you have decided to pursue negotiated work, you can eliminate certain types of businesses, such as government agencies. By narrowing your geographic range to a 50-mile radius, you can identify specific counties or towns to target.

You should also narrow down your secondary and tertiary target audiences in the same manner. If your secondary target audience was interior architects, you might assume that any-sized interior architect would have clients in the range of office space users you are pursuing, so you would not qualify those companies by their size.

You realize, however, that there is more than one decision-maker within each firm, so you decide to include all the architects in each firm. If, when quantifying your target audience, you determine that the total number of interior architects is too big to reasonably pursue, you can narrow your audience to senior architects.

In summary, for the above example, the demographic profile of your target audience might be as follows:

Primary target audience
- Office space users
- Private-sector companies of between 5 and 100 employees
- People with the title of office manager
- Companies in Sonoma, LaJola, and Bristol counties

Secondary target audience
- Interior architecture firms
- All interior architects within those firms
- Companies in Sonoma, LaJola, and Bristol counties

Tertiary target audiences
- Commercial brokerage companies
- All brokers within those companies
- Commercial real estate developers
- Project management positions within those companies
- Base building general contractors
- Estimators within those companies
- Project managers within those companies

Narrowing your target audience in this way is important for at least two reasons. First, it keeps you from wasting your time chasing prospective clients that aren't a good match for your services. Second, when it comes time to develop or purchase a mailing list of prospects, you have quantifiable parameters that a mail list seller can use to build a list for you.

Psychographic characteristics.   Once you have a clear understanding of your target audience in demographic terms, you should try to determine their psychographic characteristics. Psychographics are not as important in business-to-business selling as they are in consumer selling. Often, when buying consumer products, people make purchasing decisions based on how a product satisfies some emotional need. In business, people buy the benefits derived from a product. Sometimes those benefits are emotional, like stress reduction. But more often than not, the benefits are business details, like saving time or money.

So, you may develop a loose description of the psychographic characteristics of your target audience based upon your competitive advantages. If you feel that your honest dealings with clients and high level of customer service are your strengths, your target audience's psychographic characteristics might include an appreciation for being taken care of or for someone who is straight-forward. Although you will not find these kinds of details when purchasing mailing lists, knowing your target audience's psychographic characteristics will help you find the best words with which to communicate to them.

Speaking to one person

After describing your target audience's demographic and psychographic characteristics in some detail, you must try to narrow your definition to *one person*. This is an exercise that will prove tremendously valuable when you begin developing your marketing materials.

Why identify one person when even your carefully defined target audience still contains a diverse mix of people? Imagining one person helps you learn how to communicate more effectively with him or her.

Think of it this way: Which is a more effective way of communicating: speaking to a group of 1000 people or speaking with one person? When you speak to one person, you can establish eye contact and get to know that person's likes and dislikes. You can get feedback on what you are saying about your company, and you can establish a relationship. In this way, you can customize your presentation to suit individual needs.

When you speak to a crowd, you cannot make this type of connection. You can only make general assumptions about their needs. And, unless they are throwing tomatoes at you, you rarely get a sense of whether or not you are saying things that will have much impact.

If you think of your target audience in the same way, when you create your marketing materials—your advertisements, brochures, newsletters, or Web site—you can write as if you are speaking to one person. You can be aware of his or her needs. You won't talk to the masses but rather to one individual who can react to your message.

Try the exercise of identifying your target audience by one person. Are you speaking to a man or a woman? How old is he? Is she single or married? What is her job title? Is he an "influencer," someone who likes to be the first to buy things, then tell friends and colleagues about his purchase experience? How long has he held his position? What are her hobbies? Does he play golf or tennis? How tall is she?

By doing this, you will begin to see one person in your mind, and all of your communication efforts will become more personal and effective.

Developing a Creative Strategy Statement

By now, you should have established your communication objectives, determined your competitive advantages and stated them in terms of their benefits, identified your key selling point, and defined your target audience by their demographic and psychographic characteristics. In other words, you should have a pretty good sense of what you want to accomplish, what you want to say about your company, and to whom you want to speak.

Now, it is time to decide *how* you want to say it. Before you create your marketing materials, you need to develop a sense of the look and feel of the materials so they can best convey your competitive advantages and speak to your target audience.

The best way to do this is to develop a *creative strategy statement*. This statement simply explains what you want to accomplish, what you want to say, to whom you want to say it, and how you should communicate it. How you communicate will take some thought. The creative strategy statement does not ask you to decide what specifically to say and how to design materials. It simply asks you to determine the overall look and feel of your effort. For the interior contractor example mentioned in this chapter, a creative strategy statement may sound something like this:

> To establish preference for our services and generate sales opportunities by communicating that we are the best interior contractor for private companies with 5 to 100 employees, particularly if they need work done in occupied spaces. We will speak primarily to people within these companies who have little experience in retrofitting their office space. Because they are probably very anxious about this process, we will create marketing materials that communicate how choosing us will ease their minds. We will try to use visual images and text that show how we will make their life easier if they choose us.

You can see from the above example that, while specific visual images and text have not been decided, the people who create the marketing materials will have a very good sense of what you want to accomplish. In fact, ideas for visuals and copy may pop into your mind as you read it.

Writing your creative strategy statement may be a challenging task, and it may take several revisions. However, once you do it, you will have a clear idea of how you will conduct your entire marketing communication campaign.

Producing a Creative Platform

Now, it is time to put all the information you have gathered into a format that can be used by you and whoever else might help you develop your marketing communication materials. You will now create a document called a *creative platform*. The creative platform is essentially an outline that lists your

- Marketing goals and the communication objectives that will achieve them
- Competitive advantages, stated in terms of their benefits
- Key selling point
- Other selling points

Figure 8.5 Creative platform of a hypothetical interior construction company.

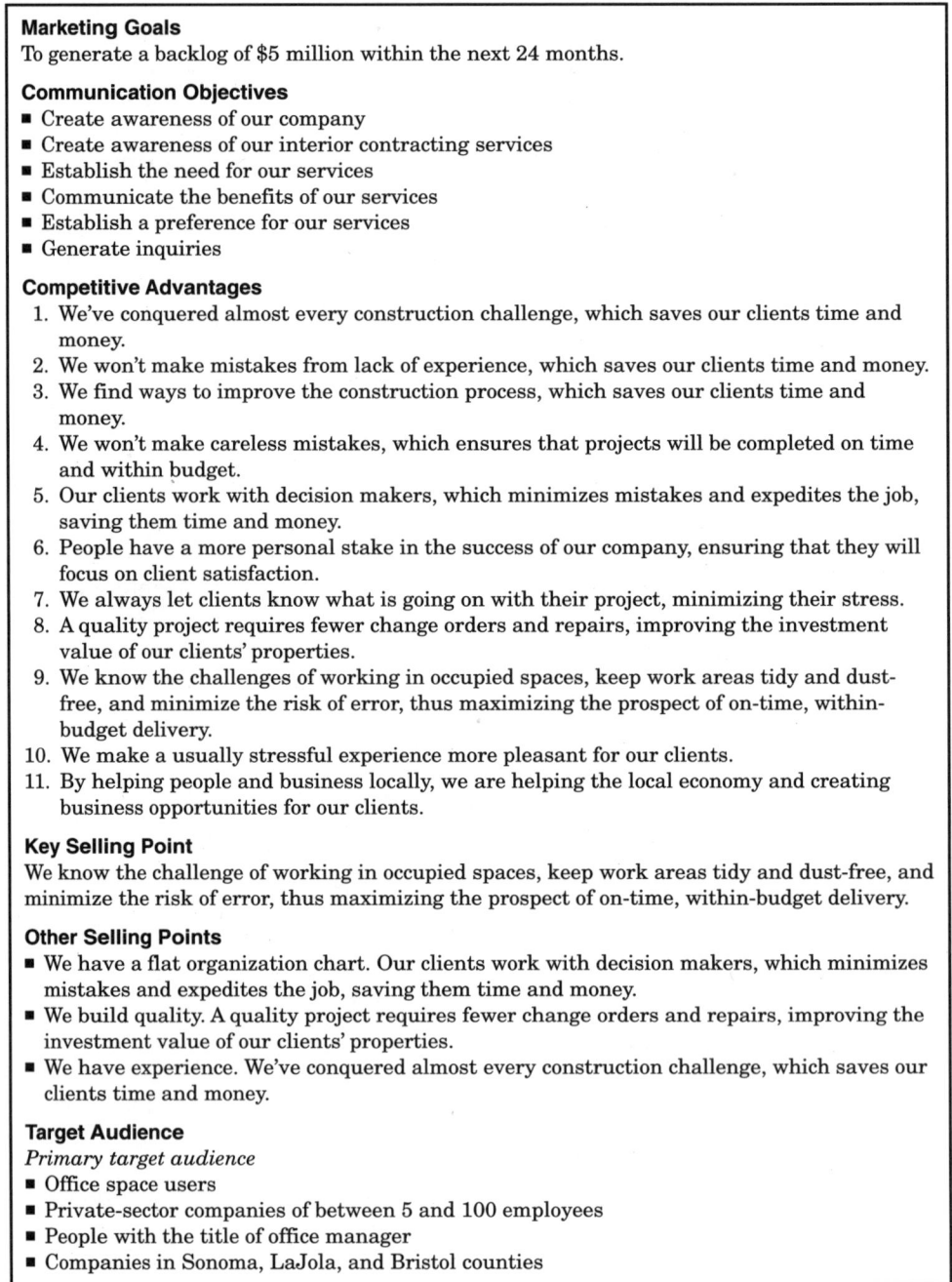

Marketing Goals
To generate a backlog of $5 million within the next 24 months.

Communication Objectives
- Create awareness of our company
- Create awareness of our interior contracting services
- Establish the need for our services
- Communicate the benefits of our services
- Establish a preference for our services
- Generate inquiries

Competitive Advantages
1. We've conquered almost every construction challenge, which saves our clients time and money.
2. We won't make mistakes from lack of experience, which saves our clients time and money.
3. We find ways to improve the construction process, which saves our clients time and money.
4. We won't make careless mistakes, which ensures that projects will be completed on time and within budget.
5. Our clients work with decision makers, which minimizes mistakes and expedites the job, saving them time and money.
6. People have a more personal stake in the success of our company, ensuring that they will focus on client satisfaction.
7. We always let clients know what is going on with their project, minimizing their stress.
8. A quality project requires fewer change orders and repairs, improving the investment value of our clients' properties.
9. We know the challenges of working in occupied spaces, keep work areas tidy and dust-free, and minimize the risk of error, thus maximizing the prospect of on-time, within-budget delivery.
10. We make a usually stressful experience more pleasant for our clients.
11. By helping people and business locally, we are helping the local economy and creating business opportunities for our clients.

Key Selling Point
We know the challenge of working in occupied spaces, keep work areas tidy and dust-free, and minimize the risk of error, thus maximizing the prospect of on-time, within-budget delivery.

Other Selling Points
- We have a flat organization chart. Our clients work with decision makers, which minimizes mistakes and expedites the job, saving them time and money.
- We build quality. A quality project requires fewer change orders and repairs, improving the investment value of our clients' properties.
- We have experience. We've conquered almost every construction challenge, which saves our clients time and money.

Target Audience
Primary target audience
- Office space users
- Private-sector companies of between 5 and 100 employees
- People with the title of office manager
- Companies in Sonoma, LaJola, and Bristol counties

Figure 8.5 *(Continued)*

Secondary target audience
- Interior architecture firms
- All interior architects within those firms
- Companies in Sonoma, LaJola, and Bristol counties

Tertiary target audiences
- Commercial brokerage companies
- All brokers within those companies
- Commercial real estate developers
- Project management positions within those companies
- Base building general contractors
- Estimators within those companies
- Project managers within those companies

Creative Strategy Statement
To establish preference for our services and generate sales opportunities by communicating that we are the best interior contractor for private companies with 5 to 100 employees, particularly if they need work done in occupied spaces. We will speak primarily to people within these companies who have little experience in retrofitting their office space. Because they are probably very anxious about this process, we will create marketing materials that communicate how choosing us will ease their minds. We will try to use visual images and text that show how we will make their life easier if they choose us.

- Target audience, defined in demographic and psychographic terms
- Creative strategy statement

The creative platform is a guide that takes you from where you are to where you want to be. It is the guide that tells the sailor how to let the wind catch the sails to get from the dock to an island hideaway. It will give your creative team the information it needs to prepare marketing communication materials that convey the right message, delivered to the right audience.

The platform is especially helpful if more than one person or consultant is working on your marketing materials; it allows them to work from the same strategy and create materials that will work together to accomplish the same results. For example, if you decide to create a new logo, an advertisement, a Web site, and a publicity campaign, you may use different people to perform each of these functions because they utilize different skills. The creative platform will ensure that all of these marketing efforts will have the same look and feel and convey the same important points.

The creative platform and the project manager of a construction company have similar functions: defining a project's goals and objectives and marrying a diverse group of skilled craftspeople to achieve those goals. Figure 8.5 shows what the creative platform of the hypothetical interior contractor we have created in this chapter may look like.

Give the creative platform to each of the creative people who will develop your marketing materials, and you will see that each of the elements work together. If you do not want to prepare the creative platform or any of its elements yourself, make sure that your in-house creative staff or outside consultants do it for you *before* they prepare any of your marketing communication materials. Make them show it to you. Use it as a guide. If at any time there are questions about which direction to take, refer to the creative platform. Use it as a litmus test to ensure that your marketing materials are consistent and stay on track.

Ensuring Ease, Efficiency, and Effectiveness

Taking the time to determine your communication objectives, define your competitive advantages, isolate a key selling point, identify your target audience, prepare a creative strategy statement, and produce a creative platform will save you untold marketing dollars and time as you prepare your marketing communication materials. It will also ensure that everything you prepare will be as effective as it possibly can be.

Congratulations. You have graduated with honors from Marketing Communication 101. From here, you can move with confidence onto developing your marketing communication materials and other marketing communication efforts.

Marketing Materials

Introduction

Once you have created your marketing communication plan as described in Chap. 8, you will have all the information you need to decide what types of promotional materials you should produce to achieve your communication objectives.

It is imperative that you go through the process of developing a marketing communication plan first. All too often, people decide they want to produce advertisements, a brochure, or a Web site before they determine their objectives and identify their target audience. Yet, your objectives and your target audience play key roles in determining which promotional materials will be most effective.

Advertisements, for instance, are effective for accomplishing the objective of establishing awareness but are not effective for generating action. A brochure is good for creating desire and action, but it is wasted if your objective is creating awareness. Web sites, too, are good for creating desire and interest but not effective for creating awareness.

The size and geographic diversity of your target audience also affect which marketing materials to develop. For example, publicity is effective for reaching a broad target audience, whereas direct mail is best when you have a small and clearly defined target audience. These are but a few of the considerations you will need to make before deciding which marketing communication materials (often referred to as promotional materials) to create for your marketing program.

This chapter discusses a wide variety of promotional materials and gives you guidance about which will be the most effective and cost efficient for meeting your goals and objectives. After describing each, you will learn the steps involved in producing marketing communication materials, and will read recommendations about what you can do yourself and for what you may need help with.

The Role of Marketing Materials in the Decision-Making Process

To sell a product or service, you must take a prospective buyer through the decision-making process. There are four steps to the process, and a prospect must go through each step every time he or she buys. The steps are

- Awareness
- Interest
- Desire
- Action

When you develop a strategy for achieving your communication objectives, you should make sure that the marketing communication materials you select take prospects completely through the process.

You may feel, for instance, that you have already achieved awareness in the geographic area that you are targeting. Your strategy, then, would focus on developing materials that take prospects through the interest, desire, and action steps.

On the other hand, if you are entering a new geographic area or are introducing a new service to your existing geographic market, you will need to develop materials that take prospects through the entire decision-making process. Table 9.1 shows how different marketing communication materials are used to help people through the decision-making process.

Marketing Communication Materials

There are dozens of different kinds of marketing communication materials you can use to accomplish your objectives. Generally, the ones used most often in the construction industry are

- Advertising
- Collateral material
- Direct mail
- Publicity
- Internet
- Exhibits
- Presentations

Each of these satisfies different objectives and helps you reach different target audiences. What becomes clear after learning about each of their strengths and weaknesses is that in most cases you will have to execute a strategy that includes more than one type of marketing communication material to accomplish your objectives. Table 9.2 will give you a quick understanding of the advantages and disadvantages of some of the primary forms of marketing communication.

TABLE 9.1   Fitting Promotional Tools into the Purchase Process

	Awareness	Interest	Desire	Action
Advertising				
Magazines and newspapers	X	X		
Directories (e.g., Yellow Pages)			X	X
Radio and TV	X	X		
Outdoor (billboard)	X			
Collateral Material				
Brochures		X	X	
Video and audio tapes		X	X	
CD-ROMs		X	X	
Software		X	X	
Article reprints		X	X	
Signs	X			
Bumper stickers and decals	X			
Direct Mail				
Stand alone	X	X	X	X
Group (e.g., deck cards)	X	X		
Exhibits				
Trade shows		X	X	X
Public shows		X	X	
Internet				
Web sites		X	X	X
Web directories		X	X	
Web advertising	X	X		
Premiums (virtually all types)	X			
Presentations				
Personal selling			X	X
Speeches (public meetings)	X	X	X	
Seminars and conferences	X	X	X	
Public relations				
News releases	X	X		
Industry representative role	X	X		
Telemarketing (out- and inbound)		X	X	X

Reprinted with permission of The Aberdeen Group.

TABLE 9.2   **Advantages and Disadvantages of Various Marketing Communication Materials**

	Advantages	Disadvantages
Advertising	Reaches a broad audience.	Cannot target small groups.
	Low cost-per-thousand (CPM) impressions.	Can be expensive overall.
	You control the message.	Limited credibility.
Collateral materials	Reaches a targeted audience.	Not good for introducing company.
	Excellent way to showcase your work.	High CPM.
	You control the message.	Limited credibility.
Public relations	Reaches a broad audience.	Cannot target small groups.
	Low or no CPM.	Risk of no return on investment.
	Maximum credibility.	Cannot control what gets published.
Direct mail	Reaches a targeted audience.	Can miss some prospects.
	Little risk of wasted investment.	High CPM.
	You control the message.	Limited credibility.
Internet	Reaches a broad audience.	Cannot target small groups.
	Low CPM.	Can be expensive overall.
	You control the message.	Limited credibility.
Exhibits	Reaches a targeted audience.	Can miss some prospects.
	Little risk of wasted investment.	High CPM.
	You control the message.	Limited credibility.

Advertising

Advertising is one of the most common forms of promotion. It can take many shapes. The most commonly used are print advertising in magazines and newspapers; directory advertising, which includes Yellow Pages advertising; broadcast advertising on radio or television; and outdoor advertising, typically on billboards.

Although advertising is quite common in consumer sales and some business sales, contractors and subcontractors do not often use it. Perhaps this is because it is hard to measure whether or not advertising can be credited for generating big-ticket purchases like concrete formwork or an office building. Nevertheless, advertising has long proven its effectiveness for establishing awareness, interest, and desire and should be factored into most company's marketing communication efforts.

Advertising is a *broadcast* medium. It reaches a broad audience. You cannot specify who will see or hear your advertisement. You can narrow the audience who will be exposed to it by selecting *media* that are targeted to a specific audience, such as neighborhood newspapers, trade association magazines, or news-format radio stations, but you cannot preselect who from within those large audiences will see or hear your advertisement. This can be advantageous if you are trying to reach a broad audience and want to establish awareness fairly quickly. It can be disadvantageous if you are trying to reach a small target audience.

When producing advertisements, you can control the message you are presenting to your target audience. Because you write and design it, you will be certain of the message it presents to your target audience. This is obviously important when you have a specific message you are trying to send.

Advertising messages are usually fleeting; in other words, as soon as they are seen or heard, they are gone. This is particularly a problem with electronic media like television or radio. You cannot ponder over the message or review it to clarify facts. Print advertising allows the prospect to reread a message. Publications with longer "shelf lives," like magazines and directories, afford even better opportunities for people to see your advertisement more than once.

However, because people who see or hear your advertisement know that you wrote it, advertising generally has limited credibility. You can enhance the credibility of your advertisements by including third-party testimonials. People believe what others say about you more than what you say about yourself. That is why so many commercials for consumer products show endorsements of people who use the product.

Another advantage of advertising is its low cost-per-thousand (CPM) impressions. No other marketing communication vehicle is less expensive per person reached than advertising, with the possible exception of media publicity. This can be a great advantage if you are tying to reach a large audience that can be reached via one newspaper or magazine. However, the overall cost of placing advertisements can sometimes be too expensive for companies. Even though a publication may reach a large number of prospects, companies often think that the overall cost is too high and choose a more direct, less expensive approach.

According to Daniel P. Anderson, president of The Aberdeen Group, which publishes five construction magazines, advertising is good for creating awareness and interest. Most contractors agree that they place advertisements to generate name recognition but do not expect to make a sale off of their ads. The possible exception is directory advertising, such as the Yellow Pages, because prospects are predisposed to buy when they use such directories. They have already established interest in a product, and they will review directory ads to identify who supplies products and learn about their competitive advantages through their advertising message.[1]

Typically, however, advertising supports other marketing materials. It is used to create awareness and reach a broad group of people within a target audience. Other forms of promotion should then be used to create desire and generate action.

Selecting the right advertising media

Deciding where to place your advertisements takes a combination of common sense, creativity, and research. Once you have defined your target audience (see Chap. 8), think about all the potential media that influence them. Ask yourself:

- Which magazines and newspapers do my prospects read?
- Are there any general-interest business publications in my city or region?
- Are there newspapers that focus solely on real estate?
- What trade or professional associations have members in our target audience? Do they have publications that accept advertising?
- Are there industry directories that prospects use to buy our products or services?

Once you have identified some of the obvious advertising vehicles, try to think of some of the less obvious ones:

- Are there any cable television shows that our target audience might be likely to watch?
- Are there any radio stations that reach our target audience or any segments of radio programs targeted to them?
- Are there any regional issues or sections of large newspapers, like our daily newspaper, that offer discounts if ads are placed only there? For instance, it costs less to place a regional advertisement in *The Wall Street Journal* than in some local business newspapers.
- Are there any private schools that produce publications that may be read by people in our target audience?
- Are billboards a viable option?
- Is mass transit something to consider, particularly in routes heavily populated by our target audience?

Look around where you live. Ask friends or clients who are a part of your target audience what media influences them. You might be surprised by what your learn.

After you have expended your personal resources to generate ideas about selecting advertising media, you can do research that will help you identify sources. In most libraries you will find *Standard Rate and Data Service,* a two-volume directory that is considered the bible for advertising media buyers. This directory is divided into business and consumer publications and is organized by subject. Each listing gives a brief profile, editor's name, and complete contract information, and it lists media by subject and location. If you want to find publications that focus on facility management, for instance, you can find a comprehensive list there.

Within the guide you will discover the mission of the publication, advertising costs and deadlines, editorial calendars when available, and contact information. You can call media representatives to get more detailed information about the readers to determine if they fall within the definition of your target audience.

Buying ad space

Identifying the right media in which to advertise is half the battle. The other half is determining how often you should run your ads and purchasing the ad space or time, depending on whether you are using print or broadcast media.

Complicated media buys—ones that involve multiple geographic markets and a wide mix of print and broadcast—might best be left to experienced advertising agencies or media buying consultants. Ad agencies have the expertise to both create the ads and identify and purchase the ad space. Typically, they will charge for their creative services and purchase the ad space for you. Most media remit a 15 percent commission to ad agencies, which the agencies use to pay for their time identifying media, creating a media plan, and buying the media space for each advertisement.

However, most often, contractors make fairly simple media buys and can pay for the cost of the advertisement directly. Often, the media will give you the agency commission. In fact, you should insist upon it if you are doing their work for them.

Frequency discounts are given when an advertiser commits to multiple ad purchases within a contract period, which is usually a year. Purchasing a single ad costs more than purchasing three or more, which is more expensive than purchasing six, and so on. If you plan to run more than one ad, take advantage of frequency discounts by signing an *advertising contract* for multiple insertions. If you cannot honor your multiple ad insertions, you will only be held responsible for paying the difference between the ad rate you contracted for and what you would have paid for fewer insertions. This is called a *short rate* and does not usually carry a penalty.

You can often negotiate with the media to get a lower rate than what they normally offer. Unsold ad space is like an unsold hotel room or airline seat. Once the publication is printed, unsold ad revenue is lost. So typically, the media will negotiate by selling you an ad at a lower frequency rate than you will actually use.

If you decide to purchase your own advertising space, talk to the *sales representatives* at the print or broadcast media in which you are considering placing advertisements. These people are generally friendly and knowledgeable and willing to work with you to figure the best specific issues to advertise in, for print media, and to help you get the best rates.

Just make sure to gather information on all the potential media in which you may advertise. Make your selection based on which has the readers, listeners, or viewers that best match your target audience, the most credibility, and offers the lowest CPM. Try not to be persuaded to select one media over another only because of a good sales pitch from a media representative. It happens sometimes.

Creating effective print advertisements

Like any creative skill, writing and designing effective advertisements takes a great deal of training and experience. You would probably be best served to find a good, creative ad agency or marketing communication firm that understands your business and the needs of your clients and then negotiating a fair price to produce your advertisements or campaign. However, if your budget does not allow it, or if you just want to be able to judge whether the advertisements produced by your consultant are effective, you should understand the basics of creating effective advertising.

First and foremost, effective advertisements must meet your communication objectives to generate the desired results. They don't necessarily need to win awards for creativity. After you have created your advertisement, ask yourself if it will satisfy your communication objectives. If not, refine it until it does.

Writing the ad: The headline, visual, and body copy.   There are three main parts of any effective advertisement: the *headline,* the *visual,* and the *body copy.* They must work in concert to communicate your key selling point, which is the most important one of your competitive advantages.

In most ad agencies, a designer, being more visually oriented, will work with a copywriter to develop ideas for headlines and visuals. The visual and the headline must work synergistically to clearly convey your competitive advantage. If you are creating the ad yourself, make a simple list with two columns, one for the visual and the other for the headline. If you are more visual than verbal, start with ideas for photos or illustrations and then try to develop a headline that works with them. If you are better with words, start with a headline. You should come up with a few options from which to choose.

Key Selling Point
We know the challenges of working in occupied spaces, keep work areas tidy and dust-free, and minimize the risk of error, thus maximizing the prospect of on-time, within-budget delivery.

Other Selling Points
- We have a flat organization chart. Our clients work with decision makers, which minimizes mistakes and expedites the job, saving time and money.
- We build quality. A quality project requires fewer change orders and repairs, improving the investment value of our clients' properties.
- We have experience. We've conquered almost every construction challenge, which saves our clients time and money.

Visuals	Headlines
1. A construction worker wearing a hardhat and dressed in a maid's apron, dusting a desk.	1. We'll keep your office so clean you'll think we're really maids.
2. A pristine office with yellow construction caution tape all around.	2. Just because you need to renovate your office doesn't mean it has to be a mess.
3. A busy office environment with construction workers all around.	3. We have so much experience building in occupied spaces, you'll hardly notice we're around.

Figure 9.1   Visual and headline ideas for a hypothetical interior contractor, Certified Construction, Inc.

Figure 9.1 shows a list of ideas for visuals and headlines for the hypothetical interior contractor from Chap. 8. The contractor's *key selling point* is: *We know the challenges of working in occupied spaces, keep work areas tidy and dust-free, and minimize the risk of error, thus maximizing the prospect of on-time, within-budget delivery.*

Good ideas for headlines. The best headlines are those that grab attention. The same is true of visuals, as well. But if an ad headline only grabs attention, it is only doing half its job. A headline must also convey your key selling point.

People only scan advertisements. Unless your headline gives them a compelling reason to read your ad by telling them something that relates to their needs, they'll turn the page. If your headline is provocative, but doesn't say anything about your selling point, chances are that people won't read further.

If your headline grabs attention because it is either clever or written powerfully and is able to convey your key selling point, you have already won more than half the battle. If all the reader does is read your headline, understand your selling point, and see your company's name on the ad, you will have succeeded. If readers study the visual and read the body copy, you've hit a home run.

There are many different kinds of headlines. Some of the more common are

- *Factual.* The headline powerfully states your selling point. *Certified Construction has experience renovating occupied office space.* It may not be clever, but it is clear.

- *Analogy.* The headline takes your selling point and uses an analogy that makes for an intriguing statement. *We'll keep your office so clean you'll think we're really maids.*

- *Testimonials.* Because people are more inclined to believe things that others say about you more than what you say about yourself, headlines that use testimonials are very effective. *"Certified kept our office so clean during renovation, we hardly knew they were there."—Deborah Carper, American Marketing Association.*

- *Play on words.* These types of headlines take an aspect of your selling point and throw in a double meaning or pun to capture attention. *Certified finished our workspace while we finished our work.*

Good ideas for visuals. Many people see the visual part of an ad before they read the copy, so an attention-getting visual is crucial to the effectiveness of your ad.

Visuals are attention-getting when they have a twist of some kind that makes them unexpected. If you don't want to use a visual with a twist, you should at least have a visual that is bold and graphic. A powerful photograph, a collage, or an illustration can be nice.

Consider the publication(s) in which the ad will appear. You don't want your ad to look like all the other ones. If you are placing an ad in a real estate publication, avoid a straightforward photo of an office building, for instance. There will probably be 25 other such photos, and yours will probably not stand out.

No matter what kind of visual you use, it has to help communicate your key selling point. A common mistake contractors make is to show a picture of completed work in their ad rather than illustrate their key selling point. For instance, if their selling point is how well they work in occupied office space, rather than use a visual that supports it, they simply show a picture of a pretty, completed office interior. It may be pretty, but it doesn't convey your selling point, and it could make your ad a waste of money.

Coming up with an idea for an attention-getting visual takes time. It is time well spent. A good visual is probably the most important part of your advertisement.

Body copy: The lead, body, wrap, and call to action. The body copy of an advertisement—the words that follow the headline—is generally divided into four parts: the *lead, body, wrap,* and *call to action.*

The *lead* is the first sentence or two. The lead acts as a transition from the headline to the body of the ad by explaining the headline. A lead sentence for the first headline in Fig. 9.1, "We'll keep your office so clean you'll think we're really maids," might be *Despite what you may have heard, at Certified, we are really builders. It's just that we keep the offices we renovate so clean that our customers sometimes think we are maids.* See how the lead sentences explain the headline?

Following the lead sentences is the *body* of the ad. It works to explain your other selling points to the reader. It can be as simple as a few sentences, each dedicated to describing one of your competitive advantages in descending order of importance. If you wrote your other selling points in sentence form, you can practically plug them in here.

Body copy for the lead sentences above might be

> In addition to a clean office space, we build a top-quality office. Quality construction means fewer repairs, which improves the value of your valuable office investment.
>
> Certified has been in business over 30 years. We've conquered almost every construction challenge, which saves our clients time and money. And when you work with us, you will work with top decision makers in our company, which minimizes mistakes and expedites the job, saving you time and money.

You can see how this body copy is only slightly modified from the way the interior contractor's *other selling points* were written. They flow nicely and let you get a few more competitive advantages across to the reader without being boring.

The *wrap* is a sentence or statement that ties the copy back into the headline and acts as a conclusion that reinforces the key selling point. For our example, it may read something like this: *We think you'll like doing business with Certified—the interior contractor that's maid to order.*

A little play on words was used to tie back to the headline and leave the reader smiling.

The final part of the body of the ad is the *call to action.* It is a statement, either part of the body or separate, that tells the reader what to do. Call us today, or For more information contact, are simple calls to action.

Your advertisement's signature.   Make sure that your company name and/or logo, phone number, and Web site address are always on your ad. The name and/or logo allows readers to make a quick connection if they only read the headline and look at the visual. They will at least be able to identify the main selling point with your company.

The phone number, obviously, gives readers vital information they need to contact you. Your Web site address, just like your phone number, should be on your advertisements and every other marketing communication material. The Web site will provide them with as much information as a brochure. If the reader visits your Web site as a result of reading your ad, and learns all kinds of valuable information about you, you will have hit a grand slam.

Designing the ad.   As we mentioned, the visual images you select should make an impact on the reader. Likewise, the type styles you use for the text and the way in which you lay out the elements of the ad must be contemporary and appealing.

David Ogilvy is the dean of contemporary advertising thought. He founded Ogilvy & Mather, the renowned firm that created memorable ads for companies like Shell, Sears, IBM, Merrill Lynch, and other major companies. Most of the design and writing decisions Ogilvy makes are based on research—what works versus what does not work. He says people view ads from the top down. For that reason, he believes the visual should be at the top, with the headline beneath it, and the body copy below. He feels the visual should have a caption because most people read them. The text should be in a serif type style, flush left and ragged right, because most people are used to reading books and newspapers, which are almost always printed that way.[2] The example in Fig. 9.2 utilizes this format with a slight variation to add interest.

If you simply follow his advice, you won't go wrong. You can make variations to help your ad stand out, but being too cute will make it difficult for people to read and understand your ad. And, if you develop a headline and visual combination that work together to convey your key selling point, and write copy that succinctly conveys the benefits of your competitive advantages, you will create an advertisement that is likely to be effective.

Advertising for radio or television

Writing and designing advertisements for radio and television is obviously different than for print. However, the steps taken before creating a print ad should be followed before preparing an advertisement for the electronic media.

Few contractors use electronic media, however, so we will not put as much emphasis on it in this chapter. Television is usually prohibitively expensive because, with the exception of some local cable channels, it reaches such a large audience. For business-to-business advertisers, your money would be wasted advertising on TV because you will reach too many people who would never need your products or services.

Figure 9.2 People typically look at the photo in an ad, then read the text. Your ads should be laid out to help them.

Radio advertising has similar disadvantages for most contractors. Most radio stations have tens of thousands of listeners at any given time, many of whom have no need for your services. You must thoughtfully consider the cost for reaching a true prospect before deciding which is the best media outlet in which you should advertise.

Writing for radio.   Effective radio advertising demands good copywriting skills. Because there are no visuals in radio commercials, all of the images you need to communicate must be done via the written word and sound effects.

The most important part of a radio ad is the beginning. It must grab your attention. People tend to use the radio as background, either while they are driving, cleaning the house, or at work with the volume turned down. Moreover, in the car, people often switch radio stations during commercials.

For that reason, your radio commercial has to have a beginning that makes a listener stop and pay attention. Sometimes it is done with a dramatic statement, and other times it is done with sound effects.

Most radio spots are 60 seconds. Thirty-second radio commercials are rarely run because most radio stations charge about the same for them as for 60-second spots.

The most effective radio spots are ones that use dialog between two people. Listeners pay more attention to these. If you use this approach, choose two voices that are sufficiently different from one another.

If you are a local supplier of a nationally distributed product, you may be able to place what is called a *donut* commercial. Visualize a donut to understand what a donut commercial is. It is a studio-produced commercial that has a prerecorded beginning and end. It may be distributed to all the local suppliers by the national headquarters. The middle 10 seconds or so is left blank for you to customize with a localized message. If you listen to the radio, you will hear these often.

Because the greatest benefit of radio is its immediacy—if you want to get a radio commercial on air within an hour, you can usually do it—you may use it to increase last-minute attendance at a special event or lobby for legislation that is about to be voted upon.

Radio commercials are often relatively inexpensive to run, so you should thoughtfully consider how it might fit in to your overall marketing communication program.

Television commercials.   Television commercials can be the most effective type of advertising because they combine words with animated visuals. It is for this reason that the Web is attracting so much attention as well.

Although radio taps the particular skills of the copywriter, it is the visual images that usually determine whether a television commercial is successful. Television commercials are similar to print ads to the extent that they rely upon words and images working together to tell a story. The difference is that with television commercials, you tell a story with a sequence of visuals and words, whereas with print ads you only have one set of images and words to work with.

Television commercials are created using *storyboards*. Storyboards are sheets of paper, each with about eight illustrations that look like television screens above boxes for text. Once you have come up with a concept for your commercial, which you do in a similar way to a print ad, you put a drawing in each TV-screen-like box to depict each scene in your commercial. Under each, you write the text that would accompany the visual. Most storyboards for television commercials have only about a dozen scene changes.

The storyboard is used to present the idea for the commercial. Once the idea is approved, the storyboards are handed over to a producer who finds the talent for the commercial and hires a director to create the spot.

There are alternatives for companies with smaller budgets. Most television studios let advertisers use their studio to film ads, and some companies use one of their own people as a spokesperson rather than use talented actors. Just beware if you do that. We have all seen cheesy commercials that make us wince, and you can actually damage your reputation if you run a television commercial that makes you look stupid. So if you have a small budget and want to place a television commercial, get the TV station to help you create a tasteful spot that uses nice pictures of your products and a professional voice-over.

If nobody else in your market is using television, you may really stand out by trying a commercial on a cable station that may appeal to your target audience. As with radio advertising, it is something for you to consider when planning your marketing communication program.

Collateral Material

The term *collateral material* applies to a broad group of marketing communication materials that are used to support the sales process. They are sometimes referred to as sales support material because they are often delivered or sent to a prospect by a sales or business development representative.

Items in the category of collateral material include such things as

- Brochures
- Newsletters
- Project or product sheets
- Site signs
- Article reprints
- Letters of commendation
- Video and audio tapes
- CD-ROMs
- Software
- Coffee cups, tee shirts, and other giveaways

The more commonly produced items among the list, brochures, newsletters, and project data sheets, are intended to move the prospect from awareness and interest to desire and action.

These items are wasted if they are sent cold to prospects. First of all, there are far more cost-effective ways to create awareness. And second, most people will not take the time to read a brochure or newsletter if they do not already know something about the company that sent it. The best strategy would be to establish awareness first, through advertising, public relations, or even a letter, before sending a brochure to a prospect.

Following is an explanation of how to best produce these elements. Your own personal taste for design and copywriting style will of course come into play when you create these materials. However, an understanding of the basics of each item's strengths and weaknesses, and some dos and don'ts about how to produce them, will help you create more cost-effective materials that achieve your intended results.

Brochures

At one point or another, just about every contractor finds a need for some type of a corporate brochure. Perhaps you have met someone at a luncheon who expresses interest in your firm's services. You've described some of the projects you have completed and they seem applicable to the prospect's upcoming project. The next thing you know, the prospect asks you to send your corporate brochure.

Corporate brochures can take many forms. They can be printed and ready to take off the shelf, or they can be customized by using a variety of preprinted or computer-generated individual sheets. There are advantages and disadvantages of either method.

Preprinted brochures are easy to use because they are always ready to pull out of a box and send, whereas brochures that need to be assembled require someone's time whenever one is sent out (see Fig. 9.3). If you send out more than a couple of brochures a week, customizing brochures can get tedious if you are busy.

Customized brochures offer more flexibility, however. If you get enough different inserts prepared, you can customize your brochure to suit the needs of each prospect or proposal. It also does not become easily out of date because you can print individual pages that contain information that changes often, like names of proposed project staff and your completed project list.

However, this flexibility comes at a price. The per-unit cost of individually assembled brochures is much higher than preprinted brochures, which are often purchased in quantities of at least 1000. Nevertheless, if you have a tight budget, the total cost of producing 50 customized brochures is less than the cost of 1000 printed brochures. If you don't send out more than 10 brochures a month, customizing will probably make more sense for you.

Typically, contractors use a combination of both: a preprinted brochure that focuses on selling points that are not likely to change over a few years and

Figure 9.3   A preprinted corporate brochure.

individual sheets that profile projects or project types, which gives them the flexibility to update information as their services change.

Preprinted brochures. Preprinted brochures have a cover that typically includes the company's name and an eye-catching graphic element like a photo of a notable project or an artistic image of a construction site at sunset (see Fig. 9.4). Whatever is chosen for the cover, it must attract attention and begin to convey the selling points you want to convey.

Most brochures are organized using an outline that includes an introduction, the body, which is usually a presentation of your competitive advantages (which you identified when you prepared your creative platform; see Chap. 8), and a conclusion. The length of the brochure is driven by how much space you need to present the items in your outline.

The brochure developed by the hypothetical interior contractor mentioned earlier might have an outline like this:

1. *Introduction.*   An overview describing the company's services and competitive advantages.

2. *Experience.*   A description of the benefits of over 30 years of office renovations.

3. *Project management.*   How our flat organization chart, service-orientation, and reputation for being nice people to work with saves our clients time and money and minimizes stress.

4. *Quality.*   The beautiful projects we build and the numerous awards we have won support our reputation for quality.

5. *Conclusion.*   If you need an interior contractor who understands all the special requirements of renovation work and want to protect your building investment, we're the right contractor for you.

Developing an outline that conveys your competitive advantages is only one of the many ways to organize a brochure. As an alternative, you may want to organize it by product or project type. For example, a masonry contractor may want to show commercial office buildings, schools, industrial projects, and health care and educational facilities on which it has worked. The text can still focus on selling points. Some brochures focus less on a formal organization and story telling and prefer to let the pictures of their products or projects tell the story. They use dramatic photos supported by only a little text (see Fig. 9.5).

How you decide to organize your brochure is largely subjective. It is based on your personal opinion about how prospects make decisions about selecting a contractor and how best you can portray your competitive advantages. Whatever you decide, there are certain rules that should be followed to produce a successful brochure. They are

Plan.   Once you have developed your marketing communication plan, you will have a good sense of your target audience and competitive advantages.

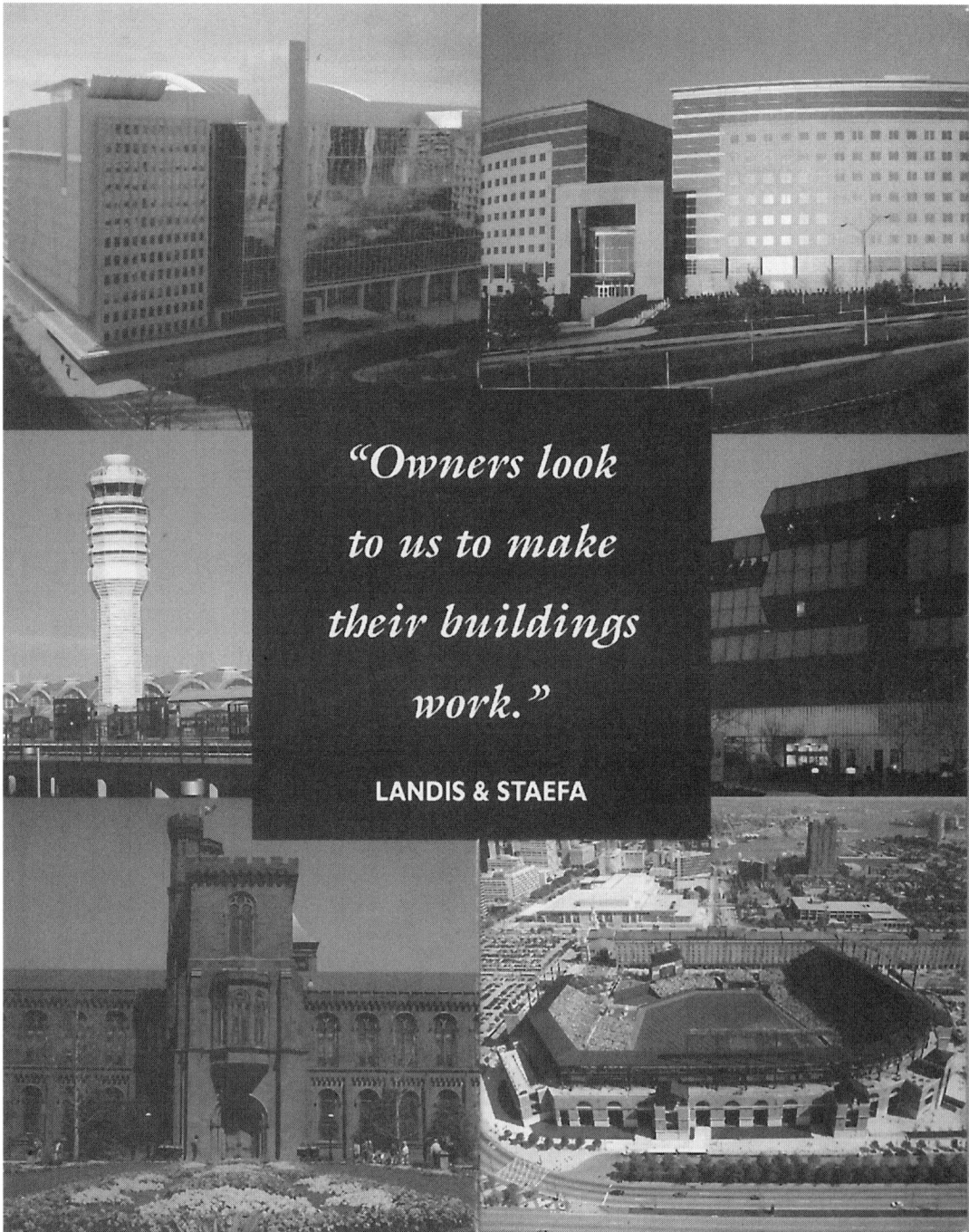

Figure 9.4 Preprinted brochures typically include the company's name and an eye-catching graphic.

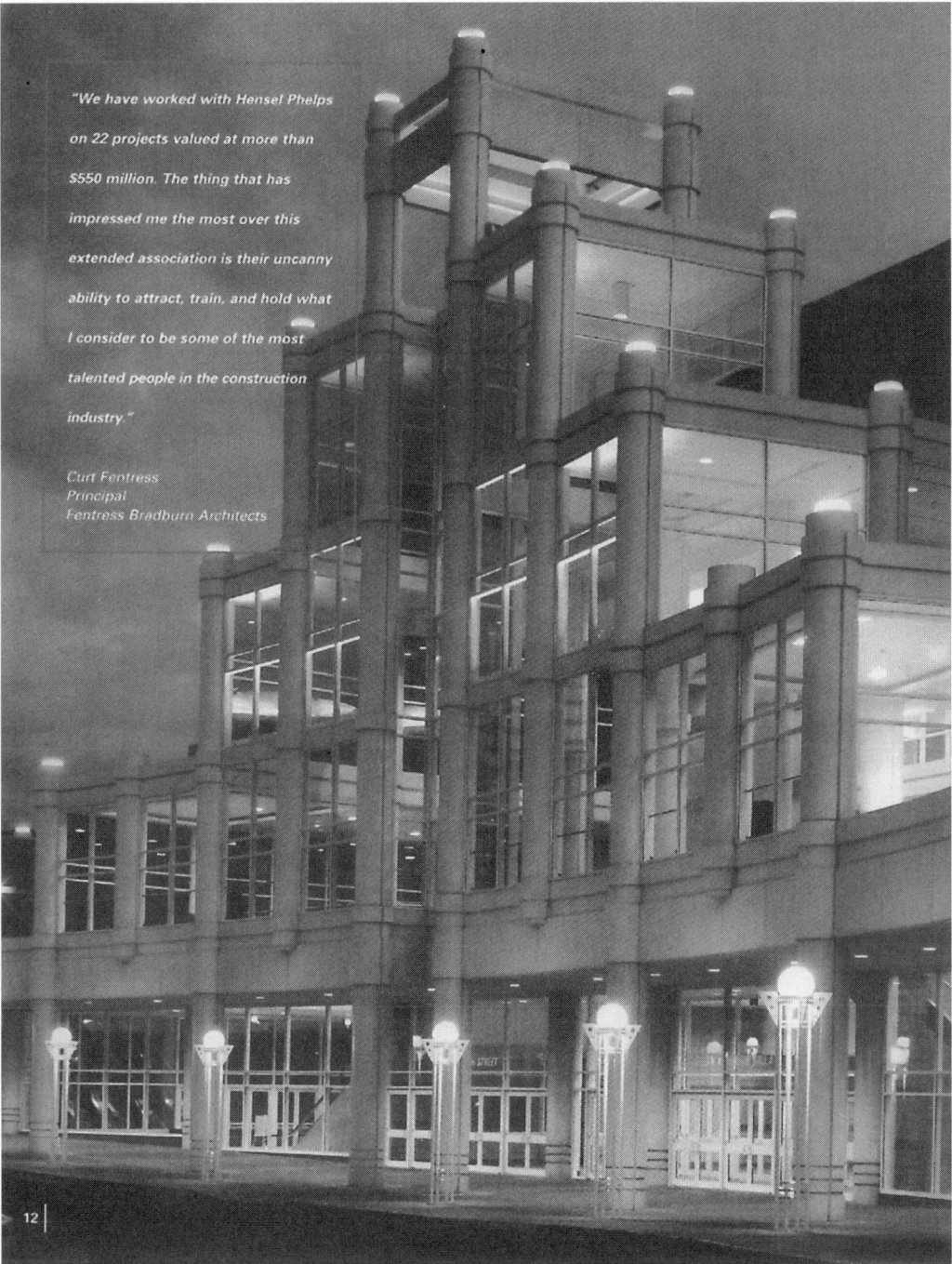

"We have worked with Hensel Phelps on 22 projects valued at more than $550 million. The thing that has impressed me the most over this extended association is their uncanny ability to attract, train, and hold what I consider to be some of the most talented people in the construction industry."

Curt Fentress
Principal
Fentress Bradburn Architects

12

Figure 9.5 Brochure interiors can vary in format as long as they tell the company's selling points.

Sahara Hotel and Casino
Renovation and Expansion
Las Vegas, Nevada

Aims Community College
Student Center
Greeley, Colorado

talented people

Tyson Foods Corporate Headquarters
Springdale, Arkansas

Colorado Convention Center

Denver, Colorado

The Lofts at Streetcar Stables
Denver, Colorado

13

Figure 9.5 (*Continued*)

You should know what selling points to communicate and the image you want your company to convey.

Choreograph.   Before you write or design your brochure, you should have a general sense of how you want it to look. Either make an outline or do a thumbnail sketch of which photos will appear on each page and where the text will go.

Simplify.   Often, companies want to say everything they can think of about themselves in their brochure. They want to show every variation of project or product they have created. The result is often a cramped, text-heavy brochure that doesn't effectively convey selling points. Instead, simplify. Reduce the number of photos and cut the text. A brochure is intended to make a positive impression, not close a sale. If you get your selling points across, you will have ample opportunities to meet with a prospect and convince him or her that you are the best choice.

Maximize drama.   People want to be entertained. Your brochure is one of your best opportunities to entertain prospects. Use big, dramatic photographs and short, punchy text. Wow them.

Minimize input.   Once your company has agreed upon the concept for your brochure in a general sense, minimize the number of people who have to be involved. Everyone has different tastes and opinions, and decisions by committees often produce camels.

Proof it.   You should have several chances to proofread your brochure before it is printed. Make sure you do so very carefully. Little mistakes get by the best proofreaders, so have more than one person do it. Pay particular attention to people's names, captions, phone numbers, and fax numbers.

Manage it.   The last place you want to see mistakes happen are at the printer. Once your brochure is printed, it's too late to correct errors. Invest the time to go to the printer when the brochure is on press and inspect press sheets. Make sure the color in the photos is exactly how you want it and that the quality meets your expectations.

Customized brochures.   If you decide to customize your brochures each time you need to send them out, there are options for how to produce them. You will probably either print quantities of individual sheets and choose the ones that enable you to customize, or you will design individual sheets but keep them in the computer for additional customization if needed. When you want to send a brochure, you will print the sheets you need (see Fig. 9.6).

In the past the only way to reproduce good quality color photos was to print them on a press. Color printing can be very expensive, though, and is not at all cost effective on a unit basis with quantities of less than 500. If you only want 80 individual color sheets at a time, or even 50 to ensure that what you print is never out-of-date, there are now far more cost-effective alternatives than printing on a press.

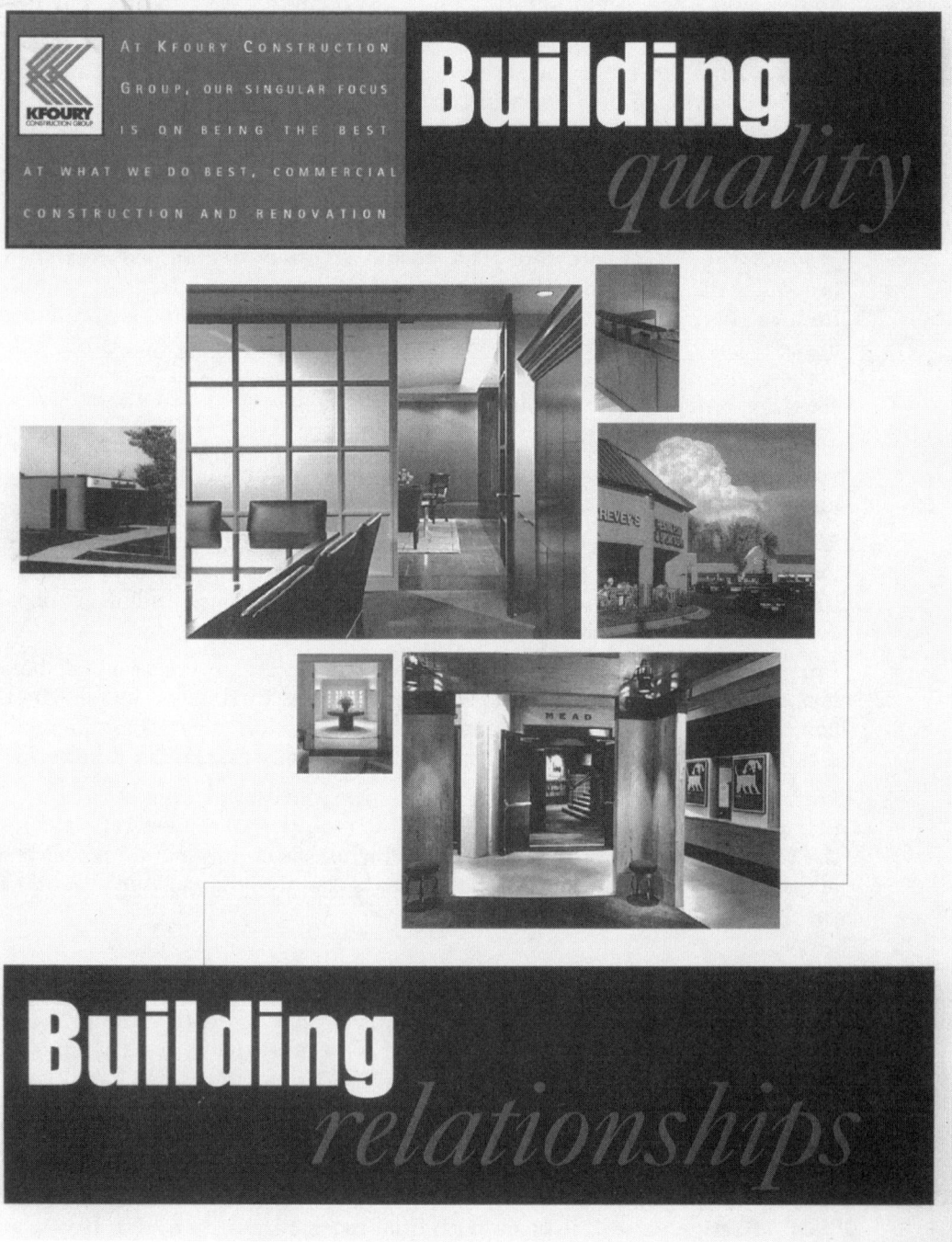

Figure 9.6 Customized brochures use preprinted individual sheets that can be assembled when needed.

Building
quality

Since 1982, our mission has remained unchanged—to be the best—not biggest—full service general contractor in the Washington DC area. We promise to build both quality and relationships.

We strive to build meaningful relationships with our clients and subcontractors. We want to be fair, open and contributing members of the project team.

Building relationships is only half the challenge. We are also driven to build the best quality space. Quality means staffing all our projects with senior level managers and field supervisors, and hiring only the best subcontractors. You will not find a more experienced project team working directly with clients than at Kfoury Construction Group.

After 625 completed projects, we are proud to say our formula for success has been confirmed time and time again. Select Kfoury Construction Group, and you will see how we live up to our promise of building quality—building relationships.

Building
relationships

We staff all our projects with senior level managers and field supervisors

Figure 9.6 (*Continued*)

Construction Capabilities
from start to finish

Kfoury Construction Group offers a professional level of general contracting services throughout the entire construction process.

BEFORE CONSTRUCTION

We like to get involved early in a project – before the design stage is completed. That's is when we can put our knowledge and experience to greatest impact. We can assist the architect, engineer and owner by offering the following pre-construction services.

- Detailed budget estimating
- Value Engineering
- Project scheduling
- Site survey/evaluation
- Lead-time research
- Permit assistance
- Subcontractor qualification

DURING CONSTRUCTION

The construction business is fast paced and our clients must have ready access to decision makers. That is why our project managers and superintendents work directly with clients. And why our project managers, estimators, accountants and superintendents' computers are networked in order to share access to critical job information. We maintain a fast-paced job site by having the technological capability to download changes transmitted to us electronically, produce revised estimates, and e-mail revised drawings to construction crews.

Kfoury Construction Group is experienced in working in occupied spaces, as well without disrupting our clients' operations. We do this by careful planning and maintaining clean and organized job sites.

AFTER CONSTRUCTION

Most companies occasionally need minor modifications or small repairs. To meet these everyday needs, we formed Prosum Service Corporation in 1995. Prosum is staffed to handle small scale repair and maintenance projects. Fast, responsive, courteous, and clean service is their hallmark – the perfect fit for the hectic pace of most businesses today.

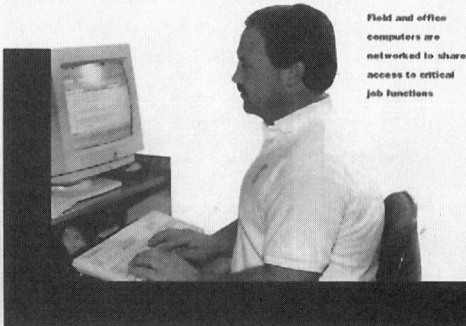

Field and office computers are networked to share access to critical job functions

Fast, responsive, courteous and clean service on office repair projects

Figure 9.6 (*Continued*)

Calibrated laser image (CLI) prints are color prints produced on a high-quality color laser printer. The quality is nearly as good as traditional color printing on a press. And yet, the total cost is far less at small quantities (see Fig. 9.7).

To produce these CLI prints, you design what is to be printed just as you would design something to be printed on press. Once you design the individual page layouts on the computer using whatever desktop software you like, give it to the firm that creates the CLI prints. The image is transmitted straight from the computer to the laser printer, not printed and then copied on a flatbed copier. The quality is a result of the high-quality computer equipment, high-resolution laser printer, and the ability to calibrate, or control, the colors.

CLI prints can be designed and printed in as few as 3 days. So if you have the opportunity to submit a proposal for a stadium, but have no individual sheets that show your stadium experience, you can design and print enough CLI prints to satisfy your needs.

If you keep individual sheets on your computer and print them out when you need them, rather than using preprinted sheets, just make sure that the quality of your color printer is sufficient to create the type of quality that befits your company.

Regardless of how you develop your brochure, the rules given above apply. Make sure your brochure clearly communicates your selling points and does so in an attention-getting, memorable fashion.

Newsletters

Newsletters are perhaps the most misunderstood, and yet potentially the most valuable, of all collateral material. Unfortunately, when most people think of newsletters they think "junk mail." That is because they are thinking about some of the newsletters they receive, ones that use most of their space bragging about the latest conquests of the firm sending it. Or worse, it devotes space to the outcome of the company's softball team's games, recent weddings, or new births. It's not that these subjects are unimportant. They are important to the employees of the firm sending the newsletter, but they should only be distributed within an internal publication.

The truth is, most people outside your company don't have the time or interest to read about your successes or personal matters. They do, however, have time to read information that will help them do their jobs better—and that is where good newsletters succeed.

Good newsletters make you want to read them. Think of the newsletters you have read. They usually aren't flashy. In fact, slick designs probably make you suspicious about the editorial value of what is within.

The newsletters you like have interesting and important content. Maybe it's financial advice, like what is found in the *Kiplinger Letter*. Or maybe it's thought-provoking vignettes, like those found in *Bits & Pieces*. They succeed because they don't waste your time telling you about things that aren't of

NATIONAL MARITIME INTELLIGENCE CENTER

SUITLAND, MARYLAND

© Max MacKenzie

Owner

U.S. Navy, Department of Defense
Suitland, Maryland

Architect

HOK/RTKL/DMJM (joint venture)
Washington, D.C.

General Contractor

The George Hyman Construction Company
Bethesda, Maryland

DYNALECTRIC

This 700,000-square-foot office and data center is located on the grounds of the Federal Center Complex in Suitland, Maryland, and consists of 600,000 square feet of office space and a 100,000-square-foot parking garage. Dynalectric installed the facility's power distribution system, which originated at unit substations located in the central plant and fed four transformer vaults. The project also included a sophisticated cable management system, which provided all necessary information regarding the complete cellular deck system. The emergency power backup system consists of six 1000 kw emergency generators with sequential paralleling switchgear. In addition, four 750 kva UPS systems augment the emergency power backup system. The project entailed the preparation of extensive, highly-detailed drawings, which were all generated by Dynalectric's in-house Coordination Department. Recognizing the company's outstanding efforts on this project, the Washington Building Congress presented Dynalectric with a Craftsmanship Award for its power distribution systems.

Figure 9.7   Color-calibrated laser image prints have nearly the quality of printed sheets but can be purchased in small quantities.

interest to you. They tell you things that help you do your job better or make you happier.

If you send out a company newsletter, it can, and must, do the same thing. If it does, it will be read and saved. And if that happens, it will become an ongoing source of business because you will have kept your name and information about your competitive advantages in front of your prospects regularly.

To create a successful newsletter, you have to understand your prospective customers' needs. If you took the time to develop a marketing communication plan (Chap. 8), you will already know their needs.

Once you understand their needs, you must satisfy them. If you are an interior contractor and most of your prospects are end users, give them advice about how to save money by doing their own minor repairs. Or tell them how to figure out how much space they need when they are looking for new office space. Give them a checklist of what they need to do when they move. Give them unit costs for tenant work (see Fig. 9.8).

You can think of many things that will help prospects. Most companies are afraid to do this because they think they are "giving away their work." Perhaps you are, but a small amount of good will goes a long way. People will save your helpful newsletters. They will remember your name. And when it comes time to hire a contractor, they will already know your name and some of your competitive advantages.

Writing and designing newsletters.   Writing a newsletter is less like writing an advertisement or brochure and more like writing for publicity (see Chap. 10). Ads and brochures demand brief, concise copywriting and rely more on making impressions than on providing details about your selling points. Publicity writing, like guest articles in newspapers, demands a well-thought-out and complete explanation of your selling points, starting with the most important and ending with the least important.

Headlines are all important for newsletters, like they are in newspapers. As readers scan the newsletter, you need to grab their attention with a headline. Article subheadings and captions are also important because people will usually read them before they read the articles. If you can get your selling points across in your headlines, subheadings, and captions, you will have accomplished much.

Designing a newsletter is more like designing a magazine. The design has to be interesting enough to capture attention but should not detract from the articles. The design needs to help the reader focus on the articles (see Fig. 9.9).

If the newsletter is mailed folded, a summary of what is inside it should be printed on the outside portion that people see first. That will help them decide to read it if they are scanning their mail and deciding what not to throw in the trash.

Overall, the design elements like paper, colors, and type style, should be consistent with the design elements used on your other marketing communication materials.

infocus:

relocation

Company Relocations Take Planning, Commitment and Teamwork

Moving into new office space is an enormous endeavor. It requires detailed planning and the total commitment of both management and employees.

"Tenants need to begin planning *at least* a year in advance," says Peter Berk of Smithy Braedon/ONCOR International of Fairfax, Va. "At that point, you need to assess your present *and* future needs and ask yourself, 'Will growth require more space, or will technology lessen the number of employees we'll need?'"

Berk added, "After the evaluation, typical tenants need about six months to research the market, evaluate proposals and negotiate a lease. The interior fit-up of your space can add two to four months."

Large space users and tenants with special needs face even greater challenges. "There are only a handful of large blocks of space available today," Berk said, "and many of the ones about to come onto the market aren't published. You need to work with a broker to set up a timeline and evaluate all the possibilities for space."

The planning process for Kfoury Construction Group's upcoming move to its own building in Reston, Va., took about 18 months. After deciding that moving into a new space would better suit the company's present organizational structure and future needs, the first step was hiring the necessary external support, like an architect, engineer, and other consultants. "They set the foundation with proper documents," said Maureen Dwyer, Kfoury's Director of Continuous Improvement.

After the documents were completed, Dwyer, who is trained as an architect, began developing the interior scheme. The new 10,000-square-foot space was designed with open floorplans to accommodate its project team structure (see *profiles* inside).

Meanwhile, Kfoury employees compiled a comprehensive list of nearly 250 items that would have to be taken care of before the move. The list included obvious items like address and phone number changes, critical ones like cabling for the computer system and notifying police and fire departments, and mundane details like buying paper towels and coat hangers.

"In the end, it has to be a seamless move as far as our clients are concerned," said company Vice President Tim Reese. "We can't be consumed with moving for an entire week. We have ongoing jobs that can't be interrupted."

Kfoury also produced an internal newsletter to inform employees about what they needed to do as move day approached.

A move of this magnitude takes the commitment of upper management, plus two or three key people to perform the wide range of tasks required. "This is truly a team effort," expressed company President Jorge Kfoury. "One person simply couldn't do it alone." ∎

> **" This is truly a team effort... One person simply couldn't do it alone. "**
>
> *Jorge Kfoury*

Figure 9.8 Newsletters are most effective when they provide information that is of value to the reader.

profiles

Kfoury Relocates to Better Serve Clients and Employees

> One of the advantages of having an open plan is that people in the team can hear what's going on with each project.
>
> *Jeff Martello*

On May 22, 1995, Kfoury Construction Group will move to larger offices in a building it will own in Reston, Va. The reason for the relocation is twofold, explains company President Jorge Kfoury: "We need more space to accommodate our project team structure and we want to be property owners to show our long term commitment to the business and to the community."

The move allows the company to create spaces specifically geared to the needs of the project teams, with ample room for team members to work close to one another. Kfoury's project teams include a senior project manager, two project managers, a project coordinator and a project accountant

Jorge Kfoury

"The perception is that teams require less office space," said Kfoury. "But in reality, we need more because teams require additional open space, separate conference rooms and support areas for each team."

"One of the advantages of having an open plan is that people in the team can hear what's going on with each project," continued Vice President Jeff Martello. "That's the real beauty of the team concept. Clients don't have just one access point on a project but three or four."

The new office will be divided into three connected "pods." The first pod includes the main entrance to the building, a reception area, executive, preconstruction and general administrative areas, as well as a room where subcontractors can review plans. The other two pods include the project teams and accounting staff. The total space is approximately 10,000 square feet.

The project team areas will have built-in custom millwork stations with tables for reviewing plans, a return area for a computer and a work surface. Work stations are laid out identically so if people switch teams they will feel comfortable in their new space. Each project team area also includes a conference room, an administrative area with a printer, copier and fax machine and a kitchenette. "It's like three separate companies, with all the required technology centered within each team," explained Martello.

Most office spaces have windows and the openness of interior areas will allow an ample amount of natural light to flow throughout.

shorts

KFOURY
CONSTRUCTION GROUP

■ Kfoury Construction Group recently promoted Michael Willard to chief estimator and Iris Boulware to project estimator. Willard has 21 years of construction experience in the Washington area. His new duties will include the management of all Kfoury estimates as well as subcontractor relations. Boulware, who joined Kfoury in 1993 as an estimating assistant, will now manage all activities necessary to develop estimates for specific projects.

■ Kfoury Construction Group led a group of volunteers who renovated the home of an elderly woman as part of *Christmas in April* on April 29, 1995. Kfoury's 30-person team was joined by Jefferson Millwork & Design of Sterling, Va. and Smithy Braedon/ONCOR International of Fairfax, Va. The home was in desperate need of carpentry, painting, plumbing and electrical renovation. *Christmas in April* is a nationwide effort to renovate the homes of elderly, disabled or poor homeowners.

■ Work began recently on a major renovation for Georgetown University Medical Center at 3833 North Fairfax Drive in Arlington, Va. The project encompasses the entire 50,000-square-foot building and includes substantial sitework, electrical and mechanical system upgrades, exterior facade renovation, window replacement and new interior finishes. Designed by the architectural firms of Elefante Mallari and Hansen Lind Meyer, the new facility will offer a broad range of medical services.

Figure 9.8 *(Continued)*

New Headquarters

checklist

The company will also upgrade all of its office equipment and systems when it moves. This includes a UPS system that will allow the company to function even during a power outage. New upgraded office equipment will include five plain paper fax machines hooked in sequence to accommodate high volume in a short time frame. All computers will be networked so project managers can take a lap top computer into any conference room and pull up data. The office will also be networked to all job sites, where most superintendents are already trained and equipped with computers.

Kfoury Construction Group's new offices will be located in the Sunset Hills Professional Center, a two acre office park located at the intersection of Sunset Hills Road and Wiehle Avenue, just off the Dulles Access Road. "When we were looking for a site, we wanted easy access to our clients. The new building is right off a main artery and is easy to get to," said Kfoury.

One of the greatest attractions of the new site is the surrounding amenities. With Reston Town Center less than a mile away, there are dozens of restaurants and shops, as well as jogging trails, an ice skating rink, golf courses and other amenities for employees. ∎

A partial list of things to consider if you are planning an office relocation:

☐ Begin the process *at least* a year in advance by assessing your future needs for space. Ask yourself questions like, "Will growth require more space or will technology lessen the number of employees you need?"

☐ Hire a broker who can help you establish a timetable and acquaint you with the current inventory of space.

☐ Assign a group of two or three employees to spearhead the move. Make sure they get the full commitment of upper management.

☐ Hire an architect, engineer and other consultants to establish a good foundation with suitable designs and proper documents.

☐ Make a list of everything you need to take care of prior to the move. Don't forget the smallest details like turning in old keys and finding out when the new landlord allows moves to take place.

☐ Find a qualified general contractor to build your new office space.

---

for more information ...

KFOURY
CONSTRUCTION GROUP

☐ Please send me additional information about your general contracting services.
☐ Please add my name to your mailing list.

Name_____

Title _____

Company _____

Address _____ Suite_____

City _____ State_____ Zip_____

Telephone _____ Fax_____

Mail this form to **Kfoury Construction Group**, 102 West Jefferson Street, Falls Church, VA 22046-3444.
Or fax form to (703) 241-2411.

Figure 9.8   (*Continued*)

Photos: Richard Anderson

Martin Marietta

Maintenance Services: Branching Out

First impressions. So often it is our initial reaction to a place or a person that is remembered months, even years, after the first contact. At Chapel Valley Landscape Company, we know how important it is for our clients' properties to make a favorable impression on their clients, employees and visitors. We also know that maintaining a positive impression is important.

While Chapel Valley is best known for quality landscape installation, we are committed to better serving our clients through complete landscape management services. Over the years additional specialties have been developed in design/build, water management, residential, and maintenance.

The Virginia and Maryland Maintenance Branches at Chapel Valley have grown tremendously over the years. Our success is apparent in the projects we maintain—**EDS Headquarters, Martin Marietta Corporate Headquarters, Rockledge Executive Center, Owings Mills Corporate Campus, L'Enfant Plaza,** and **Tysons International** to name a few.

Quality improvement is key to our maintenance approach. Each project is monitored with consistency and flexibility. While strict attention is paid to the specifications of a given project in order to uphold schedules, budgets and productivity, our field managers and foremen can offer suggestions to avoid problems that can occur when nature does not cooperate.

Chapel Valley is dedicated to on-going staff training at all levels. Many of our project foremen are Certified Professional Horticulturists and all are provided with extensive training. Each month, maintenance crews are involved in quality control checks, training through our mobile training unit and on-site assessments.

Chapel Valley has tailored its staff structure and procedures to better meet the needs of our maintenance customers. Each of our customers benefits from

Accurate Appraisals. Even before a project is accepted, careful consideration is given to every detail of a job site. The initial appraisal is crucial in meeting a client's expectations and in helping to anticipate problems before they happen. Our team of estimators and horticulture experts works with each client to produce the best possible maintenance program.

Client Communications. Regardless of the maintenance project, a written status report is provided to our clients by the branch manager on a monthly basis noting any changes in schedules or specifications. Our foremen and field managers are always available to our clients, and they make personal visits to the client a part of each site visit.

Special Projects. Chapel Valley employs a full-time specialty field manager who deals

CONTINUED ON PAGE 6

Chapel Valley Landscape Company

Figure 9.9   The design of a newsletter should grab attention but not take the focus of the reader away from the articles.

Project or product sheets

For most companies, a primary way to demonstrate their experience is by showing examples of products or completed projects. These are typically printed on $8\frac{1}{2}$- × 11-inch paper and are often called project data sheets, project profiles, or product sheets (see Fig. 9.10).

What you show depends on what your prospects want to see. If you are a contractor, architect, or engineer and are targeting developers, architects, or general contractors, it is likely that they want to see finished work. Even if you

A Residence in Potomac
Grand Award, Total Residential Contracting   1.

Tysons International
Grand Award, Commercial Landscape Maintenance   2.

Dorsey's Search Village Center
Grand Award, Commercial Installation   3

A Residence in Potomac
Grand Award, Residential Landscape Maintenance   4.

Photos: Scott Sanders

1. The naturalized beauty of this residence is attributed to the combined efforts of the owners and landscape contractor to create a setting sensitive to the existing flora and topography. Upon initial inspection, however, it was determined that the soil was in poor quality; therefore, Chapel Valley carefully regraded the site, literally improved what was poor soil, and assisted the owners in the management of all subcontractors.

2. Tysons International's landscape contains more than 5,000 perennials that provide colorful, year-round interest. The intensive maintenance routine involves periodic division of plants, proper fertilization, and pruning techniques which maximize the plant's blooming effect.

3. This shopping center's courtyard, which is surrounded by Bradford pear trees and filled with colorful perennials, is the focal point of the site throughout the year. Chapel Valley successfully overcame a difficult construction schedule, which required installation of plant material during the hottest summer months.

4. The lively composition of grasses, perennials, and flowering evergreen shrubs creates a symphony of color, texture, and motion at this Potomac residence. It also keeps the maintenance team busy throughout the year.

Chapel Valley Wins Four Grand Awards

The Landscape Contractors Association of MD•DC•VA (LCA) recently awarded Chapel Valley Landscape Company four Grand Awards for its quality landscape installation and maintenance in Maryland and Virginia. A Grand Award is the highest honor that LCA awards to industry professionals.

Chapel Valley was the only company that won four grand awards in four separate categories—commercial landscape installation, commercial maintenance, total residential contracting, and residential maintenance. The projects included **Dorsey's Search Village**

Center in Columbia, Md. (landscape installation), **Tysons International** in Tysons Corner, Va. (commercial maintenance), and a residence in Potomac, Md. (total residential contracting and residential maintenance).

Each project had a unique set of challenges and a variety of opportunities for creative problem-solving. All of the more than 55 projects entered in the LCA competition were judged first from slides and narrative of each project, and then by actual site visits.

Chapel Valley would like to thank

CONTINUED ON PAGE 3

2.

Figure 9.9   (*Continued*)

are a specialty subcontractor or a mechanical or electrical engineer, it is often best to show how your work fits into the overall project. You may also want to show a smaller detail shot of your work product, particularly if it is unique or complex. By and large, however, a prospect is interested in seeing if you have relevant experience in projects similar to the one they are planning to build. So showing a completed arena instead of concrete seat bends will quickly convey your related experience.

Branch Updates

Maryland Maintenance

Chapel Valley's involvement with **The Washington Home**, a long-term care nursing facility located in Washington, D.C., began in the fall of 1990 when its two courtyard gardens were installed. The inauguration of one of the courtyard gardens, The Barbara Bush Garden, was scheduled for April of 1991.

The recently installed Barbara Bush Garden was in excellent condition; however, the property surrounding it was in need of major renovation. The Maryland Maintenance Branch was approached in early April to evaluate the property with yearly maintenance in mind along with providing a "face-lift" for the entire property by the April 22nd ceremony.

Existing conditions that needed to be addressed in less than two weeks included: removal of the existing shredded bark from all beds, establishment of a crisp edge along all planting beds, cleaning of many beds with existing weed problems, pruning and shaping of all shrubs, and turf maintenance.

Two crews were immediately assigned to the job, and after more than 160 man hours, 100 cubic yards of mulch, and each specific concern addressed, the site was in top condition for the inauguration and a personal visit from Mrs. Bush.

Virginia Maintenance

A metamorphosis has occurred on the 18 medians between 15th and 22nd streets on **Jefferson Davis Highway** in Crystal City, Va. Originally, the county-owned property was unkempt and had virtually no landscaping—a company came in once a month to bushog and mow. Now, the Virginia Maintenance Branch maintains the property, which contains approximately 7,000 vibrant annuals, along with perennials, trees, shrubs and turf!

The challenge has been to maintain the municipal highway project as though it were a private office park. Due to high pedestrian and vehicular traffic, Chapel Valley pays a great deal of attention to employee safety and environmental concerns. Because it is a non-irrigated site, the medians require daily attention to monitor soil moisture. To meet the demands of this project, two full-time employees provide water on a daily basis. This has been accomplished by hand-watering and the use of a 500-gallon capacity water truck with 2,000 feet of hose. Chapel Valley also has a four-man maintenance crew on-site each week.

Thanks to the diligent efforts of **Charles E. Smith Company** and Chapel Valley there is now a lush oasis where only an asphalt desert existed.

Maryland Landscape

Our design/build project, the new **McCormick & Company Headquarters** in Hunt Valley, Maryland, has been at the forefront of the Maryland Landscape Branch's operations recently. Working closely with the **Gaudreau Architectural Firm**, Chapel Valley's and the architect's goals have been to preserve as many of the mature trees from the site as possible, while adding to the feeling of serenity that this site offers.

Chapel Valley Landscape Architect Eric Rains designed the plan that blends 6" and 7" caliper Honey Locusts, 12' Blue Spruce, 20' White Pines, and 10' Kousa Dogwoods into the landscape. Many more deciduous and evergreen trees and shrubs were added to help smooth the transition and balance the scale of the new building and natural trees.

One of the site's focal points is an aquatic pond, that will be surrounded by color throughout the spring and summer. More than 16,000 square yards of wildflowers, 40,000 narcissus, 2,800 daylilies, 250 rhododendron, and other flowering plant material adorn the site.

Whenever possible, we try to preserve beautiful, natural environments rather than razing sites and starting over with new, smaller plant material. The cost that was budgeted for clearing can be used to purchase larger additional plants.

Water Management

Working in a retail-oriented environment requires careful coordination so that the flow of customers is not interrupted. The **Montgomery Mall** was especially challenging as Chapel Valley worked with as many as six subcontractors without disturbing the normal operations of the mall.

CONTINUED ON PAGE 8   6.

"Whenever possible, we try to preserve beautiful, natural environments..."

Editor's Note: Branch Updates is a regular feature of *Landscope* which highlights projects from Chapel Valley's six branches. The Virginia Landscape and Residential Branches will be highlighted in the spring issue of *Landscope*.

Figure 9.9   (*Continued*)

If you are a supplier, you may want to show a large picture of your product so prospects can compare it to competitive products; also show a smaller photo of how the product is used.

The design style of your project or product sheet can take many forms, depending on the amount of information you want to present and the design elements in your other marketing materials, with which it should be consistent. Typically, project data sheets include a photo of the completed project

Lincoln Center

Minneapolis, Minnesota

Construction of this towering 32-story office building with three levels of underground parking was enhanced in several ways by the selection of Ceco as concrete form-work subcontractor. Ceco's use of wide module construction and haunched girders reduced concrete quantities in the floor system and shearwalls of this 820,000-square-foot structure. In addition, larger bays resulted from the long spans of space created by wide modules. All horizontal and vertical formwork was performed by Ceco's able crews who used flying forms to complete a 25,000-square-foot floor plate every six days.

Developer
Lincoln Property Company
Dallas, Texas

Architect
Kohn, Pederson & Fox Associates
New York, New York

Structural Engineer
Brockette, Davis & Drake
   Associates
Dallas, Texas

General Contractor
M.A. Mortensen
Golden Valley, Minnesota

CECO

Figure 9.10 Project data sheets show pictures of projects and provide important information.

plus possibly a detail photo showing a specialty trade; the name, city, and state of the project; the primary project team members you feel are important to mention; and a description of the project. You may want to include a definition of the project scope, in size, contract value or duration, and a description of how the client benefited by working with your firm.

Product sheets include much of the same information, but the description is usually more technical in nature. Rather than listing a project team, it should list specifications. Product and project sheets must also include contact information, such as your company name, address, phone and fax numbers, and Web address.

Your decisions about design and copywriting should be based on your in-house talent and the expectations of your prospects. Most consumers of construction materials and services are sophisticated and are used to seeing professional graphic design and well-written text.

In addition to good design and writing, the quality of the photographs you show will have a big impact on how effective your data sheets are. You have options in obtaining photos. Read the section later in this chapter entitled Producing Your Marketing Communication Materials, for a discussion of the options given your needs and budget.

How you reproduce your project data sheets depends largely upon the quantity you intend to print. For quantities of 500 or more, traditional press printing is the best option. It gives you the best quality and is competitive on a unit cost basis versus other options.

If you print fewer, you should consider CLI prints (see the subsection above on brochures).

Site signs

Site signs are perhaps your best forms of advertising. They instantly associate you with your work product, are totally credible because you are making no claims about your superiority, can be highly visible to a large audience, and are essentially free.

You should have a published guide, or *style manual,* that outlines how any size or shape sign you may be able to install on a job site should be reproduced. You should be consistent with every sign you produce, in size, color, and location on the job site.

Anytime you are contracted to build, you should place the largest sign you can in as many locations around the job site as possible. Your business is unique in this regard. How many other businesses get to advertise their work in progress as blatantly?

Truck and equipment signs are also important because they maintain name recognition. You should include dimensions and locations of these signs in your style manual as well. Moreover, make sure your trucks are always well painted and clean. A contractor has many more opportunities with outdoor advertising such as these. You should always take maximum advantage of them.

Article reprints

Getting publicity for your company is discussed in great detail in Chap. 10. There, you will learn how to get publicity by sending press releases, generating feature stories, and writing guest articles in publications, among other things.

Having an article printed about you is a tremendous boost to your marketing effort. It has been said that people believe what they read in print 10 times more than they believe what you say about yourself in ads or brochures. Therefore, any time you are lucky enough to have a positive article written about you, or are prominently mentioned in an article, you should get reprints and include them in your marketing packages.

Some publications will reprint articles for you; they usually charge a fee. Some will give you permission to copy or reproduce the article. Either way, seize every opportunity to do so.

Other forms of collateral materials

Although brochures, newsletters, project data sheets, signs, and article reprints are the primary forms of collateral material used by most construction-related firms, other methods are also used, among them

- Letters of commendation
- Video and audio tapes
- CD-ROMs
- Software
- Coffee cups, tee shirts, and other giveaways that feature your company name

All of these can be effective. How you use them depends on their relevance to your type of work and the needs of your target audience.

Testimonials, in the form of letters of commendation, are one of the best ways to promote your products or services. Readers are more likely to believe what other owners, architects, or contractors say about you. If your clients don't send you letters at the end of a successful project, ask them to write a nice letter. They usually will. These should go in your marketing packages and be used in proposals.

Videotapes may also be useful if the best way to promote your products or services is by showing a process at work. Still pictures can't capture a special rigging system, for example, but a video may. Videos can show how much faster or quieter a certain product is. The only disadvantage of using video is that most prospective clients don't have video cassette players in their offices, so you will be asking them to take work home, which might be met with some resistance.

As a substitute for video, many firms are producing CD-ROMs. They have sufficient storage capacity to combine large amounts of information, including text, data, graphics, video, and audio. If you are a concrete formwork contractor, for instance, and want to show different forming methods being installed, a CD-ROM is a good option.

You may feel that the best way to illustrate your services, or provide a needed service, is by developing a software program that prospects can load and use on their computers. You may be an interior contractor, for example, and want to produce a software program that allows tenants to plug in square footage and other measurements along with certain materials to develop budgets for construction. Although the information won't be precise enough to develop contracts, it might be another way to portray your expertise and keep your name in front of prospects.

Putting your logo and a tag line on coffee cups, hats, tee shirts, and other giveaways will also keep your name in front of prospects, as long as what you send is useful and nice enough to keep out of the trash can. If prospects use coffee cups with your name on them, you will be keeping your name in front of them on a daily basis—a marketing coup.

Direct Mail

Direct mail promotion is primarily used to make the recipient take action. Although you can send just about anything directly to someone in the mail—like a brochure or other collateral material—traditional direct mail has a specific purpose and a format that is fairly consistent from one piece to the next. A standard direct mail solicitation has four elements:

1. *An attention-getting envelope* that states the key selling point in a compelling way and entices the reader to open it

2. *A very personal, subjective letter* written by the head of an organization or someone influential explaining why you will benefit by purchasing the product or contributing to the cause

3. *A very objective, well-written brochure* or pamphlet describing the selling points

4. *A response vehicle*—usually a return card—that allows you to pay for the product or contribute to the cause

Strengths and weaknesses of direct mail

The strengths and weaknesses of direct mail are quite different from those of advertising. Direct mail is a *narrow-cast* medium. Unlike advertising, which does not allow you to control exactly who sees your message, with direct mail you can target people who match perfectly with the demographic and psychographic characteristics of your target audience. You can develop a mailing list with the name, title, address, phone, fax, and e-mail address of your prospect, either on your own or by purchasing it from list brokers (this will be covered in more detail later in this chapter).

Therefore, direct mail is best if you have a small, clearly defined target audience, which is typically the case with most contractors and subcontractors. Rather than wasting money promoting your products or services to people who have no interest, as may be the case with advertising, direct mail allows you to speak only to people who need what you are selling.

Like advertising, with direct mail you can *control the message*. You write, design, and print it, so you know what it will say. However, also like advertising, direct mail has limited credibility because people who receive it know that you wrote, designed, and printed it.

Direct mail messages are *lingering*; in other words, prospects can reread your message as often as they like after receiving your direct mail. This is why in direct mail you can include more information, which people will need while they work their way through the decision-making process.

Direct mail has a *high cost per thousand*. It costs more to reach a person via direct mail than advertising. However, the people who see your direct mail are qualified prospects; with advertising, you're paying to place an ad in a media that also includes a significant number of people who probably have no need for your product or service.

There are two reasons direct mail is so popular within the construction industry. It can be targeted to certain groups, as mentioned above, and it can take a prospect through the entire decision-making process, from creating awareness to generating an order.

A good response rate for direct mail is about 1.5 percent, depending on the quality of the message and the complexity of the product or service being sold.

Creating effective direct mail

Creating effective direct mail requires the same skills as creating effective advertisements and other marketing and promotional materials. You must identify and clearly convey your competitive advantages and be creative enough to design a mailer that has quick visual appeal.

Writing for direct mail is different from writing for advertising. With direct mail, you typically have more space to explain your benefits. Although the goal of most advertisements is to create awareness and perhaps action, the goal of a direct mailer is to take a prospect from awareness through interest, desire, and action. Therefore, you need to tell a compelling story that will attract attention, explain your selling points, and force your prospects to take action.

The typical direct mail piece as described above is mailed in an envelope, but the beauty of direct mail is that you are not constrained by the limitations of a page size like you are with advertisements. You can mail anything.

The size or shape of a direct mailer can be creative, as long it does a good job of leading a prospect through the decision-making process. Boxes and tubes are excellent vehicles for direct mail. You can be assured that they will stand out in a crowded in-box of mail and be opened.

Mailing lists

Obviously, one of the keys to creating an effective direct mail program is developing a good mailing list.

Building your own mailing list. You can build your own mailing list using personal information manager (PIM) software like ACT!, by Symantec, or GoldMine, among others (see Chap. 7 for more information about software for construction marketing). This is the most time-consuming way to build a mailing list, but it is also the best way to ensure that the list contains valuable prospects and that the information is accurate. Many firms hire marketing interns to help them build a database.

There are many resources for gathering the names to put on your list. You can sometimes buy lists from trade or professional associations whose members make up your target audience. Many national associations have state or local chapters. See if your area has local offices of associations like American Institute of Architects, American Subcontractors Association, Apartment Owners and Managers Association, Associated Builders & Contractors, Associated General Contractors, Building Owners and Managers Association, Construction Specifications Institute, Design-Build Institute of America, National Association of Home Builders, National Association of Real Estate Investment Trusts, National Multi Housing Council, or many, many others. If they do not have local offices, contact the national office to inquire about the availability of lists customized for your area.

Trade and professional associations are exceptional resources for building a database of prospects. There are several directories in your library that list national and local associations by industry, geographic location, size, and budget. *National Trade and Professional Associations of the United States* (NPA) and *State and Regional Associations* (SRA), both published by Columbia Books, Inc., in Washington, are excellent resources. Most associations you belong to have member indexes you can scour for names to add to your list.

You can also look elsewhere. Most local business publications or business sections of newspapers have periodic lists of companies, like the top 25 property managers or top 100 privately held companies. You can also add to your database information about companies that are published in these periodicals.

Most database management software programs give you the option of producing mailing labels. Check the software's User's Guide to learn which size and type of label the software can print onto. If you send labels to a mailing house, they will probably ask for a specific format for their automated labeling machines.

Purchasing a mailing list. If you don't have the time or resources to develop your own list, there are outstanding resources for purchasing lists. There are dozens of companies called *list brokers* who sell complete mailing lists. They typically have their own database and have access to other sources, like subscriber lists for highly targeted magazines and newspapers.

There are many list brokers, like Dunhill International List Company, Dun & Bradstreet Information Services, The Specialists, Database America, Compilers Plus, American Consumer Lists, and others, who can be found in your Yellow Pages or other directories under Mailing Lists. Database management has become so sophisticated that you can select people by almost any demographic or psychographic characteristic, and the list broker can tell you how many people match your criteria in a matter of hours.

Most list brokers sell names for either one-time or unlimited use. It typically costs about twice as much to purchase a list for unlimited use. Most brokers recommend that you buy a new list every year to account for changes in the database. The brokers usually guarantee an accuracy rate of about 95 percent for their lists.

Mailing services. After you have produced your direct mailer, whether it is printed or assembled in a box, you will send it to a mail shop along with the labels and a check for postage. The mail shop will put the labels on your mailer, sort and bundle them to conform to postal regulations, and deliver them to the post office. They can also be found in your Yellow Pages directory under Mailing Services.

You should consult with your mail shop before purchasing lists to make sure that the lists are organized and delivered in a format which is most cost effective for your mail shop. Lists can be saved electronically on disks or printed as pressure-sensitive or Cheshire labels.

Publicity

Public relations, or PR, is a broad term that can be used to describe many functions dealing with the public—anything from how a receptionist answers the phone to how salespeople are trained to how job site construction fences are painted.

All these things do affect how the public perceives your firm, and they should be considered as part of your overall marketing efforts. For the purpose of this book, however, we will focus on *publicity,* which is how to use the media to communicate information about your company.

Chapter 10 is devoted entirely to publicity. To help you easily compare the value of publicity to that of other forms of marketing communication—like advertising and direct mail—following is a description of its advantages and disadvantages.

Advantages and disadvantages of publicity

Like advertising, publicity is seen in *broadcast* medium. Articles appearing in newspapers or magazines or stories broadcast on radio or television reach a broad audience—all those who read, watch, or listen to the media. Although you cannot specifically control who sees your articles, you can narrow your efforts by targeting media that focus on issues of importance to your target audience, like trade magazines or all-news radio stations.

Publicity's greatest strength is its *credibility*. People typically believe what they read, hear, or see in the media. Unlike with advertising or direct mail, most people believe that the information is objective and written by the media itself. This is true even though much of the information about companies in the media is generated by the companies themselves. It has been said that people find articles or news stories 10 times more credible than advertisements.

The biggest potential disadvantage with publicity is that you *cannot control the message,* as you can in advertising or direct mail. You may send a carefully worded press release to the media and they may extract only a small amount of the information. You may provide quotes to a reporter and find them used in a story in a way that does not convey the true intention of what you said.

If you are going to use publicity as part of your marketing mix, you must be prepared for the downside risk of misinformation and know how to deal with it. There will be more on that in Chap. 10.

Publicity has a very *low cost per thousand* people who are exposed to your message. Certainly you have heard publicity referred to as "free advertising." If you can get an article placed in a publication, you only pay whatever it costs you to prepare it, if anything, but not the cost of the space in the publication. The same is true for television or radio. This is an exceptional bargain, if the publicity is positive. Keep in mind, however, that your efforts are never guaranteed. You can pay tens of thousands of dollars to a public relations firm and get *no return on your investment.*

Like advertising, publicity messages are usually *fleeting.* After they are seen or heard, they are gone. This is particularly true of electronic media like television or radio. At least with print publicity, the prospect can reread an article. And certain publications, like monthly magazines, have long *shelf lives,* meaning they are left around the office a long time.

Publicity is excellent for establishing awareness, and sometimes interest, but rarely will it lead to action. That is still best left to other forms of promotion like direct mail, exhibits at trade shows, telemarketing, the Internet, and directory advertising.

Once you have finished reading this chapter, learn much more about how to incorporate publicity into your successful marketing program by reading Chap. 10.

Internet

Marketing your products or services on the Internet is discussed in great detail in Chap. 11. To help you easily compare the value of promoting your products or services via the Internet to that of other forms of marketing communication—like publicity and collateral material—following is a description of its strengths and weaknesses.

Strengths and weaknesses of the Internet

More than any other form of marketing communication, the Internet is the broadest of *broadcast* media. Once you post something on the Internet, whether it is an article, a comment on a newsgroup, or a Web site, it is available for the entire world to see. Over a billion people now have Internet access, and that number will surely grow.

Most companies are setting up in an area of the Internet known as the World Wide Web. Unlike other areas of the Internet, which are mainly text-based, the Web uses a graphic interface that offers interactive capabilities including animation, sound, and full-motion video.

There is no way to control who sees your Web site. It cannot be targeted like placing an ad in a trade publication. However, unlike all of the other forms of marketing communication mentioned here, the Internet is largely a *self-directed medium*. With the exception of e-mail transmissions, the only way a person can find your information on the Internet is to look for it or stumble across it. You can increase the likelihood that people will see your Web site, for example, by putting your Web address on all of your other marketing communication materials and registering it with Internet search engines and directories.

Internet messages are *not necessarily credible*. Like advertising and direct mail, most people view with skepticism claims of superiority that companies make about themselves. Third-party endorsements are always good to add to Web sites, brochures, and direct mail. However, also like advertising and direct mail, you can *control the message* you broadcast via the Internet because you write and/or produce it.

Internet marketing has a very *low cost per thousand* people it reaches because the number of potential viewers is in the millions. Unfortunately, few firms who use the Internet for purposes other than commerce, like most firms in the construction industry, know what kind of return on investment they are getting. Only time, experience, and modifications to accommodate users needs will tell whether it is a cost-effective marketing tool.

Another exciting advantage of the Internet is the number of times people can refer to it. If you have an interesting Web site that is regularly changed, people will come back to it often. No other marketing communication material can make that claim.

For marketers, a great advantage is that you can update a Web site whenever something on it needs to be changed. You can add and subtract key personnel, change project photos, update descriptions of your products, or modify your client list, all with relative ease.

The Web is particularly well suited for creating preference and action; however it is not a very cost-effective way to create awareness because you have no way of controlling who sees it. People who already know about you, either by seeing an ad or hearing a news story, can refer to your Web site for more information.

The Internet, particularly the World Wide Web, provides exciting opportunities for you to mix words with still pictures, animation, and audio to present your products and services in a new and dynamic way. Once you have finished reading this chapter, learn much more about how to incorporate the Internet into your successful marketing program by reading Chap. 11.

Exhibits

Companies that promote themselves using exhibits do so at either trade shows or consumer shows. The distinction between a trade show and a consumer show is simply that a trade show limits attendance to people who qualify because of their business affiliation, whereas a consumer show is open to the public.

According to Daniel P. Anderson of The Aberdeen Group, this difference is important mainly because it means that greater care must be taken at a consumer show to screen out those people who are not good prospects for your services.[3]

Like direct mail, exhibiting at a show theoretically has the ability to move a prospect through the purchase process, from awareness to action. In practice, however, shows are not very good at building awareness on any large scale. Even with a show like The Aberdeen Group's World of Concrete, which attracts as many as 50,000 people, to create awareness among just 3000 of these attendees your exhibit needs to have a measurable effect on one person every 30 seconds. That's a tall order unless you have a very large exhibit (see Fig. 9.11).

Although exhibits at shows may have a minor role in creating awareness, properly used they can be highly effective at moving people from the interest stage through desire and action. They also provide an excellent opportunity to

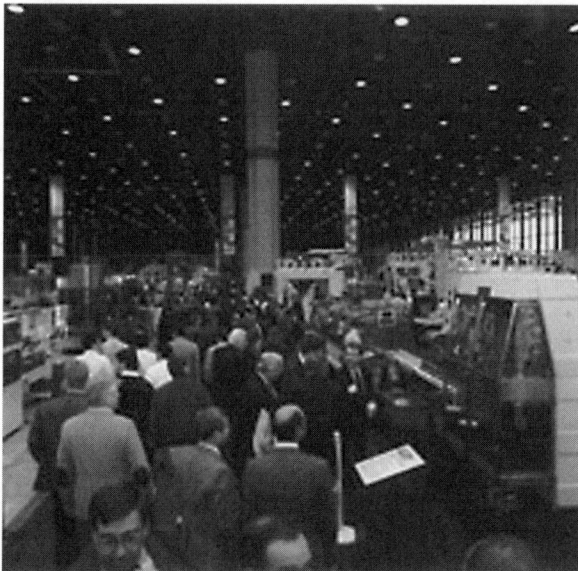

Figure 9.11 Exhibits can be highly effective at moving people from interest to action.

screen prospects and build that defined database that is essential if you need to communicate with prospects repeatedly over the long term.

However, without good planning, exhibiting at a show or industry convention can waste money faster than almost any other marketing communication tool. Success with using exhibits depends largely on the planning done before the show and the follow-up work done afterward.

Before committing to an exhibit presence at a show or convention, define your objectives precisely and know exactly what target audience you expect to reach at the show. If your objective is to attract customers and you are exhibiting at a consumer show, you may end up talking to hundreds of people and giving out hundreds of expensive brochures. But do you have any solid prospects you can contact after the show? One solution is to give out only a simple brochure at your booth but show people a more elaborate brochure that you promise to mail to them if they give you their address and phone number. Rather than waste your time talking to people who want to discuss your trade but could not possibly be considered prospects, ask a couple of simple questions at the outset to qualify them.

With respect to trade shows, the most effective exhibitors are the ones who recognize that the face-to-face time you get at a show is best used to move people who already are interested in your product or service closer to the action step. This usually involves using some combination of preshow advertising, direct mail, and/or telemarketing to generate interest and to get the right people seeking you out at the show. If you are marketing concrete pavements at an American Association of State Highway and Transportation Officials convention, for example, you know you have succeeded when someone wearing a state department of transportation badge walks up carrying the coupon you have sent out that can be redeemed for a free copy of your pamphlet, "What Every Pavement Engineer Should Know about Designing Permanent Pavements."

If you need to communicate to prospects over the long term, trade show exhibiting provides a good opportunity to add qualified people to your database of prospects and, more importantly, to gather much of the detailed intelligence about each person that you need for effective follow-up. This requires having copies of your data-gathering questionnaires in your booth and creating a situation where people will be comfortable giving you the time and attention required. One device sometimes used for this purpose is a prize drawing where people are required to provide certain information on the entry form.

Amazingly, many exhibitors don't give a thought to postshow follow-up until the show is over. As a result, either no follow-up is attempted or what is done happens long after the momentum created by the show has disappeared. Planning the postshow activity as carefully as the preshow work can prevent this.

Developing trade show exhibits

Once you have decided in which show to participate and have developed a strategy for attracting and following-up prospects, you must develop an exhibit that will generate traffic. Trade show exhibits are as varied as advertisements, direct mail, and other marketing materials. Your decision must be based on what will most effectively reach your target audience.

With the wide variety of materials at your disposal, your exhibit need not be just a static display of photos and words. Although enlargements of photos and text are inexpensive, they are not very exciting. You can rather easily include a video, PowerPoint, Internet, or interactive CD-ROM presentation that will not only catch people's attention but give them the kind of information they need to move them through the decision-making process.

Many marketing communication and graphic design firms specialize in trade show marketing, and several of the major exhibit manufacturers offer creative services. You should contact potential vendors well in advance of the show to discuss concepts and costs.

Presentations

A presentation is a form of marketing communication that is delivered in person. This can happen in a group setting such as a speech at a trade association meeting or at a one-on-one discussion with a federal procurement official.

Either way, giving a presentation can be an intimidating experience for those who aren't accustomed to it. Part of this pressure comes from the expectation that a personal presentation should result in a sale or a commitment of some type. Although the seasoned salesperson doesn't like to admit it, the probability of getting a commitment on a given sales call usually is determined not by the skill of the presenter but by the mindset of the customer. If the customer has been exposed to the right messages and has developed a positive perception of your product prior to your arrival, the rest is easy.

All presentations are made easier through careful preparation. If you have been given information about the people to whom you are presenting, you can tailor your presentation to address their needs. In any group presentation you should make a point of learning who will be in the audience and what level of background they might have regarding your subject.

If you expect to do personal presentations with any frequency, you should become familiar with one of the software packages now available for creating visually exciting presentations. (PowerPoint from Microsoft is the best-known software for this purpose. For more information about presentation software and hardware, see Chap. 7.)

Producing Your Marketing Communication Materials

After reading this chapter and using your instincts, you should have a good idea of which marketing materials should be used to achieve your goals and objectives. The next step is getting them produced.

Clearly, there are qualified consultants available who can provide the creative and production services you need to create powerful and effective materials. However, with the basic understanding of marketing communication available to you in this book, you can do much of the work yourself. Either way, by being well prepared, you will save time and money and yield more effective materials because you know your industry, your company, and your competitive advantages better than anyone.

Following are the services you will need to properly execute your marketing communication program. Some of these can easily be done internally, and for some you might be best served getting help from outside consultants.

Research

To quantify the characteristics of your target audience and figure out their needs, sometimes you have to do research. There are basically two kinds of research, *primary and secondary research*. Primary research is original research that you conduct or hire someone to conduct. It includes such things as questionnaires—conducted in person, by mail, or over the telephone—or focus groups. Secondary research is gathering data that has already been collected by others.

Developing anything more than very basic questionnaires or focus groups takes professional assistance. Gathering information from these forms of primary research can be expensive, too, so if you want the information you gather to be of value, you need to make sure the research is conducted properly. The questions on questionnaires must be worded correctly and the questionnaire must be prepared in such a way as to generate the highest possible response rate.

Focus groups require a trained moderator to ensure that the participants provide as much information as possible. Moreover, it is often better to have someone outside your company moderate so participants will give more honest answers.

Primary research is good for getting less statistical information, like people's opinions and attitudes. This can prove valuable in helping you determine whether your products or services will be beneficial to them. Although it is expensive to generate primary research, it is the best alternative for you if the information you need is very specific.

Gathering secondary research is easier to do because it is already available. Finding the specific information you need is sometimes difficult though, in which case you may need outside help. Secondary research usually results in more factual information, like getting names of prospects from existing directories or databases on company size, number of employees, and geographic location from list brokers. Secondary research can also include reading magazine and newspaper articles on the Internet to learn more about the marketplace or getting statistical information from the Department of Labor.

Gathering information is essential if you are going to base your marketing plan and marketing communication materials on facts, not guesses. It need not be time consuming or expensive, but it needs to be done.

Planning

Developing a marketing communication plan, as was discussed in Chap. 8, is something that must be done before you develop advertisements, brochures, direct mail, publicity, or any other promotional effort. It is something that you can do yourself if you follow the steps in that chapter.

If you need help, consult a firm that calls itself a marketing communication firm, preferably one with experience in your industry. If you cannot find a marketing communication firm, interview advertising agencies, design firms, or public relations firms that have knowledge of your market and ask if they provide communication planning. Either way, make sure the plan clearly identifies your goals and objectives, target audience(s), and key selling points and has a strategy (see below) that you agree will properly achieve your marketing goals.

Strategy

Preparing a creative strategy is perhaps the most challenging part of any marketing communication plan. It is where you decide how you will communicate the benefits of your selling points to the target audience in a way that will achieve your goals and objectives. If you follow the suggestions in Chap. 8, you should be able to create a functioning strategy. To make it truly special, you may want to hire a marketing communication, graphic design, advertising, or public relations firm for some help.

Remember that the creative strategy will guide all of your specific tactics, whether they are advertising, a Web site, or a publicity campaign. It will be the document that will allow you to assess whether the creative ideas you or your consultants generate are on target. Make sure you put the requisite amount of time into this activity.

Media planning

If you decide to communicate your competitive advantages to prospects through advertising, the first step is determining which medium is best suited to reach your target audience (there may be more than one). Tasks such as selecting the media, developing a schedule of ad insertions, monitoring placements, and managing payments fall under the umbrella term *media planning*.

Effective media planning requires some experience finding the right publications or electronic media, determining which of them delivers the right *reach* and *frequency* to generate results, negotiating rates, and tracking placement. There are plenty of books and resources available to help you learn media planning. However, if you have a large and diversified target audience, you might consider hiring an advertising agency or an independent media buyer to help you.

The advertising agency can help you develop the advertisements as well as provide media buying, if you need help with both. It will typically charge a creative fee to write and design the ads and then pay the media directly for the cost of placing the ads. In exchange for providing this service, it collects a 15 percent agency commission from the media. Most agencies let the commissions cover media planning and account management services. The 15 percent commission represents a discount from the published price, so, theoretically, you (the advertiser) will pay the same cost for the ad placement whether you buy the placement or you let the agency buy it.

An independent media buyer only handles media planning and will typically negotiate a fee with clients that is less than the 15 percent commission it would get for paying for the media. This may seem like the best deal, unless the ad agency provides additional services that you need and those services are compensated through the commission. Also, if you need an ad agency to create the ads, it might be easier to deal with just one consultant.

If, however, you determine that you can adequately reach your target audience using one or a few publications, you may want to handle the media planning yourself. Once you have determined the publications and created a schedule, the media plan essentially works by itself.

First, however, you have to find the right media. Again, if your target audience is either small or well defined, you may already know what publications, radios stations, or television stations reach them. If you don't know the best media to use, there are directories available that list and describe media by subject. These directories also provide ad rates and contact information. The best directory for gathering this information is *Standard Rate and Data Service*. This two-volume set is divided into business and consumer publications and is organized by subject. Each listing gives a brief profile, ad manager's name, and complete contact information. It can be found in most libraries.

You should make a list of several different media that might work for you, then contact them, and meet with an advertising sales representative to learn more about the audience and *demographic profile* (statistical information about their readers, viewers, or listeners) for that medium. Be forewarned that most advertising sales representatives are very good salespeople and can easily persuade you to select their publication. Make up in advance a list of questions you need answered, like a clear breakdown of paid readers, an editorial calendar that shows what topics are covered, and so on.

Photography

Good-quality photography is essential to selling construction services. A crisp, artistic photograph taken at dusk has far more impact than a washed-out picture shot with a 35-mm camera that makes a building look like it is falling in on itself.

If you need photographs for your marketing materials, the first, easiest, and perhaps least painful option is developing a relationship with a professional photographer. If you sell products, you will want to work with a product photographer—someone who specializes in composition and lighting.

If you want photographs of buildings, *architectural photographers* possess clearly different skills and use different equipment from publicity or product photographers. Their cameras use 4- × 5-inch or 120-mm film, which eliminates vertical distortion, also called "keying," that creates the effect of a building's tilting inward rather than going straight up. They know how to use natural light and the effects different kinds of interior light have on film.

If you want shots of people, either on a job site or in an office, you will want to find a photographer who specializes in people. Their skills include capturing people in spontaneous, rather than posed, settings. In addition, if you need photos for press releases or newsletters, you need a publicity, or "grip-and-grin" photographer.

Interview photographers in advance of needing one. Judge them as you would be judged—by the quality of their finished product. Get references to find out how easy they are to work with. If you like more than one, give each a few assignments and judge the quality and relationship for yourself. They will be representing your firm when they talk to building owners and property managers to gain access to buildings, so they should act accordingly.

Another option is buying the rights to existing photographs. If you are in a hurry to obtain a photograph, or if you need photographs of projects in different parts of the country and you can't afford to send a photographer to different cities, you may want to see if photos of the projects you want already exist.

The best way to do this is to contact the owner, architect, or general contractor of the project and ask if the project has already been photographed. If it has, you will probably be referred to the photographer. Unless the photographer has given a client unlimited rights to use the photographs, the photographer retains the rights and can sell them to anyone else for a usage fee. Typically, the fee is slightly less than the cost to hire a photographer for original work.

The only problem with this method is that it requires you to make a lot of phone calls to track down a photo, and you will not have visual consistency from your various photographs because different cameras, lenses, settings, and film will have been used. You need to weigh your needs and budget with the aesthetic impact of using this method.

Still another way of obtaining photos is using stock photographs. There are dozens of stock photography houses, like The Stock Market (www.stockmarketphoto.com), PhotoSource International (www.photosource.com), and Unicorn Stock Photo (www.unicorn.photos.com), that have catalogs you can review and choose from. These can be expensive because the stock house will charge you based on the number of times you use it, which can be prohibitive for advertising. A less-expensive alternative is buying a CD-ROM of stock photos. One supplier of stock CD-ROMs is Photo Disc, Inc. (www.photodisc.com). You typically pay about $300 for the disc. It will have several images you can use as often as you like.

Graphic design and copywriting

Graphic design and copywriting are probably the two most important resources you will need to produce effective marketing materials and are the two that you should weigh most carefully in deciding whether to do them with in-house resources or qualified consultants.

We are a visually oriented society. There is no way of getting around this. With the overwhelming assault on our senses from direct mail, advertising, television, e-mail, the Internet, and so on, and the daily distractions in all of our lives, it is increasingly important that the marketing materials you send grab attention and convey your main selling point quickly. This can best be accomplished through good graphic design and crisp, clear copywriting.

Graphic design. Basically, graphic design works in three stages: concept, design, and layout. The *concept* is the thinking stage. The designer, who thinks visually, works with the copywriter, who thinks verbally, to blend visuals and words together to convey your main selling point in the most attention-getting manner.

Some concepts are design-driven, whereas others rely more on words. The combination is somewhat subjective and depends on what your creative team thinks will work best. For instance, if your competitive advantage is saving money for your clients, you may develop a direct mailer with George Washington's face from a dollar bill peeking through a cutout in the envelope with some powerful words that compel the reader to open it. Upon opening the envelope, the reader finds a crisp dollar bill with text that cleverly tells the prospect how they can get more of these by working with your firm. That is as far as the concept has to go.

The next stage for the graphic designer is the *design*. Today, design is done on the computer using *desktop software*. Two commonly used desktop software programs that have all the tools needed to generate the electronic files service bureaus or printers need to create film are QuarkXPress (www.quark.com/quarkxpress) and Adobe PageMaker (www.adobe.com).

Desktop software allows the designer to create page layouts using any dimensions, add text boxes, which are invisible but define spaces for words, and import photographs or other graphic images like illustrations. The designer will have many typography styles to choose from, most of which can be purchased in packages.

Once you have laid out your text and graphics, the software links them together in a way that allows film to be made properly. It is important to buy desktop software that links images in this way, or else it cannot be used for printing, which renders it essentially useless except for being output on your own laser or ink jet printer.

A graphic designer will design the promotional piece and print it out, usually in full color, to show how the piece will look. If changes are needed, the designer can make them in the software program and then print out a new design.

Once final comments have been made, the designer moves to the *layout* stage. This is where all the remaining pages or pieces of the promotional material are created, based on the approved design. The desktop software then links the pages and images together in a format needed to give to the service bureau or printer.

Copywriting.   Concurrently, the copywriter has moved into the second stage as well. After the concept is approved, the text for the piece will be written. The text can be reviewed in manuscript form or be incorporated into the design or layout and reviewed in a more finished fashion.

The availability of desktop software has misled some people into thinking that it gives them design skills. It does not. Desktop software is a tool for the designer, just like T-squares and X-acto knives used to be. Because we are a visually oriented society, it is important that the design and writing of your marketing materials be as good as possible.

That being said, you might want to weigh the pros and cons of performing copywriting and design in-house for certain materials, depending on their importance. You may also want to consider having professional designers and writers establish templates for such things as newsletters and data sheets and let your in-house staff execute layouts.

Whatever you decide, weigh carefully the use of outside experts versus in-house talent to get the right blend of creativity and value in producing your marketing materials.

Programming

There is a certain amount of computer programming that is required for developing Web sites. It is explained in detail in Chap. 11. If you are considering building your own Web site, two of the most commonly used off-the-shelf Web publishing software packages are Adobe PageMill (www.adobe.com) and Microsoft FrontPage (www.microsoft.com). You can place text and visuals on each page and then the software embeds the coding needed to make the site work on the Internet.

Other more sophisticated programs can be used, such as Macromedia Flash (www.macromedia.com/software/flash), The Network Director (www.nrsinc.com/nrs.2200.asp), and Macromedia Authorware (www.macromedia.com/software/authorware), to add more of the "bells and whistles" that many of today's Web sites feature. These programs can be used to add higher-end multimedia and interactivity functions to your Web site.

Hypertext Markup Language (HTML) is the primary code that Web pages are written in. Internet browser software, which provides a graphical connection to the Internet, reads the HTML code. Web publishing software will "write" this code for you, through an interface that is very similar to desktop publishing software.

Although there are many different types of Web browsers available, Netscape Communicator (www.netscape.com) and Microsoft Internet Explorer (www.microsoft.com) are the most commonly used Internet browsers. At a minimum, someone who is developing a Web site must have a basic understanding of HTML and how it works. Often, Web publishing software will write code that is not interpreted the same by different Web browsers, and therefore someone must edit the file at the HTML level (removing or modifying tags, etc.).

Web development firms have the ability to design, develop, and maintain your Web site. Simple additions or deletions, however, can be made in-house with just a bit of HTML training and understanding of a Web publishing software package. Even a word processor or simple text editor can be used to create or edit your Web site. HTML files are, in reality, just text files with an .htm or .html extension.

If you are going to maintain your Web site in-house, you will need to have the ability to transfer your files to and from the server where they are housed. File Transfer Protocol (FTP) software, such as WSFTP, must be used to perform this task. In this way, you can simply download the original file (transfer it from the server to your hard drive using the FTP interface), open it in your HTML editing program, make your changes (add a press release, correct a problem with the coding, etc.), and upload it back to the server using FTP.

Vendor management

To ensure that you are satisfied with your marketing communication materials, you must work closely with your vendors to make sure they deliver the quality, price, and schedule that was promised. If an outside graphic design firm or ad agency is creating your material, you will need to decide whether you want to work with vendors or let your consultant do it. Chances are the consultant will have the knowledge and relationships needed to ensure success. It will probably charge a production management fee and mark up the vendor's bill, typically between 20 to 30 percent. It is usually worth the added cost to have an expert handle that aspect of the project to ensure that your entire investment is not lost due to a careless oversight.

However, if you prepare your marketing materials in-house, or if you have someone on your staff in-house with expertise in working with vendors, you may decide to work with them directly. You would save production management fees and mark-ups and establish relationships with people whose services you will depend on throughout your career.

For the purpose of this discussion, *vendors* are the people who produce the final product and use equipment rather than the creativity used by designers, writers, photographers, and other creative resources mentioned in this section. The vendors with whom you will work most often are

- Service bureaus
- Printers
- Color imaging labs
- Quick print or copy centers
- Photograph developers

Before you begin a relationship with a vendor, you should meet with a sales or customer service representative, review samples of the vendor's work, and contact references, both those given to you and the companies for whom the samples were done. Ask the vendor's customers questions like, How do you find the quality of the vendor's work? Does it deliver on time consistently? Does the final bill match the estimate? Does it charge an inordinate amount for customer changes? Is the company easy to work with? Is it flexible? Does it use the latest technology for outputting film, making plates, and proofs?

Service bureaus.   Service bureaus are often the first vendors you will use when producing a marketing communication piece. They take electronic files of your marketing piece and convert them to either film or a high-resolution print.

The printer uses the film to create printing plates. High-resolution paper prints often are used for black and white advertisements or copies or by one-color printers who make plates directly from them. The resolution, measured in dots per inch (dpi), is at least twice as good as most high-quality laser printers. If you do not have photos in your material, the 600-dpi resolution of most laser printers will probably be acceptable.

The qualities to look for in a service bureau are speed and computer skill. Given the deadline orientation of this business, you will usually want your film or print quickly. Moreover, even professional designers sometimes give service bureaus electronic files with missing fonts, bad links, or other technical errors. A good service bureau will call you and be able to fix the problems for you.

Printers.   There are several different kinds of printers. Each specializes in different types of work, usually in these categories: one- and two-color printers, four-color printers, and web-fed printers. One- and two-color printers should be used for stationery, newsletters, pocket folders, and other materials that are not printed in full color (known as four color). It is usually best to avoid asking two-color printers to print four-color pieces because to do so they have to run a job through a two-color press twice. This is often logistically difficult when trying to align the printing perfectly or to ensure that the job is in "register." For color brochures and direct mailers, it is best to use a printer with four-color presses. It is called four-color printing because all the color images are created when the paper is run through four rollers on the press, each of which has one of the four primary colors, yellow, magenta, cyan, and black. Most four-color printers actually use five- or six-color presses these days. The extra roller is used to put varnish or an additional specified color on the page.

Regardless of whether you work directly with a vendor or subcontract that function, the key issues are managing quality, time, and cost—any of which can get away from you quickly when you are producing marketing materials.

Quality is very subjective. On the back of many printers' estimates is a long disclaimer about how they follow printing industry standards. Read it carefully. It sometimes allows the printer to deliver "pleasing color," which may not really be good enough to "please" you. Understand how your printer defines this and other issues in advance. If you want near-exact matches of your photographs, for instance, you have to make that clear, in writing, up front.

To ensure quality, especially on four-color work, you should ask for a color proof before the entire job is printed. For material printed on press, the proof will be a chromalin, iris, or Fuji proof. These proofs are made from the four pieces of film generated to make printers' plates. The proofs are made using powders instead of ink; therefore the proof is not an exact reproduction of what the printed piece will look like. You can get a good sense of the richness of colors and sharpness of images, but you will not be guaranteed that the final product will look like the proof.

The best way to ensure that the final product meets your needs is to inspect the job on press. You will be given a time to go to the printer and will be shown some of the first sheets to come off the press, once the press operator feels that the colors are correct. At that point, you can make adjustments. As long as your recommendations make the printed piece more carefully match the proof or original photos, you won't be charged for these changes. If your recommendations are made to improve or change the product, you may have to pay for changes. A press check is time well spent, whether your in-house staff is doing it or you are paying a consultant.

Another essential proof is called a blueline proof, named for the color of the images on the paper. This proof is typically the last proof that you will see before a job goes on press. Although you should have carefully proofed and made any changes to your work before it goes to the printer, the blueline proof gives you one last chance to make any final changes.

You will find that managing *time* and *cost* are very much like managing the same items for a construction project. You should ask the printer to give you a detailed estimate. It is up to you to evaluate whether it includes all of the elements your project requires. If it does, managing cost is a matter of making sure that there are not unauthorized changes.

In addition, get a written schedule from the printer. Managing a printing schedule is much like managing a construction job site because unforeseen problems can crop up. Build some flexibility into your schedule to allow for such problems so that you won't be disappointed if your project comes off the press a few days late.

Color imaging labs, copy centers, and photo labs.   There are several other types of vendors who supply copies or computer laser prints. Copies might be used for reproducing documents in a proposal, whereas color laser printers would produce CLI prints used for project data sheets and other color output.

A color proof for a CLI print will be the first print itself, so the proof you see should be exactly the same as the finished product. If you are printing a large quantity of CLI prints, it is well worth it to see a proof before printing the whole job.

Seeing a color proof for CLI prints is important because it will vary from the colors on your computer screen. That is because the colors used on computer screens are different from the colors produced from powders, which is how color laser prints are made. By getting a color proof, you can either ensure that you are satisfied or make minor adjustments to the color and brightness of the printed piece.

You will also probably work with photo labs to process your photographs. Here, too, there is a wide range of quality and services. Labs with 1-hour processing typically do not produce the kind of quality you will want if you need photo prints that will be reproduced in a brochure or newsletter. A high-speed machine that does not allow for much color calibration does 1-hour processing. You should use a custom lab for making prints that will be reproduced in printed material. It will be slower, sometimes taking a full week to develop points, so allow for the extra time in your production schedule.

Notes

1. Daniel P. Anderson, *How to Promote the World's Greatest Construction Material,* The Aberdeen Group, Addison, Ill., 1998, pp. 13–14.
2. David Ogilvy, *Ogilvy on Advertising,* Vintage Books, New York, 1983.
3. Anderson, op. cit., pp. 16–17.

Getting Publicity for Your Business

Introduction

The press wants your news. That's right. Reporters who represent all forms of media—television, radio, newspapers, magazines, newsletters, and the Internet—can only stay in business if they have information about which to report. The better, more timely, and more original the news they report, the more their particular media is used by subscribers, viewers, or listeners. And, most importantly, the more people who see or hear their news, the more the media can charge for advertising or subscriptions.

So in that context, you are actually doing the media a favor—in fact, enhancing their profitability—by sending them news about your company.

Although this may seem like an oversimplification, it is essentially true. The media depends on businesses and public relations consultants for a significant amount of the information they publish. In 1980, *The Wall Street Journal* executive editor Frederick Taylor admitted that as much as 90 percent of its daily news originates from press releases. The symbiotic relationship that exists between the media and people who provide information to it should help you feel comfortable working with them.

Of course, reporters are only interested in what would be perceived as "newsworthy" by their subscribers or users, so as long as you do not try to use the media as a vessel for "free advertising," you can have great success integrating publicity into your marketing mix.

Generating publicity is so valuable that all contractors and subcontractors should include it as part of their marketing mix. The most important reason is the credibility associated with published stories. It has been said that people perceive information reported by the media as 10 times more credible than messages delivered by companies themselves in forms like advertising or direct mail. Despite the fact that the originator of the information might be the company itself, via a press release or story idea, most people assume an article is an objective reporting of information, as opposed to marketing materials, which of course are subjective.

Getting publicity about your company in the business or trade press has many benefits:

1. *Publicity is inexpensive.*   With the exception of costs you may incur to prepare a press release or feature article, getting a story in the newspaper or on the news costs nothing. If you measure the inches of space in all the newspapers and magazines that carry a story generated by your press release and compare it to the cost of comparably sized advertisements, you will quickly see the dollar value of the publicity. In fact, when considering the enhanced credibility of publicity, the dollar value, and therefore the cost savings, is much greater.

2. *Publicity acts as a "safety net" for your overall marketing communication plan.*   Other promotional efforts like direct mail, and to a lesser degree advertising, along with direct sales efforts, allow you to reach people you have already identified as prospects. But what about the people you have not identified, the people who simply didn't make your list, new companies to the area, or even decision makers from companies visiting your area? For these people, a broad-based publicity effort will ensure that you get your name in front of all prospects, both those you have identified in advance and those you have not. Our firm got one of its biggest clients after we issued a press release announcing work we had done for Microsoft. Our new client reasoned that if we could work for Microsoft, we could work for it. The client went to the trouble of looking up our number and calling us to set up an introductory meeting. This company was not on any of our mailing lists and we would likely never have contacted it, but because someone there saw our press release, we were contacted. The company fell into the "safety net."

3. *Publicity sets you apart from your competitors.*   The more prospects see your name, the more likely they will remember you when they need your services. If you can execute a strategy that enables you to have a variety of media exposures—brief mentions, feature articles, guest articles, and so on—you will make your subsequent selling efforts more successful.

4. *Publicity establishes you as an expert.*   Because people usually believe what they read in print, the more they read about you, the more they will come to perceive you or your firm as an expert in your field. This is particularly true if you are the author of a regular column in a newspaper or feature on radio or television. It is common sense to expect that your prospects will be receptive to what you have to say in a sales presentation if they already perceive you as an expert.

5. *Publicity makes people feel good.*   Publicizing the promotion or hiring of a person in your company makes that person feel good. There are few greater rewards than having friends tell you, "Hey, Jeannie, I saw your picture in the paper and read about your promotion. Congratulations!"

6. *Publicity equates with success.*   If you read often about a company starting or finishing construction projects, you assume it is busy. People often equate being busy with being successful. Generally speaking, people want to be associated with those who are successful. People assume that others are

successful because they are good at what they do. Regular publicity makes people think you are successful.

7. *Publicity allows prospects to get to know you.* An interesting feature article or a series of brief mentions about new projects, promotions, and awards lets a prospect learn about you before you ever meet them. This gives you a significant leg up on the competition at presentation time.

8. *Publicity begets more publicity.* Reporters are avid students of media other than the ones for which they report. Once you become regularly featured in the media, reporters will begin calling upon you for quotable information when they are writing stories about your field of expertise.

The sum of all the benefits mentioned above is revenue. Getting publicity will help you get business—something any construction company or subcontractor wants.

Although every size or type of company stands to benefit from increased publicity, the ability of publicity to level the playing field for small or start-up firms is perhaps its greatest asset.

Steve Heidenberger remembers well how publicity helped jump-start his business. Armed with years of experience with large interior contractors, Steve decided to branch out and start his own firm. He asked us how he could promote his business, without a budget for advertising or marketing materials. Our advice was to send out press releases about any- and everything he could think of. We gave him a simple template to store on his hard drive (several such templates are included in this chapter) and told him to send a release to a list we gave him of every media in the area that might reach his target audience—smaller office tenants, brokers, and architects.

After a few weeks, he had sent out a release announcing the formation of his company, a second one when he was low bidder on a job, and a third one about a new administrative assistant (actually a part-time receptionist) he hired. None of the stories were big news items, but they were picked up by many of the media. Typically, for press releases about new projects and new hires, the business press will print a couple of sentences regardless of the size of the project or title of the employee. Therefore, Steve's stories ran alongside mentions of much larger projects and promotions of vice presidents.

"I was at a cocktail reception soon after the press releases hit the papers and I remember being approached by a prominent competitor who patted me on the back and said, 'Gee Steve, you must be going gangbusters. I see your name in the papers all the time!'"

The moral of the story is that most people don't remember *what* they read, but they do remember seeing your name in the paper.

Publicity's Advantages and Disadvantages

Publicity, like advertising, appears in *broadcast media*. Publicity reaches broad audiences—everyone who reads the publication or is exposed to the broadcast medium in which your story appears. You cannot specifically control who sees it, as you can with direct mail, for instance.

You can narrow who is exposed to your publicity by attempting to place articles in media that focus on specific audiences. Identifying these publications is discussed in detail later in this chapter. Targeting these media, construction industry magazines like *ENR* for instance, increases the likelihood that your press releases will be published because these publications are more targeted to people interested in construction news than general-interest publications. The first question that has to be answered by an editor before your news item will be published is, "Is this of interest to my readers, today?"

The biggest advantage of publicity over other forms of marketing communication, as was mentioned earlier in this chapter, is its *credibility*. The biggest potential disadvantage is that you *cannot control the message,* as you can in advertising or direct mail. Ultimately, it is up to the editor to decide how a story will run. You may send a carefully worded press release to the media and only a small amount of the information may be published. You may provide quotes to a reporter and then find yourself referenced in a story where your quotes were pieced together in ways that don't convey the true intention of what you have said.

Stories published in the media have a very *low cost per thousand* people who are exposed to them, compared to advertising or direct mail. Certainly you have heard publicity referred to as "free advertising." If you can get an article placed in a publication, you only pay the cost of preparation but not the cost of the space in the medium. This can be an exceptional bargain. Keep in mind, however, that your efforts are never guaranteed. You can pay tens of thousands of dollars to a public relations firm and get no return on your investment.

Like advertising, publicity messages are usually *fleeting*. After the story is aired or appears in a publication, it is gone forever. This is particularly true of electronic media like television or radio. Print advertising, at least, allows the prospect the opportunity to reread an article, and weekly or monthly publications have long shelf lives that allow stories to sit around a while before they are thrown away.

Publicity is excellent for establishing awareness, and sometimes interest, but rarely will it lead to the action step of the decision-making process. That is still best left to other forms of promotion like direct mail, exhibits at trade shows, telemarketing, the Internet, and directory advertising. Nevertheless, working in concert with other forms of marketing communication, publicity is a crucial tool.

Developing a Plan

If you are going to realize the full benefits of publicity, and properly integrate it into your overall marketing communication program, you should develop a publicity plan. The plan should be roughly divided into these sections:

- Establishing your goals
- Creating a media list

- Meeting the press
- Developing a strategy

Establishing goals

In her book *6 Steps to Free Publicity* (Penguin, New York, 1994), Marcia Yudkin states,

> Effective publicity involves a match between your goals and the needs of the media. Without considering what you hope to achieve from publicity, you're unlikely to receive an optimal outcome. And unless you take into account what the media want to cover, you might as well have addressed your materials to a black hole.

She goes on to present the following checklist to help you zero in on your specific goals:

1. *Which of the following do you want most—credibility and prestige, customers, clients, donors, or attendees; changed or opened minds?* Your goals will help you determine what types of press releases and feature articles to write, or even to decide what type of speaking engagements to pursue. Most contractors are probably interested in establishing credibility and getting clients or customers. Articulating your priorities is fundamental to any successful publicity venture.

2. *Where, geographically, does it make sense for you to aim—nationally, regionally, or locally?* If you are starting a new business in Cincinnati that will only work with clients in a 20-mile radius, it is likely that business and trade press in the Cincinnati area will report on it. National trade press may also mention it, but business press in other areas are not likely to be interested. Focus your efforts where your prospects are, and don't waste your time and postage flooding media around the country with your press releases.

3. *Who, specifically, are you hoping to reach?* Focusing your publicity efforts to media that are aimed at people in your target audience—like architects, developers, property managers—will increase the likelihood that your press releases will be picked up because the information will be valuable to the readers, viewers, or listeners. (You should have already identified your target audience when you prepared your marketing communication plan, described in this chapter.)

4. *Are you hoping to sell a particular product?* If so, then persuading, cajoling and even begging the media to include contact information—your address and/or phone number—is crucial. According to Cambridge, Massachusetts, publisher Jeffrey Lant, most entrepreneurs greatly overestimate the willingness of consumers to go to a lot of trouble to order something. If getting in touch with you requires anything more complicated than a call to local directory information, most people won't bother.

5. *Will you welcome any and all publicity opportunities, or will you want to pick and choose among opportunities to maintain a certain image and focus?* Any publicity isn't necessarily good publicity. If you are a paving contractor and get a call from a Sierra Club publication preparing an article about how certain types of companies are destroying woodlands, you may want to pass on providing information.

6. *Do you have a reason to lean more to one medium than another?* Most contractors would be attracted to the tangibility and relative permanence of print, but if you have time-sensitive material to report or can manage to make a regular

appearance on a radio or television news show, you might consider making a push for electronic media as one of your goals.

Taking the time to establish your publicity goals will ensure a far greater likelihood that news about your company is reported on in media that reach your target audience. You will ultimately save time and money by carefully planning in advance.

Creating a media list

Once you have established your goals, developing a media list should not be difficult. It should flow directly from your goals, particularly your decisions about who you are trying to reach and where, geographically, you are trying to reach them. Developing a media list involves three steps: selecting the right media, identifying the proper contact person at the media, and building the list.

Selecting the media. Deciding upon the right media takes a combination of observation and research. Start with the obvious local media that influence your target audience, like your daily newspaper. It probably has a separate section devoted to business or real estate news. You may also have a business or real estate weekly in your area. Most major cities are part of American City's *Business Journal* network (www.amcity.com). Some larger metropolitan areas also have monthly magazines that focus on news of interest to your client types. Identify trade or professional associations in your area that serve your prospects. Chances are they will have publications that accept press releases.

After you have done a little creative brainstorming and built a list based on your observations, you can then look at reference materials in the library. There are several resource books that list media by the subjects they cover and their geographic area. The most commonly used are

Bacon's Newspaper/Magazine Directory. This two-volume directory lists names, addresses, and phone and fax numbers of daily and weekly newspapers, with names of section editors as well as lists of columnists and news syndicates. Daily papers are also listed by "areas of dominant influence" so that you can look up any city and find a list of newspapers in descending order of circulation.

Bacon's Radio/TV/Cable Directory. This two-volume directory lists names, addresses, and phone and fax numbers of radio, television, and many cable stations. It, too, contains areas of dominant influence listings as well as a subject index that allows you to look up relevant shows.

Columbia's NTPA (National Trade and Professional Associations) and SRA (State and Regional Associations) Directory. Two volumes. The *NTPA* directory lists more than 30,000 national associations, indexed by subject and geographically, so you can easily locate construction or real estate-related associations in your area or nationally. There are 157 listings for construction associations alone. The *SRA* directory provides similar information and is divided into states. You should use both directories to get a more complete listing of associations serving your target audiences.

Oxbridge Directory of Newsletters. Complete contact information for more than 20,000 newsletters. It states that newsletters constitute 30 percent of all publications.

Standard Periodical Directory. Very brief listings of more than 75,000 magazines, yearbooks, and directories, with the name of the editor in chief and contact information.

Standard Rate and Data Service. This bible for advertising media buyers is divided into directories for business and consumer publications and is organized by subject. Each listing gives a brief profile, editor's name, and complete contract information.

Working Press of the Nation. These essential directories are divided into four volumes and contain similar information as *Bacon's.* Volume 1 lists newspapers, volume 2 lists magazines and internal publications, volume 3 lists radio and television, and volume 4 lists feature writers, photographers, and professional speakers.

Most major libraries carry these or other directories and will help you find the resources you need to develop your media list.

Figure 10.1 is a listing of more than 50 national and regional construction trade publications, most of which were compiled by the Associated General Contractors and the Construction Writers Association. They are contained in an informative and targeted workbook entitled *PR Handbook for Contractors.* For information about the handbook, contact the Associated General Contractors in Washington, D.C.

Identifying the proper contact. Although these directories list contact information, it would be wise to call each of the media on your list to verify current addresses and phone and fax numbers. Generally, you should not include specific names of reporters or editors who cover real estate or business on your media list. Rather, when you contact the media, you should ask for the title of the editor who handles business or real estate.

This may seem to be an impersonal way to address your press releases, but there is a good reason to do it. Generally, reporters change positions within their media, or change media altogether, with some frequency. Even editors move about fairly often. Most reporters or editors are put off when they receive press releases addressed to people who have preceded them. To avoid this embarrassment, simply address press releases to the appropriate editor. The editor will decide which reporter to assign the story to, if it is considered news-worthy.

The exception to this rule is your list of most important media—the ones that have the greatest influence on your target audience. You should include the name of the appropriate editor or reporter on your list, along with his or

Figure 10.1 A sampling of national and regional construction publications, most of which were compiled by the Associated General Contractors of America and the Construction Writers Association in their book *PR Handbook for Contractors*. This list gives each publication's name, phone number, and Web address, if available.

ABC Today
(703) 812-2063
www.abc.org

Architectural Record
(212) 512-3104

Architecture
(201) 536-6221
www.architecturemag.com

Automated Builder
(805) 642-9735
www.autbldmag.com

Better Roads
(214) 827-4630

Builder
(202) 736-3377
www.builderonline.com

Builder Insider
(214) 871-2913
tx.bin.net/insider

Building Design & Construction
(847) 390-2129
www.bdcmag.com

Building Operating Management
(414) 228-7701
www.facilities.net

California Builder & Engineer
(909) 328-1920
www.acppubs.com

Civil Engineering
(703) 295-6213
www.pub.asce.org

Construction (Connecticut)
(860) 523-7518

Construction
(757) 988-1045
www.cmdg.com

Construction Bulletin
(612) 537-7730

Construction Business Review
(703) 734-0017
www.constructionnet.net

Construction Digest
(317) 293-6860
www.cmgnews.com

Construction Dimensions
(703) 534-8300
www.cd@awci.org

Construction Equipment
(847) 635-8800
www.coneq.com

Construction Equipment Distribution
(630) 574-0650
www.aednet.org

Construction Equipment Guide
(215) 885-2900

Construction Link
(626) 932-6175

Construction Monthly
(205) 969-0088
www.constmonthly.com

Construction News
(800) 766-2611
www.cmgnews.com

Construction Specifier (The)
(703) 684-0300
www.csinet.org

Constructioneer
(215) 942-2389
www.cmdg.com

Constructor
(202) 383-2768
www.agc.org

CraneWorks
(816) 254-8735

Dixie Contractor
(770) 417-4119
www.acppubs.com

Dodge Construction News
(312) 616-3253

ENR (Engineering News Record)
(212) 512-3249
www.enr.com

Figure 10.1 *(Continued)*

Equipment Today (920) 563-1657 www.equipmenttoday.com	*Ohio Contractor* (614) 846-8761
Equipment World (205) 349-2990 www.equipmentworld.com	*Pacific Builder and Engineer* (425) 486-8553 www.acppubs.com
Hard Hat News (518) 673-2381	*Professional Builder* (847) 390-2135 www.probuilder.com
Heavy Equipment News (205) 444-8112	*Rental Product News* (920) 563-6388 www.rentalproductnews.com
Louisiana Contractor (504) 292-8980	*Roads & Bridges* (847) 298-6622 www.roadsbridges.com
Intermountain Contractor (801) 972-4400	*Rock Products* (330) 497-6034
Michigan Contractor & Builder (313) 962-3337 www.acppubs.com	*Rocky Mountain Construction* (303) 295-0603 www.goldenpress.com/~joes
Michigan Roads and Construction (517) 332-7600	*Southwest Contractor* (602) 258-1641
Midwest Contractor (816) 561-3300 www.cmgnews.com	*Texas Construction* (512) 458-1343
Nation's Building News (202) 822-0427	*Texas Contractor* (972) 271-2693
New England Construction (978) 433-5742 www.cmdg.com	*The Western Builder* (414) 453-7700 www.acppubs.com
New Hampshire Highways (603) 224-9399	

her e-mail address. You must track these publications closely. If the reporter changes positions and a new person is assigned to cover the subject, you should immediately change the name on your mailing list. These are the people you will want to get to know.

Building a mailing list.   The rest is simple. Once you have gathered your names and addresses, build your media list using whatever *contact management software* you use (see Chap. 7 for more information on contact management software). Most of the more commonly used software programs will print lists or labels, allow you to keep notes about your contacts (such as what subjects the reporter has already covered), automatically fax releases to them, and automate your e-mail.

Take good care of your mailing list. It will grow to become a valuable asset as you integrate publicity into your marketing mix.

Meeting the press

Much to some people's surprise, reporters are actually people. In fact, generally speaking, they are very friendly people, interested in talking to you and learning more about the subject areas on which they report. This is said "tongue in cheek" because most people have a misconception that reporters are difficult to get along with. The experiences of most PR practitioners show that reporters are "people-people." Their jobs depend on learning and understanding things about people, politics, and human nature. They are generally very interested in what you have to say.

One of the most beneficial things you can do when you launch a publicity plan is to spend a day meeting with the key reporters in your area—the people on your list who you have identified by name. Call each reporter or editor on the phone, introduce yourself, and tell him or her a little about your company. Explain that, from time to time, you will be sending out information about your company—new projects, new hires, promotions, awards, and so on—and that you thought it made sense to meet for 10 or 15 minutes during a slow time of the week. You are likely to get a positive response. The next step is to prepare for your meetings with the media.

Press kit.   Prepare a press kit in advance to give to the reporter when you meet. A press kit is simply a package of information that reinforces your claim of being worthy of publicity.

The information should be contained in a pocket folder. If you have preprinted folders with your company's name and logo on the front and your address information and Web site address on the back, use them. If not, buy a dozen attractive two-pocket folders from an office supply store and put a label with your company name or logo on the front. In the long run, it would be wise to invest in printed custom pocket folders. Two hundred and fifty standard 9- by 12-inch folders with an imprinted or foil stamped logo on the cover should cost you less than $750.

Inside the kit, include the following information:

A company brochure.   If you have a brochure, include it. If you don't have a brochure, a description of your company, which you probably prepared for a previous proposal, will be fine. Make sure the information describes your mission, the products you sell or services you provide, your competitive advantages (see Chap. 8), and the geographic area in which you work. If they are not included in your brochure, throw in a few black-and-white, 8- by 10-inch photographs of projects you have worked on.

Biographies of key people.   Biographies should be interesting descriptions of accomplishments that support your claims about your key people being industry experts. They will be used to describe you and others when you are quoted in articles. The biographies should be written in paragraph form, not in lists like resumes. They should be written in descending order of importance and not necessarily chronologically.

Following is a sample, fictitious biography:

> Pete Johnson is vice president and director of preconstruction services for Andrews Construction Co., a $150 million-a-year commercial general contractor in Saddle River, N.J. Johnson has worked in the construction industry for more than 20 years, since graduating from Virginia Polytechnic Institute in Blacksburg, Va., with a master's degree in civil engineering in 1978.
>
> Johnson began his career as a field engineer laying out highways and bridges, and later worked as a project manager, coordinating the construction of such notable area projects as the Westbridge Center 3 office and retail center in Rochester, N.Y., and the Park Regency on Central Park in Manhattan.
>
> His experience in the field and his significant training as a writer and speaker have given Johnson a blend of skills that enables him to be an excellent presenter of information in sales or speaking situations. He has authored several articles about bridge and highway construction in notable industry publications like *ENR, Building Design & Construction,* and *Civil Engineering Today.* He has also spoken about engineering, construction, and marketing at national conferences sponsored by the American Society of Civil Engineers and The Urban Land Institute, and has been a guest lecturer at Virginia Polytechnic Institute.

You will notice a few things about the sample biography. It is brief and succinct, only three paragraphs long. It also refers to Pete Johnson by his last name after the first reference, which is how he would be referred to in the media. Formatting issues like these will be discussed in more detail later in this chapter, where you will be strongly urged to purchase one of the journalistic style books that reporters use to format their articles.

Other items to include in your press kit are

Photographs of key people. Black-and-white 5- by 7-inch photos of your key executives, particularly those who will interact with the media, are important to include in your press kit. On the back, put the full name and contact information, either on a label or written with a felt-tip pen. The pose of your photos should reflect the image you want to portray. You should invest in having the photographs shot by a professional in a studio. You will have many uses for these photos, including proposals, when someone is selected to the board of directors of an association, and when one of your key people is invited to speak at various functions.

Fact sheet. A fact sheet is a list of interesting information about the area in which you are involved, not about your company. Reporters will use it as a valuable resource when they are writing about your industry. It should be printed on your letterhead with your contact information in case it gets separated from your press kit. For the bridge and highway contractor, it may include information about the interstate highway system, the condition of infrastructure in the United States, the modern technology behind bridge building, and the complexities of rebuilding highways already in use, among other things.

Client list.   To the extent you feel comfortable revealing this information, a client list establishes your credibility, particularly if you have worked for major companies. It also is a valuable resource for the reporter if he or she is researching an article and needs to contact someone at one of your clients' companies.

Other material.   There are several other things you can include in your press kit if you have them available. Reprints of previously published articles about you or your company can serve as good background information. Recent press releases you have issued can also be valuable to reporters. If you have testimonials from clients, particularly those who are well known, they will lend credibility to your claims of quality.

The meeting.   When you meet with editors or reporters, briefly review the information you have gathered and tell them you are giving them this information as background on your firm, in case it is ever useful. You should offer to be available if they have questions of a general nature about your industry, perhaps as background information for stories they are writing. You will be positioning yourself as an industry expert.

You should take one current press release with you, either announcing the establishment of your business, your promotion to director of business development, a major award, or a new project. This will give you a specific reason to visit in addition to the introduction, prolong the meeting, and increase the likelihood that you will see a brief article soon after your meeting.

Keep the meeting brief, no more than 10 or 15 minutes. Reporters are almost always either researching information or writing a story on deadline. When you call, make sure you set up a meeting at a time that is not near deadline. They will appreciate a brief meeting but will get anxious if it runs too long.

These meetings will be beneficial for both you and the reporters. If they know you, the reporters will be more likely to run your press releases, as long as they have news value, and will be predisposed to run feature articles about some aspect of your firm. They will come to consider you an expert in your industry and call you for information when they are writing general articles about your industry.

The relationship benefits the reporters because the largest portion of their job is gathering information. If they know you, it will make the job that much easier. You can also refer them to others who can provide them with information.

Just remember that you are developing a professional relationship, not a friendship. You should never ask them to do you a favor by running an article or going easy on your firm if a negative issue arises, just as they would not ask you to lie about a subject to get a quote that supports the theory of an article. Your job is to try your best to get positive publicity for your company. Their job is to gather information and report facts. You can work cooperatively with reporters to achieve your goals.

Developing a strategy

After you have established your goals and created a media list, you should develop a strategy that will accomplish your goals. It should identify what type of information you will try to get published or reported upon and how you will go about doing it. There are several levels to a publicity strategy, depending upon your personnel resources and comfort level with working with the media. It could include any or all of the following.

Press releases.   Press releases are the staple of any publicity plan, and they should be the very least that any company does. The press release format, discussed later in this chapter, is simple and straightforward and provides the reporter with enough information to either write a brief story or prepare the outline of a story idea that he or she can present to an editor. More on writing press releases, including templates for almost every occasion, is provided later in this chapter. Following are standard subjects that any contractor or subcontractor can write press releases about:

1. *New projects.*   Any time you get a new project you should send a press release announcing it. If it is a publicly awarded contract, and there is no stipulation in the contract about restraining from issuing publicity, you should send it out after the contract is awarded. If it is a privately awarded contract, you should ask the client for permission. The benefit of publicity will not offset the loss of a client if it is upset about your issuing a press release.

2. *Promotions or new hires.*   Whenever executives or management-level personnel are hired or promoted, you should send a press release. These announcements are covered in almost every daily newspaper, weekly business newspaper, and trade publication. If you can afford it, you should include a professional photograph. If you cannot afford a professional photo, do not include snapshots that someone in your office has taken. There is nothing more unprofessional than a mug shot of someone standing against the kitchen wall with a green face and red eyes. Even if your budget is limited, try to include photos of yourself plus vice presidents or your president if they are the subjects of a release.

3. *New products or services.*   Any time you launch a new product or service it is newsworthy. Often, if it is a major announcement and something that is unique to your marketplace, it will warrant a feature article, which is discussed in more detail later in this chapter. If you are announcing a new product, you should include a photograph.

4. *Awards.*   Winning awards clearly demonstrates your abilities. Any time you receive an award for innovation, quality of construction, management, or whatever, you should issue a press release. It is imperative that you issue the release as soon as possible after you receive the award (or before if the award

is not supposed to be a surprise). Include a photograph if the award is for a project you built.

5. *Company relocations.*   Most publications will print brief stories about companies that move or expand.

6. *Special events or seminars.*   If you schedule a special event, like a ribbon cutting ceremony, a trade show, or a job fair, or if you or one of your key employees is hosting or speaking at a seminar or conference, it is worthy of a press release. Certain special events may warrant special media attention. Event publicity is also discussed in more detail later in this chapter.

7. *Topped-out and completed projects.*   If you have the time, it is worthwhile to prepare and send press releases about topped-out or completed projects, even if you have already issued a press release announcing the start of the project. You may only get a small mention in the press, but publicity is cumulative, so any positive mention you get will add to your name recognition and credibility. You may want to announce top-outs in advance because sometimes they make interesting photo opportunities for the media, particularly if the project being topped-out is significant or well known.

8. *Articles of yours which have been published.*   It never hurts to send a press release announcing published articles that you authored. It may be picked up in trade press and will help establish your credentials as an expert in your field.

Among these topics, you should easily be able to generate a different press release every month or 6 weeks. That would yield 8 to 12 press releases each year, which should give you a good amount of publicity.

One major national general contractor issues as many as 80 press releases a year, all of which are newsworthy. This inundation of press releases not only gives the company a broad, balanced media presence, it initiates a variety of calls from major publications, including *ENR*, that are interested in writing feature articles about it. It allows the company to establish valuable relationships with the media and generates frequent calls when reporters need opinions or background information for articles that they are writing.

There is really no downside to issuing regular press releases. They may never generate a large feature story about your firm, but they will ensure that your name and services stay in front of prospects, establishing that critical first step in the decision-making process—awareness.

Feature articles.   A feature article goes into much more depth than what will be extracted from a press release. As its name implies, it usually is written about an interesting subject rather than an item of news, as is the case with the type of article that is generated from the press release. You should try to schedule one feature article idea each year. Writing and pitching feature stories to reporters is covered in more detail later in this chapter. Good ideas for feature stories might include

A new service you are providing.   If you are providing a service that is new to your market, or it solves a problem in an innovative way, it may be worthy of a feature article.

An acquisition of a new company.   If the acquisition gives you the ability to provide services heretofore not available in your market, a reporter might want to develop it into more than just a short mention in the Business Notes section.

A change in leadership.   The announcement of a new president or the purchase of a company by an employee or group of employees that leads to a leadership change may generate interest. If the leadership includes a minority or woman, the press will be even more likely to take notice.

How you have overcome obstacles to be a leader in your field.   Human interest stories, like how a woman overcame obstacles to build a thriving construction company in a male-dominated industry or how an owner of a masonry contractor emerged from bankruptcy in 1990 to become an industry leader, will usually attract the attention of the media.

A profile of a particularly difficult project and how you found innovative ways to manage it.   Stories like these are particularly interesting to trade press because readers like to learn how to overcome challenges they may also be facing.

A feature on your non-industry-related activities.   Darrel Rippeteau is a Washington architect who has a particular flair for designing retail spaces. It was not only his designs, however, that attracted considerable media attention in 1998, including a feature on the Washington CBS television news affiliate and a cover story in the *Washington Business Journal,* each within 6 months of the other. It was his ties. Darrel's hobby is making ink drawings of city scenes and landmark buildings, of which there are many in Washington. He parlayed this hobby into a flourishing tie and scarf business called Archifeti.

You should try to include one feature article every year or 18 months. Any more would be difficult to sustain. Coming up with ideas for feature stories, then pitching and writing them, is time consuming and somewhat tiring. If you can generate any feature articles, you will add immensely to your overall marketing communication efforts.

Publishing articles.   At some point, you have probably been reading your favorite business or industry trade publication and been impressed to see someone you know is the author of an article. Or perhaps you have even read an article written by an expert and went to the trouble to contact that person to discuss how he or she could provide services for your firm.

Publishing articles is not a big mystery. It is one of the best, most enduring forms of publicity, yet few people make the effort to try to get published.

As Marcia Yudkin says,

> Getting published on your own is the fastest way to establish yourself as an expert. It helps you reach out to your target market; and it keeps on working for you forever when you include copies of your publications in your press kit or send them out to potential clients. In comparison with courting the media through press releases, you have more control of your message through article writing. You'll also have the opportunity to develop your ideas in more depth.[1]

You will learn how to write your own articles later in this chapter, but for planning purposes, you should focus upon what might be of interest to the readers of the newspapers, magazines, and newsletters in which you might be published.

If you are an interior contractor, for instance, you might want to consider publishing articles like these:

- The 10 essential things you can do yourself to maintain your office space.
- What every architect needs to know about building codes for renovations.
- How to cut 6 months off your search for your next office space.
- A guide to planning your office move.

You will notice that in each example, the article will give the reader valuable information. This might encourage them to clip out the article and save it—extending its publicity value for you.

These articles will also help establish you as an expert in your field. When readers need information on these matters, they will turn to you.

"Advertorials."   Even if you are not successful getting feature stories placed in your target media, or do not get much interest in your efforts to publish articles in your local business or trade publications, you can still achieve your goals of getting publicity. Advertorials are hybrids of advertisements and editorials. You have surely seen them dozens of times, whether you are aware of it or not.

Advertorials appear to be articles in publications. They are laid out very similarly to the publication, with very similar typefaces for the headline and body copy and similar column widths. Indeed, advertorials are meant to fool you into thinking that they are news columns. What differentiates advertorials from actual editorials (news stories) is that a border typically surrounds them that has the word "advertising" printed in small type along it.

Despite the warning that the advertorial is actually advertising, many people are drawn to advertorials because they look like editorial content. The border surrounding them, which is placed there by the publication, also draws attention to it.

Advertorials are paid advertising. They will cost the same as a traditional display ad of the same size as your advertorial. However, because more people are drawn to them, the value of your advertising investment is often greater than it would be for a display ad.

Writing and publishing advertorials is discussed in more detail later in this chapter, but for your planning purposes, you should plan advertorials that help support your main selling points that were established in your marketing communication plan (Chap. 8). Select subjects that would be of interest to the readers of the publications in which you advertise. If the advertorial is self-serving or boring, you will waste your money.

Use advertorials only as a contingency if your other publicity efforts fail. You may want to set aside some of your advertising budget for an advertorial if you need more publicity than you are getting.

Speaking.   Just the mention of speaking before an audience evokes dread in most people. Yet, with the right preparation and the confidence that the audience will benefit from what you have to say, speaking can be one of your best publicity tools.

Speaking can establish credibility faster and more completely than any other form of publicity. Listeners assume you have been chosen to speak because you are a leader in your field. Although you may indeed be an expert, many organizations actively seek potential speakers and are delighted to find credible volunteers who may or may not be experts.

In addition to establishing credibility and expertise among the people listening to you, speaking engagements also generate additional publicity. Often, the organizations sponsoring your presentation publicize the event in advance, by running an article in their own publication, sending a notice to area media, or advertising. You could also do pre-event publicity yourself by sending out a notice to the media.

Also, whether invited or not, media representatives may attend your speech and write a follow-up article. They will take notes during your presentation and follow-up either immediately afterward or by phone. You may also initiate postevent publicity by sending a press release outlining your presentation.

Preparing for a speech or seminar appearance will be discussed in more detail later in this chapter, but if you are willing to give it a try, you would be well served by adding at least one speaking engagement to your publicity plan this year.

Selecting the audience is very much like selecting media for your press releases. Trade and professional associations are probably the best place for you to make your foray into public speaking. One of the major missions of associations is education, and most good ones have regular seminars, panel discussions, and conventions that need qualified speakers. Identify associations to which people in your target audience belong. Contact the executive director, or director of member services, to discuss your interest in speaking. You will more than likely get a very enthusiastic response.

Events.   Events are opportunities to recognize significant accomplishments. Award ceremonies, groundbreakings, and charity functions are examples. If an event you stage generates publicity, you should consider it a bonus. Staging an event with the sole purpose of generating publicity is a risky business that, more often than not, is a waste of money.

If you have a worthwhile reason to stage an event, it will likely attract your target audience directly. For example, if you have a ceremony celebrating the completion of a major construction project, you are likely to attract architects, engineers, brokers, and others who may not even be part of the project team. If the media is attracted to the event because it may be of interest to their readers, you get the added benefit of generating publicity for your efforts.

Conversely, if you stage an event for publicity, it is probably because you don't have a good reason to attract members of your target audience. Press conferences to announce unspectacular news, celebrity appearances, and presentations set up with the sole intent of attracting media usually fail because the media see them for what they are—attempts to generate publicity and not newsworthy events.

If events are a part of your overall marketing effort, create the most interesting, topical, and fun events you can. Then send out a few invitations to the press. If you succeed in making the event worthwhile for those who attend, you will probably succeed in getting some press coverage as well.

Your PR strategy, like your overall marketing communication plan, should be revisited frequently, perhaps every 6 months, so you can update ideas for press releases and feature articles to include news events that have happened to your company over that time period. Once you have finalized your plan, it is time to execute it.

Writing for Publicity

There is an art to writing acceptably for the media. It takes an understanding of journalistic techniques, a command of the English language, and the ability to think and write logically. For the most part, journalistic writing is not like creative writing. Although you need to be creative in coming up with ideas for stories that will be picked up by the media or, in the case of publishing articles, will be interesting to readers, writing for the media is more about clearly presenting information than flowery prose.

Fortunately, if you follow a few basic rules, you can write an effective press release that will be read by the media. Templates for most situations that warrant a press release are including in this chapter. If you put these templates onto your computer, you will have the beginnings of a good, basic publicity program. From there, if you think you need help writing press releases and pitching feature articles or for writing articles to publish, you can probably find a publicity consultant in your area with knowledge of the construction industry.

All press releases and articles you write should follow basic journalistic rules and be formatted according to style books used by the media. If you are going to write your own press releases, you must own the *Associated Press Stylebook and Libel Manual,* published by the Associated Press, the University of Chicago's *Manual of Style,* or a comparable style book.

These books are both style books and reference works and are organized like dictionaries. If you need the acronym for a government agency, look under the agency's name. If you need to know whether to capitalize a word, check the word itself or the *capitalization* entry. They show you how to handle titles, which should always be lowercase, except when immediately preceding the name. They show how to abbreviate states, how to cite academic titles, and so much more.

The anatomy of a press release

The press release is the basic tool for getting publicity in the media. It presents information in a way that allows the reporter or editor who reads it to quickly discern if the information you are providing is newsworthy.

Your press releases should be written in the same style that reporters write news stories. Whether the press release is a simple announcement of a new project or part of your pitch to generate a feature article, it should follow the same basic format, called the *inverted pyramid*. This simply means that the most important information with the broadest impact is at the beginning, and as the story progresses, the information becomes focused on details of narrower importance (envision an upside down pyramid with the broad information at the top and the narrow information at the bottom).

Reporters write articles this way so that readers can glean the most important information at the outset and continue reading if they want more. Also, this style gives editors more ease in cutting articles that are too long. They simply chop off enough of the end of the article to fit their space needs. If you follow this style when writing your press releases, you will help the reporter decide if the information should be developed into a story.

The Five W's.   Because of this inverted pyramid style, the first paragraph, called the "lead paragraph," is the most important part of the story. Every journalism student learns on day 1 of classes that the lead paragraph must include the Five W's: who, what, where, when, and why (or how). These essential bits of information must be written into the lead paragraph as concisely as possible. A good lead paragraph should not exceed 35 words, as follows:

> Andrews Construction Co. has just begun construction of a 12-story mixed-use development that will add much-needed office space and hotel rooms to the rapidly growing Airport Corridor in Rochester.

In 32 words, the above fictitious example addressed

- Who: Andrews Construction Co.
- What: Construction of a 12-story mixed-use development
- Where: In Rochester
- When: Just begun
- Why: To add office space and hotel rooms to the rapidly growing Airport Corridor

The second paragraph should elaborate on the premise of your release, usually providing additional detail:

> The development, to be called Airport Plaza, will include 115 hotel rooms atop 120,000 square feet of office space, with a two-level retail mezzanine of nearly 30,000 square feet. Construction began in March and will be completed within 24 months, according to Project Executive Richard Kline of Andrews Construction.

The third paragraph should provide additional information of lesser importance and may begin with a quote to attract interest:

> "Vacancy rates in the Airport Corridor are hovering around 4 percent," stated Phyllis Brock, a project manager for the building's owner, Fleet First Investment Trust of Rochester. "Airport Plaza will give major tenants a chance to occupy large blocks of space and provide much needed hotel facilities for business travelers." ADI & Partners of Albany, N.Y., is the project architect.

The final paragraph should provide information about your firm and should become boilerplate for the end of each of your press releases:

> Andrews Construction Co., headquartered in Saddle River, N.J., is a commercial construction company serving the northeast United States. Established in 1962, its annual volume of construction projects exceeds $150 million.

Even though it appears at the top of the press release, it is best to write the headline last—after you have solidified your lead paragraph. The headline should capture the attention of the reader with the most important information. This is sometimes referred to as a *hook*. The lead paragraph of the press release serves to elaborate on the hook in the headline:

> Construction of Much-Needed Office and Hotel Space Begins near Airport

Succinct and to the point, the headline conveys the most important information, which in this case is that the office and hotel space is needed. You will notice that the name of the contractor is not in the headline, because the fact that they are building the mixed-use facility is not as important as the fact that it is being built to satisfy a need.

When you write press releases, remember that reporters just want the facts—the five W's. They receive dozens of press releases every day, and they don't have the time to search through each to find the salient facts. Make their job easier by presenting the facts up front, in the lead paragraph, and provide only enough additional information to explain your lead paragraph. If they need more information, they will contact you.

Formatting your press release. There is a very specific style you should follow in preparing your press release. This is to ensure that editors can get all the information they need exactly where they are accustomed to seeing it. The press release crafted above is shown in Fig. 10.2 in its proper format.

First line. Either print your press release on your stationery or provide the business name and address on the top of the release. Some reporters prefer

Andrews Construction Co.
1200 Hudson Street
Suite 200
Saddle River, NJ 10028

Contact: Alton Fryer, (206) 555-1212

FOR IMMEDIATE RELEASE

CONSTRUCTION OF MUCH-NEEDED OFFICE AND HOTEL SPACE BEGINS NEAR AIRPORT

Rochester, N.Y., November 1, 1999—Andrews Construction Co. has just begun construction of a 12-story mixed-use development that will add much-needed office space and hotel rooms to the rapidly growing Airport Corridor in Rochester.

The development, to be called Airport Plaza, will include 115 hotel rooms atop 120,000 square feet of office space, with a two-level retail mezzanine of nearly 30,000 square feet. Construction began in March and will be completed within 24 months, according to Project Executive Richard Kline of Andrews Construction.

"Vacancy rates in the Airport Corridor are hovering around 4 percent," stated Phyllis Brock, a project manager for the building's owner, Fleet First Investment Trust of Rochester. "Airport Plaza will give major tenants a chance to occupy large blocks of space, and provide much needed hotel facilities for business travelers." ADI & Partners of Albany, N.Y., is the project architect.

Andrews Construction Co., headquartered in Saddle River, N.J., is a commercial construction company serving the northeast United States. Established in 1962, its annual volume of construction projects exceeds $150 million.

###

Figure 10.2   Sample new project press release.

releases on plain white paper rather than stationery. Perhaps that is because a company's letterhead automatically signals that the release is for self-promotion and may put the reporter on the defensive.

Second line.   Provide the name and phone number of someone who can provide additional information about the project. It should be someone prepared, and perhaps trained, to speak with the media, like your marketing director or communication director and not the job's project manager or superintendent who may be uninterested in spending time on the phone with a reporter. You will be best served if you choose one person in the company to handle this function. If that person does not know the answers to every question, the reporter won't mind if he or she promises to call back with the answers as soon as possible.

Third line.   FOR IMMEDIATE RELEASE should be printed in capital letters. This lets the media know that they can publish the information immediately. If you are presenting information that should not be released right away, like results of an award ceremony on March 19, write, For Release March 20, 1999.

Headline.   Capitalize and center the headline. Use bold type to ensure that the headline is the first thing that the editor or reporter who receives the press release sees. If the headline carries over onto two or three lines, break the headline where a reader would logically pause. "Design" the headline so each line has maximum impact on the reader.

Dateline.   This is the city and state where the news takes place and the date the material is released. Make sure that you mail your releases quickly so that the date is close to the time the releases are received by the media.

Format.   Double space your press release and leave a 1-inch margin all around for easy editing. Keep your sentences brief and to the point, and restrict your paragraphs to one thought at a time—usually no more than two or three sentences. Try to keep your release to one page. If you need to include a second page, put the word *more* in parentheses, centered at the bottom of page 1, and put *continued,* in parentheses, flush left just above where the text continues. On the top left corner of page 2, write an abbreviated form of your headline, such as Construction Near Airport Begins, for the above sample. Underneath it, put the date of the release, then on the next line, write Page 2 of 2. Each of these elements help the reporter keep the information together if the pages get separated.

Bottom.   ### signifies the end of the release.

When writing your press release, you may need to modify it slightly to suit the needs of particular media. An engineering-oriented publication is targeted to different readers than a real estate publication, which is different still from a neighborhood newspaper. By customizing your press release to suit the needs of a publication's readers, you will increase the likelihood that it will be published.

How to write the release.   Not everyone who wants to write a press release has the benefit of taking journalism classes, but there are a few basic rules that will help make your press release more readable and therefore more likely to be used by the media:

1. *Write economically.*   Remember, this is not twelfth grade creative writing class. You don't need to find the most eloquent way to convey your thoughts. Write clear, simple sentences that use the fewest, shortest words needed to complete an idea. Become a student of how stories are written in major dailies like *The Wall Street Journal.* Despite reporting on complex subjects, its reporters write so clearly that their material is easy to digest and pleasant to read.

2. *Use active phrases.*   Use active verbs instead of passive ones. Active phrases say directly who does what, rather than passive phrases that say

something happened to someone. Active phrases are livelier and keep the reader's interest.

3. *Use positives.* Negative words like *don't, not, failed, avoid,* and so on, make readers uneasy and force them to work harder to understand your meaning. Whenever you feel the need to use negative words, try to figure out how to keep the meaning of the sentence while using positive words.

4. *Jargon.* Keep in mind that a reporter looking at your press release may have little or no knowledge of construction. Even commonly used terms like *general contractor, tilt-up, precast, top-out, fit-out,* and *rebar,* may be unfamiliar and cause a reporter to become confused and disinterested. If you need to include industry terminology, be sure to explain what it means.

5. *Grammar.* Always follow the rules of good grammar. If you don't already have one, get a book like Strunk & White's *Elements of Style* and reference it whenever you have questions about word usage.

6. *Edit.* Edit your press release for two reasons: to make sure you are saying as much as you can in as few words as possible and to make absolutely sure there are no misspellings or typographical errors.

Sending press releases to the media. Fold the press releases as you would any correspondence and stick it in the envelope. You don't need to include a cover letter. If you are sending a photograph with your release—people who are promoted, renderings of a building you are constructing, an award you have received—put a large label on the back that clearly describes it. The label serves two purposes: to identify the photo as going with your release if it gets separated, and as the basis for a photo caption.

There is no need to follow up press releases with a phone call. Most reporters are too busy to field calls and sort through the dozens of releases they get each day to find yours. Also, many releases go to a central mail room and are distributed directly to appropriate editors who may not be the ones to whom you addressed the envelope. If your release is well written and contains information that is valuable to the people who read, watch, or listen to the media, chances are it will be used on its own merits, without the need of a cover letter.

Template press releases for all occasions

Following are templates you can use to prepare press releases for several situations that arise for most construction industry businesses. Simply type these into your word processing software in a folder titled Template Press Releases and fill in the blanks when you are ready to issue a release.

For each example, there is also a sample press release to show you how your press release should flow (the sample press release for a new project was shown in Fig. 10.2). Don't hesitate to copy these samples word for word except for replacing relevant information. Template releases include

- New projects (or topped-out and completed projects)
- New hires
- Promotions
- New products or services
- Company relocation
- Company expansion
- Awards

Figure 10.3 is a template press release for new projects (or topped-out and completed projects). Remember, if you use industry jargon like "topped-out," be sure to explain the term soon after you use it.

Figure 10.4 is a sample press release for a new hire, and Fig. 10.5 is the corresponding template.

Figure 10.6 is a sample press release for a promotion, and Fig. 10.7 is the template for this release.

Figure 10.8 is a sample press release for a new product or service, and Fig. 10.9 is the corresponding template.

Figure 10.10 is a sample press release for a company relocation, and Fig. 10.11 is a template for that release.

Figure 10.12 is a sample press release for a company expansion, and Fig. 10.13 is a template for this release.

Figure 10.14 is a sample press release for company awards, and Fig. 10.15 is the corresponding template.

Writing feature articles

Feature articles are not always written in the same style as news articles. They often lead off with an aspect of the story that is not the most important but is intended to capture the attention of the reader or viewer. They are written more like stories with a beginning, middle, and end. Unlike news stories, the end of feature stories is often the most important part.

Feature articles typically offer far greater publicity value than a small story or mention generated from a press release because they are longer and go into more detail describing your company's services and competitive advantages. You can derive value indefinitely if you make copies of the feature article and include them in marketing packages.

The first step in getting a feature article written about your company is coming up with a good idea. You have already developed a marketing communication plan (Chap. 8), so you know what competitive advantages you want to communicate about your company. Now you have to think like the editor or reporter who will be on the receiving end of your pitch for a feature story.

(Your company's name)
(Your company's address)

Contact: (Name of media contact), (area code and phone number of media contact)

FOR IMMEDIATE RELEASE or FOR RELEASE (date to be released)

**(HEADLINE THAT SUMMARIZES MAIN POINT
OF LEAD PARAGRAPH)**

(Location of news item), (date of release)—(Your company's name), has just begun construction (or topped-out construction or completed construction) of a (brief description of project, including physical dimensions—number of stories, square footage, hotel rooms, number of retail stores, linear feet of pipe, contact area of concrete, tons of steel, or whatever best describes the project) in (location of project).

The project, (name of project), will include (a more detailed description of the project, expanding upon the brief description in the lead paragraph). Construction (or work—whatever term best describes what you do: electrical engineering, painting, rehabilitation, drywall work, concrete forming, etc.) began in (when work started) and will be completed (broad time frame of work completion, either "within 15 months," "in fall 2004," "in early 2003," and so on), according to (title and name of a company executive, often the president, who will benefit from the added exposure. This does not have to be the same as the contact person.).

"(A quotation that describes the relevance, complexity, or particular interest of the project)," stated (name of person quoted), (title of person quoted) for the (project type—building, hotel, power plant, highway)'s (affiliation—owner, architect, engineer, supervisor), (company or organization name of person quoted) of (city and state or country if outside United States of person quoted). "(Continuation of quote if enough relevant information is available—no attribution of the quote is needed. If it is in the same paragraph as the first quote, it is assumed to be from the same person)" (names and locations of key team members; at the least, they should include your client. You may also want to include any or all of the following: the owner, architect, engineer(s) and general contractor).

(A boilerplate paragraph that describes your company, including your location, the type(s) of business(s) in which you are engaged, your years in business, and so on).

###

Figure 10.3   New project template.

The first question a reporter or editor is going to ask when he or she sees your article idea is, *Is this of interest to my readers (viewers or listeners) today?* Media only want to publish interesting stories to keep their "customers" buying their product. So before you pitch a story idea, ask yourself the same question.

The first critical part of the reporter's question is *my readers (viewers or listeners).* For a feature story to be picked up, it must be relevant to the audience of the media you are pitching. One of the reasons that many privately held companies have a difficult time getting press coverage in *The Wall Street*

Andrews Construction Co.
1200 Hudson Street
Suite 200
Saddle River, NJ 10028

Contact: Alton Fryer, (206) 555-1212

FOR IMMEDIATE RELEASE

**PRYOR JOINS ANDREWS CONSTRUCTION AS
VICE PRESIDENT OF FINANACE**

Saddle River, N.J., November 1, 1999—Allison Pryor has joined Andrews Construction Co. as vice president of finance.

In this capacity, Pryor will oversee financial management of the company, including investments, accounts payable and accounts receivable, payroll and taxes, and office management.

"Allison Pryor is an experienced financial strategist who will help propel Andrews Construction into its next level of growth," stated Andrews Construction President, Anthony P. Andrews.

Pryor has nearly 15 years of experience in finance, including 6 in the construction industry. After graduating in 1984 from Columbia University in New York, with a bachelor's degree in business administration, she spent 9 years with the accounting firm of Arthur Andersen in New York, during which time she earned a CPA. In 1993, she was hired as controller of James P. Lewis & Sons, a masonry and drywall contractor in Ithaca, N.Y.

Andrews Construction Co., headquartered in Saddle River, N.J., is a commercial construction company serving the northeast United States. Established in 1962, its annual volume of construction projects exceeds $150 million.

###

Figure 10.4   New hire press release.

Journal is because the publication primarily covers publicly traded companies. The better job you do of identifying publications targeted to people who would be interested in your story idea, the more likely you will be to get a positive response from a reporter.

The second critical part of the reporter's question is *today*. Timeliness is key to getting stories published, particularly in print where magazines and newspapers have to compete with television and radio, which can deliver news much faster. Ask yourself if your story idea could be published 1 month or 6 months from now without having to be changed. If so, you need to find a way to make it relevant to events taking place in the industry today. If you can make your story a statement about a trend, it will capture the editor's interest.

Once you have come up with a timely, topical idea for a story, the next step is gathering enough information to present to the media. Collect a group of people who are knowledgeable about the subject you are going to present and have a brainstorming session. Have the group offer as many ideas as it can. Remember that brainstorming sessions are for free exchanges of information. Ideas presented should not be evaluated for their relevance until all ideas are on the table. Once the brainstorming is completed, test your ideas against the key question, Is this of interest to my readers (viewers or listeners) today? After you have done this, you will have the beginnings of an outline of ideas for your story.

Pitching a feature story.   Pitching a feature story includes three steps: writing a press release that summarizes your idea, writing and sending a pitch letter, and following up with a phone call.

The *press release* you write is much like other press releases for new projects, promotions, and so on. It should be concise but complete. It should not be written as a complete story but rather the basis for a story. It must be written well enough to capture the editor's attention and interest and provide enough support information to lend credence to the value of the story. The editor or reporter assigned to the story will want to write an original piece, not merely edit what you have written.

Figure 10.16 is a press release that was submitted to pitch a feature article about a large commercial construction company that added a real estate division.

The *pitch letter* is a brief statement about the contents of the press release and the relevance of the story and contains enough supporting information to

---

(Your company's name)
(Your company's address)

Contact: (Name of media contact), (area code and phone number of media contact)

FOR IMMEDIATE RELEASE or FOR RELEASE (date to be released)

(NEW HIRE'S LAST NAME) JOINS (COMPANY NAME) AS (TITLE)

(Location of news item), (date of release)—(New hire's full name) has joined (your company's name) as (title).

In this capacity, (new hire's last name only) will (job description).

"(Quotation that describes person's positive contribution to the firm)," stated (company name and title of person who would be best qualified to render judgment of the new hire's contribution to the company).

(Overview of new hire's experience in their field, including relevant experience within your industry, if applicable. Relative experience adds credibility to the new hire). (Description of new hire's education, including undergraduate and graduate degrees and any professional designations, plus work history. You may not want to mention names of other companies the new hire has worked for if they are competitors. If the other companies are well-respected firms, mentioning them will add credibility to the new hire).

(A boilerplate paragraph that describes your company, including your location, the type(s) of business(s) in which you are engaged, your years in business, and so on).

###

Figure 10.5 New hire template.

Andrews Construction Co.
1200 Hudson Street
Suite 200
Saddle River, NJ 10028

Contact: Alton Fryer, (206) 555-1212

FOR IMMEDIATE RELEASE

RAMIREZ PROMOTED TO PROJECT EXECUTIVE
AT ANDREWS CONSTRUCTION

Saddle River, N.J., November 1, 1999—Manuel Ramirez has been promoted to project executive of Andrews Construction Co.

In this capacity, Ramirez will oversee the field and office management of several commercial construction projects being built by the firm. He will supervise several job teams, which include project managers, project superintendents, and field managers. He will join two other existing project executives in this capacity.

"Manny Ramirez has consistently demonstrated his management and leadership skills since joining our firm, and he is an ideal candidate to assume this important position" stated Andrews Construction President, Anthony P. Andrews.

Ramirez has 22 years of experience in the construction industry, including 14 with Andrews Construction. After graduating in 1977 from the University of Georgia in Athens, Ga., with a bachelor's degree in civil engineering, he spent 8 years working in positions of increasing responsibilities with construction firms throughout the southeast United States.

In 1985, Ramirez joined Andrews Construction as a project manager. In that capacity, he has worked on several of the firm's major construction projects, including the 15-story Mercury Center in Albany, N.Y.; the 135-room Riverside Suites hotel in Jersey City, N.J.; and the 650,000-square-foot, three-phased Albany Convention Center complex.

Andrews Construction Co., headquartered in Saddle River, N.J., is a commercial construction company serving the northeast United States. Established in 1962, its annual volume of construction projects exceeds $150 million.

###

Figure 10.6   Promotion press release.

make the recipient want to read the press release. It should not beg, but rather make a compelling argument for why you feel the information would *be of interest to the particular media's readers (viewers or listeners) today.* Include that very statement in your letter.

Figure 10.17 is a letter that was sent with the Fig. 10.16 press release.

Notice that the letter states that Davis will *follow up with a phone call* the following Tuesday. He knows that this is not a deadline day for this particular reporter. Notifying her of this allows her to prepare for the call, which encourages her to read the material.

Several things could result from the phone call. There is a chance you will get a "thanks but no thanks" response, in which case you should press a little by asking, "What might make a story like this relevant or interesting to your readers? Perhaps there is an angle we haven't thought of." This line of questioning might take you in a different direction, but it that might also lead to a feature story.

Another scenario is that the press release will be used but as a smaller story than what you had hoped for. A little gentle pressing similar to what was just mentioned may help expand the reporter's interest, so long as it does not push her away from writing anything at all.

Have contingency plans in place. If one publication does not want to run your story, there are probably others that you can pitch. Just don't pitch more than one at a time. You could embarrass and alienate an editor or reporter if two competing media run the same story at the same time.

If the editor or reporter is interested, you must be ready to act fast. You can offer to meet the reporter at the newsroom, but the environment in most newsrooms is crowded and noisy and not the best place to conduct an interview. Chances are, the reporter will recommend an interview at your office. If so, you should know the schedules of every person in your company who will participate in the interview. If trying to organize a lot of people's schedules

---

(Your company's name)
(Your company's address)

Contact: (Name of media contact), (area code and phone number of media contact)

FOR IMMEDIATE RELEASE or FOR RELEASE (date to be released)

(PROMOTED PERSON'S LAST NAME) PROMOTED TO (TITLE) AT (COMPANY NAME)

(Location of news item), (date of release)—(promoted person's full name) has been promoted to (new title) of (your company's name).

In this capacity, (promoted person's last name only) will (description of the promoted person's new position).

"(Quotation that describes person's positive contribution to the firm)," stated (company name and title of person who would be best qualified to render judgment of the promoted person's contribution to the company).

(Overview of promoted person's experience in his or her field. Relative experience adds credibility to the person). (Description of promoted person's education, including undergraduate and graduate degrees and any professional designations, plus previous history. You may not want to mention names of other companies the new hire has worked for if they are competitors. If the other company's are well-respected firms, mentioning them will add credibility to the new hire).

(Description of promoted person's work history with your firm, including major projects on which he or she worked and any previous promotions, awards, or honors that person has received.)

(A boilerplate paragraph that describes your company, including your location, the type(s) of business(s) in which you are engaged, your years in business, and so on).

###

Figure 10.7 Promotion template.

Andrews Construction Co.
1200 Hudson Street
Suite 200
Saddle River, NJ 10028

Contact: Alton Fryer, (206) 555-1212

FOR IMMEDIATE RELEASE

ANDREWS CONSTRUCTION
ADDS INTERIORS DIVISION

Saddle River, N.J., November 1, 1999—Andrews Construction Co. has added a division that will build interiors of office buildings and retail establishments.

"With the increased demand for office space in New York and New Jersey, the need for interior construction services has grown," stated Andrews Construction President Anthony P. Andrews. "The unique demands of interior construction prompted us to set up a wholly independent division with its own staff to manage these projects."

Andrews Construction Vice President Bruce Bentson will head the new division. He has been employed at Andrews for 5 years. Prior to that, he spent 14 years with Blunt Construction, Inc., a New York firm that specializes in interior construction. He left that firm as senior vice president.

According to Bentson, Andrews Construction's Interiors Division will build office interiors for a variety of businesses. The average size will probably range between 5000 and 50,000 square feet. The firm will also build interiors of retail spaces, like restaurants, shopping centers, and malls. Most of the firm's clients will be the tenants, or occupants, of the spaces.

Andrews Construction Co., headquartered in Saddle River, N.J., is a commercial construction company serving the northeast United States. Established in 1962, its annual volume of construction projects exceeds $150 million.

###

Figure 10.8   Press release for a new product or service.

hampers your ability to set up a meeting at the media's convenience, don't include everyone. It is more important to be responsive. You can always follow up a reporter's question after the interview.

If the reporter suggests a telephone interview, make sure that whoever will act as your spokesperson is responsive and prepared.

Preparing for the interview.   When you are interviewed for a story, particularly one you have pitched, you better be prepared. You have worked very hard to get to this point, starting when you first developed a marketing plan, refined it into a marketing communication plan, and then developed your plan into an idea for a feature story.

Write down the essential points you want to get across and read them over and over again. Be open and honest when answering questions asked by the reporter, but also be sure you weave your main points into enough of the answers to ensure they get into print.

Study how political candidates answer questions in debates. Usually, they briefly answer the question asked and then launch into a speech about one of their ongoing campaign points. Henry Kissinger once opened a press conference by asking, "Does any reporter have any questions for my answers?"

Publishing articles

Getting articles you have written published requires a different approach to the media. Once you have determined a good story idea, or a series of stories that will provide valuable, useful information, you should prepare a pitch letter similar to what would be prepared to pitch a feature story.

In this letter, however, you will be pitching the idea of publishing the article under your name as a guest writer. The value to the media is the credibility you lend to the article. They know that a mix of articles they write and ones written by guest, or contributing, writers will keep their publications interesting to readers.

(Your company's name)
(Your company's address)

Contact: (Name of media contact), (area code and phone number of media contact)

FOR IMMEDIATE RELEASE or FOR RELEASE (date to be released)

(COMPANY NAME) HAS ADDED
(NEW SERVICE OR PRODUCT)

(Location of news item), (date of release)—(Your company's name) has added (brief description of service or product your company is adding).

"(Quotation that explains why the service or product is being added—one that explains the benefit you will offer clients)," stated (your company's name, title, and name of individual who is qualified to render an opinion on the benefits of the service or product). "(A continuation of the quote, if it is needed to further explain why you are well qualified to render this new service or provide the new product)."(Supporting information that further establishes your credentials for adding this service or product. If it is a service, describe the qualified person who will provide the service. If it is a product, explain who will oversee the operation and distribution, or why your firm is qualified to add the product).

(A paragraph describing how the service will be provided or how the product can be used).

(A boilerplate paragraph that describes your company, including your location, the type(s) of business(s) in which you are engaged, your years in business, and so on).

###

Figure 10.9 New product or service template.

Andrews Construction Co.
1200 Hudson Street
Suite 200
Saddle River, NJ 10028

Contact: Alton Fryer, (206) 555-1212

FOR IMMEDIATE RELEASE

ANDREWS CONSTRUCTION
MOVES OFFICE HEADQUARTERS

Fort Lee, N.J., December 1, 1999—Andrews Construction Co. has moved from Saddle River, N.J., to Fort Lee, N.J.

The company's new headquarters address is 200 Prosperity Plaza, Suite 500, Fort Lee, NJ 08295. The offices include 52,500 square feet of space on three floors.

"We moved to be closer to New York City where much of our work is located and because the size of our new building allows us room for much-needed expansion," stated Andrews Construction President Anthony P. Andrews. "It is with very mixed emotions that we moved, because we enjoyed working in Saddle River, which had been our home since our founding."

Andrews Construction Co., headquartered in Fort Lee, N.J., is a commercial construction company serving the northeast United States. Established in 1962, its annual volume of construction projects exceeds $150 million.

###

Figure 10.10 Press release for company relocation.

(Your company's name)
(Your company's address)

Contact: (Name of media contact), (area code and phone number of media contact)

FOR IMMEDIATE RELEASE or FOR RELEASE: (date to be released)

(COMPANY NAME)
MOVES OFFICE HEADQUARTERS

(Location of news item), (date of release)—(Your company's name) has moved from (where you were), to (where you are now).

The company's new headquarters address is (complete postal address of new offices). (Description of new offices, if you feel it is helpful or if it helps position you in the way you want to be perceived by clients and prospects).

"(Quotation explaining why you moved)," stated (your company's name, title, and name of individual who is qualified to render an opinion why you moved). "(A continuation of the quote, if it is needed to further explain why you moved or adds relevant information about your company)."

(A boilerplate paragraph that describes your company, including your location, the type(s) of business(s) in which you are engaged, your years in business, and so on).

###

Figure 10.11 Template for company relocation.

Andrews Construction Co.
1200 Hudson Street
Suite 200
Saddle River, NJ 10028

Contact: Alton Fryer, (206) 555-1212

FOR IMMEDIATE RELEASE

ANDREWS CONSTRUCTION
ADDS ROCHESTER, N.Y., DISTRICT OFFICE

Fort Lee, N.J., December 1, 1999—Andrews Construction Co. has expanded to include a district office in Rochester, N.Y. The company's new district office is located at 45 Brown's Ferry Road, Rochester, NY 06825.

"We established a district office in Rochester to better manage our construction projects in that area," stated Andrews Construction President Anthony P. Andrews. "This is our first corporate expansion and is an exciting milestone for our company."

Jerome Williams, a senior vice president of Andrews Construction, will manage the new district office. Williams has worked at Andrews Construction since 1987 and has served as a project superintendent and project executive, overseeing several projects in both the New York City and Rochester areas. He has spent the last 3 years exclusively managing projects near Rochester.

Andrews Construction has already built or begun work on many notable projects in the Rochester area, including an eight-story office building at 200 State Street, the renovation of the 205-room Excelsior Hotel at Cambridge and Leigh Avenues, and the Cineplex Eight, an eight-theater movie complex near White Marsh Mall.

Andrews Construction Co., headquartered in Fort Lee, N.J., is a commercial construction company serving the northeast United States. Established in 1962, its annual volume of construction projects exceeds $150 million.

###

Figure 10.12 Press release for company expansion.

Select publications whose readers will benefit most from the information you have to present. Then, send a pitch letter to the editor of the appropriate section of the publication (assuming you are not pitching guest spots on radio or television). In the letter, outline the types of articles you would like to publish, why you feel they would be *of interest to their readers today* (remember that all-important question), and your credentials or those of the person writing the articles. You may, indeed, want the articles to be authored by someone else who will most benefit your company by this added media exposure. If so, it is all right for you or someone else to write the articles for them.

Figure 10.18 is a sample pitch letter to an editor from a fictitious interior contractor who wants to publish an article in a weekly business newspaper's monthly commercial real estate feature section. Use your company letterhead or a standard letter format with your name and address in the upper right corner.

(Your company's name)
(Your company's address)

Contact: (Name of media contact), (area code and phone number of media contact)

FOR IMMEDIATE RELEASE or FOR RELEASE (date to be released)

(COMPANY NAME)
ADDS (LOCATION) (TYPE OF) OFFICE

(Location of news item), (date of release)—(Your company's name) has expanded to include a (type of office, if applicable—branch, district, regional, etc.) office in (city and state of new office). The company's new (type of office) office is located at (postal address of new office).

"(Quotation explaining why you expanded)," stated (your company's name, title, and name of individual who is qualified to render an opinion why you expanded). "(A continuation of the quote, if it is needed to further explain why you expanded or adds relevant information about your company)."

(Name of person), a (title) of (your company), will manage the new (type of) office (this will help establish name recognition for the individual, which will help make it easier for him or her to establish contacts in the new market). (Background information about the person that establishes credentials and qualifications for heading the new office).

(A paragraph describing work you have already performed in the new location, if applicable. This information adds a comfort level to new prospects because they will know you have already managed work and hired workers and other trades in their area.)

(A boilerplate paragraph that describes your company, including your location, the type(s) of business(s) in which you are engaged, your years in business, and so on).

Figure 10.13   Template for company expansion release.

Another strategy for publishing articles is sending a letter pitching an idea, rather than sending a completed article. This takes less time and gives the editor an opportunity to give you feedback on your ideas and direction about what exactly to write. If you choose this direction, your letter might be like that shown in Fig. 10.19.

Submitting letters to the editor is another way to get published, which is often overlooked by marketing people. Business newspapers, trade magazines, and even dailies publish articles or editorials dealing with real estate issues. There are several ways you can respond with a letter to the editor. You can agree or disagree with an article or editorial, correct something inaccurate, correct an oversight if your company is not mentioned but should be, or comment on an article that you feel is accurate and informative.

Letters to the editor typically begin with "To the Editor" and include a reference to a specific, dated article or editorial. Below your signature, include your name, title, company, and phone number. Keep the letter brief, usually to between 250 and 300 words.

Advertorials

Advertorials are paid advertising; however, they are written just like an article. Because you are paying to place the article, you can be a little more self-serving than you would be allowed to be in an article published by the media. Nevertheless, to get people to read it, it must provide information that helps them do their jobs better.

Also, because you are paying for space, you will want to write as economically as possible. Most publications either charge for advertising by the

---

Andrews Construction Co.
1200 Hudson Street
Suite 200
Saddle River, NJ 10028

Contact: Alton Fryer, (206) 555-1212

FOR IMMEDIATE RELEASE

ANDREWS CONSTRUCTION RECEIVES
THREE AWARDS FOR SUPERIOR CRAFTSMANSHIP

Fort Lee, N.J., December 1, 1999—Andrews Construction Co. has received three awards from the New Jersey Building Congress for its work on two office buildings and one hotel in New Jersey.

The awards were received for the masonry exterior of Lexington Place, a 55,000-square-foot office building at 100 Hudson Blvd. in Fort Lee; the concrete structure of a 90,000-square-foot office building at 210 Grace Street in Fort Lee; and the marble façade of the Marriott Suites extended-stay hotel at 1001 Briardale Road in Saddle River, N.J.

The awards were presented November 29, 1999, at the New Jersey Building Congress' Forty-First Annual Craftsmanship Awards at the Regents Country Club in Fort Lee. Each year, awards are presented for craftsmanship in several construction trades. Industry experts judge nominated projects. This year, the Building Congress presented 23 awards. Two firms each received three, which were the most given to any one company.

"This is a tremendous recognition of the skilled crafts people employed by our firm," stated Andrews Construction President Anthony P. Andrews. "It is a testament to our commitment to providing the absolute best-quality construction services to our clients."

Andrews Construction has participated in the Craftsmanship Awards for nearly 20 years, and in that time, it has received 32 awards. The individual craftsman honored included Herbert Perry and Richard Nightingale for their work at Lexington Place; Vladimir Dubkoff and Cecil Cooper for their work on the office building at 100 Hudson Blvd.; and Melody Snowden, Heather Cross, and Antonio Cocovelli for their work on the Marriott Suites extended-stay hotel.

Andrews Construction Co., headquartered in Fort Lee, N.J., is a commercial construction company serving the northeast United States. Established in 1962, its annual volume of construction projects exceeds $150 million.

###

Figure 10.14 Press release for company awards.

(Your company's name)
(Your company's address)

Contact: (Name of media contact), (area code and phone number of media contact)

FOR IMMEDIATE RELEASE or FOR RELEASE (date to be released)

**(COMPANY NAME) RECEIVES AWARD(S)
FOR (DESCRIPTION OF WHAT AWARD RECOGNIZED)**

(Location of news item), (date of release)—(Your company's name) has received (quantity) award(s) for its work on (brief description of what was awarded) by (name of organization issuing the award).

The award(s) was (were) received for the (a complete description of what was awarded).

The awards were presented (date awards were presented), at the (name of organization issuing the award)'s (name of award program) at (location of awards presentation). (Information about the award process, the history of the awards, and any newsworthy information about your award, if applicable).

"(Quotation expressing your reaction to the award)," stated (your company's name, title, and name of individual who is qualified to render an opinion). "(A continuation of the quote, if it is needed, to illustrate how receiving the award is consistent with the company's goals and mission)."

(This paragraph can provide additional details about the awards, people who were involved in receiving them, and your company's history of receiving such awards).

(A boilerplate paragraph that describes your company, including your location, the type(s) of business(s) in which you are engaged, your years in business, and so on).

###

Figure 10.15 Template release for awards.

column inch or in predefined sizes, like sixth-page, quarter-page, half-page, and so on. An advertisement that is 12 column inches can either be three columns wide by 4 inches deep or two columns wide by 6 inches deep. Consult with the publication in which you are interested in placing the ad, or refer to *Standard Rate and Data,* which publishes advertising rates and specifications for most publications.

Advertorials should probably only be used if you are not able to get articles written or published because it will be expensive to place one that is as large as an article.

Speaking

The dreaded S word. Many people will just flat out refuse to get in front of an audience and speak, but the benefits, as discussed earlier, are so great that you should try hard to get someone from your company onto the speaking circuit in your marketplace.

Figure 10.16 Press release that pitches a feature article.

Horton Construction Co., Inc.
455 East Alamo
Austin, Tex. 78714

Contact: Andrew Davis, (512) 555-7890

FOR IMMEDIATE RELEASE

HORTON ADDS SERVICE DIVISION TO HANDLE SMALL CONSTRUCTION AND REPAIR PROJECTS FOR OFFICES AND RETAIL SPACES

Austin, Tex.—Horton Construction Co., Inc., has added a service division to provide repairs and small construction for office and retail tenants in the Austin area. Horton, which was founded in 1985, has built only complete interior spaces until now.

"A majority of the construction work that companies need is small-scale items such as moving doors, painting walls, and replacing carpets," explained Audrey Horton, president of Horton Construction. "Typically, businesses do not have a reliable source for these services, which we are now offering, backed by the reputation for quality that we have established over the years."

Last year, Horton was ranked the third-largest interior construction company in Austin by the Mid-Texas Building Industry Association. The Austin Chamber of Commerce ranked it as the largest woman-owned business in the city. Horton's construction volume was $45.6 million.

"We did not establish the service division to significantly increase our revenue," said Horton. "We did it because we perceived a void in the marketplace for high-quality, cost-effective repair and small construction work. We think it is an excellent way to keep our clients happy."

Nate McCallum, a senior vice president of Horton who has been with the firm 13 years, will head the service division.

Initially, the service division will work out of Horton's main office. It will be staffed by an entirely different group of people from those managing the company's interior construction projects. This will allow the service division to respond rapidly to customer needs.

"Our intention is to eventually spin off a separate company, with a small office and minimal overhead to keep costs down," explained McCallum. "It will be managed by experienced construction professionals, though, to maintain quality."

Building Off a Solid Reputation

"The key to the success of the service division will be keeping our expenses down and trading off our excellent reputation," said Audrey Horton.

Horton Construction has won over two dozen major building awards from various national and local building industry associations. Its reputation is attested to by many of its clients.

"Horton listens, then comes back to you with creative solutions," stated Brant Aylway, senior partner at Odom, Strunk and Aylway, an Austin law firm. "When we built our two-level offices next to the State Capitol, Horton reviewed our architectural drawings, listened to our needs, and made recommendations that improved the quality of our building and reduced the cost. What more could you ask for?"

(Continued)

Figure 10.16 (*Continued*)

> Most businesses believe the cost to construct their office or retail space is
> an investment. As such, this space needs to be maintained.
>
> "Our store takes on a lot of wear and tear," said Barbara Butler of the
> Cotton Factory, a 18,000-square-foot clothing store in Austin. "We've needed
> periodic repairs to our space over the years. Until now, we've had to accept
> less-than-quality work, which ultimately affects the way people perceive us.
> The need for a firm that provides quality work on smaller projects is acute."
>
> "Service is so important in this industry," said Horton. "As long as we
> remember that we are a service company first, and that our product is
> construction, we should succeed."
>
> ###

The first thing any novice should do is sign up for a formal speaking program like Toastmasters or Dale Carnegie. Either will give you considerable public speaking skills that will make you far more comfortable in front of a crowd. But even seasoned speakers and professional performers get nervous when the lights come up. They have learned, as you should, to channel the nervousness into energy and imbue your presentation with enthusiasm.

The second thing new speakers should realize is that they don't have to be perfect. If you come prepared, know your subject, and can be conversational, you will capture the audience's attention and give a very effective presentation.

To give a really polished presentation, use these public speaking tips:

1. *Start from an outline.*   Organizing your thoughts is essential. Develop an outline first. Make sure that you completely cover your topic and that you provide information that you think the audience will not already know. Do research to support your arguments, if necessary.

2. *Use the tried and true format.*   Organize your presentation by telling the audience what you are going to say, saying it, and then summarizing by telling them what you said. This enables the audience to know what to expect in advance, which should keep them more attentive. It also enables you to reinforce your main points at the end when most people will remember them. Some people prefer to write entire speeches; others only prepare notes. It is best to only use note cards once you reach the dais, however. Reading from a completely written speech can be disastrous if you lose your place. It also minimizes eye contact between you and the audience.

3. *Get the audience's attention with a story or a joke.*   You've heard the old adage; "You never get a second chance to make a first impression." Try to start with a topical joke or an interesting story. If you capture the audience's attention at the beginning, it is more likely that you will keep it throughout your presentation.

4. *Include audience participation.*   Let's face it, it's not easy to keep people's attention for 20 minutes, unless you are Jim Carrey. Find a way to stop once or twice during your presentation and get the audience involved. Studies show that when people participate in a presentation, the material becomes at least three times more memorable than if they merely listen to a lecture. Ask ques-

Ms. Maryann Lewis
The Austin Business Post
1150 Capitol Street
Austin, TX 78714

Dear Maryann:

A recent study by the National Building Owners Association reported that over 40 percent of the dollars spent on construction were for repair work. Of that amount, over half was for interiors—both offices and retail space.

Companies that do not aspire to the high-quality standards found in the new-construction trade do much of the small-scale work. So their clients, who have invested hundreds of thousands of dollars building their spaces, end up having them repaired in a low-quality manner.

Horton Construction recognizes the need for high-quality, responsive repair and retrofitting needs. We have therefore recently established a division to provide these services, and we believe that we are the first major construction company in the area to do so.

We think that this information would be of interest to your readers. I would like, then, to discuss the possibility of your running an article in the *Post* about our service division. Audrey Horton, the company president, and Nate McCallum, the head of the new division, are available at your convenience for interviews.

I will call you next Tuesday to follow up. If you have any questions in the meantime, please call me. Thank you for your consideration.

Sincerely,

Andrew Davis

###

Figure 10.17 Pitch letter that accompanied the press release in Fig. 10.16.

tions, have the audience perform an amusing exercise that illustrates your point—anything you can think of to get them involved.

5. *Use visuals sparingly, if at all.*   Visuals only aid a presentation when they simplify a complex idea. They should be used sparingly. They can disrupt eye contact between you and the audience because they divert attention from your face to a screen. They also typically need to be viewed in the dark—a very dangerous practice, as you will remember from your 8:00 a.m. classes in college.

6. *Practice, practice, practice.*   You can't be too prepared for a speech. Practice your speech out loud—to a friend, the dog, or even yourself. Consider tape recording the speech and playing it back to hear how you are pacing the presentation and whether you are slurring any words. If you are speaking early in the morning, practice the night before your talk, go to sleep, and then relax in the morning. Your speech will have literally soaked into your mind over night. It is best not to practice within 1 hour before you talk because you will tend to remember only the last part of the speech.

7. *When you're speaking, go slow.*   The best way to lose an audience's attention is to speak so fast that they can't understand you. Write, GO SLOW in big bold letters at the top of each of your note cards. Your nervous energy will

make you speak faster than you think, so force yourself to slow down and converse with the audience. Force yourself to take frequent deep breaths. They give the audience a break and slow your heart rate down so you can relax and be more conversational on stage.

8. *Project from the diaphragm.*   Stage actors are taught to speak from the *diaphragm,* the muscular partition between your chest and your abdominal cavity. During normal conversation, the air we use to speak with comes from the throat. To drive your voice across the room, use the diaphragm to push air from the bottom of your lungs. To practice, make short exhale bursts from your stomach area. If you feel muscles in your stomach area contract, you will know you are using your diaphragm. Now, speak using those bursts of air and see

---

February 20, 2001

Ms. Anne Marie Cooper
Real Estate Editor
Denver Business Journal
1000 16th Avenue
Denver, CO 60026

Dear Anne Marie:

   I have read with great interest the informative articles contained in the *Commercial Real Estate Monthly.* I always find something in them that helps me do my job better. This led me to write an article that I thought may be of interest to your readers.

   As a builder of office interiors, tenants often ask us how they can best prepare for an office move. Because this seems to be a concern of many companies planning to relocate, I wrote the enclosed article, "Twenty Essential Ways to Prepare for an Office Relocation."

   The article explains how to prepare for a sequence of events that begins more than 1 year ahead of the move when you prepare an audit of your current and future space needs. It covers things you must not forget on the day of the move, like making sure to disconnect your phones, and items you must consider several weeks after the move, like inviting clients and friends to an open house.

   I feel well qualified to write this, having worked in the construction industry for nearly 20 years and authored several articles in local professional association publications. They include, "Cutting Costs on Office Maintenance," and "How to Get the Most Out of Your Limited Office Space."

   I hope you will find the subject of my article interesting and will consider publishing it in the *Denver Business Journal.* I will follow up with you before next Tuesday to make sure you have received this and answer any questions you might have.

Sincerely,

Justin Brothers Construction Co.
Ron Grieve

Figure 10.18   Letter pitching article submission.

February 20, 2001

Ms. Anne Marie Cooper
Real Estate Editor
Denver Business Journal
1000 16th Avenue
Denver, CO 60026

Dear Anne Marie:

I have read with great interest the informative articles contained in the *Commercial Real Estate Monthly.* I always find something in them that helps me do my job better. This led me to think of several ideas for articles I could write that might also be of interest to your readers.

As a builder of office interiors, tenants often ask us for advice on how they can save time or money. Because these seem to be concerns of many companies, I thought of the following story ideas:

- *Twenty essential ways to prepare for an office relocation.*   This article would be a step-by-step guide of how to plan and execute an office move, from preparing an audit of present and future space needs to christening the new office with a gala open house.
- *Ten easy things you can do to maintain your office space.*   Companies spend thousands of dollars a year; some of it is wasted on office repairs and maintenance. This article would list ten things any tenant can do itself to maintain its space, from repairing carpet seams with an iron to selecting the right paint for the office kitchen.
- *How to cut six months and 10 percent off your next search for office space.* This article would give companies inside advice on how to act as their own real estate broker. They will learn how to find available space that may not be published and how to lower their rental rates by not paying a commission.

I feel well qualified to write these articles, having worked in the construction industry for nearly 20 years and authoring several articles in local professional association publications. They include "Cutting Costs on Office Maintenance" and "How to Get the Most Out of Your Limited Office Space."

I hope you will find these ideas interesting and will consider publishing them in the *Denver Business Journal.* I will follow up with you before next Tuesday to discuss these ideas in more detail.

Sincerely,

Justin Brothers Construction
Ron Grieve

Figure 10.19 Letter pitching several ideas for articles.

how far you project your voice. This is not to be confused with yelling. Projecting from your diaphragm puts no strain on your vocal chords.

9. *Stay within the allotted time.*   Even a great speaker can turn off an audience by speaking well beyond the allotted time. Everyone has somewhere to be, so don't alienate your audience by forcing them to sit longer than they are supposed to. Practicing your presentation will ensure that you keep it within time limits.

Publicity on the Internet—The New Frontier

The Internet has opened new channels for publicizing the products and services your company offers. Not only does the Internet let you reach reporters easier, it provides access to people through entirely different avenues from the traditional print and broadcast media, advertising, and direct mail.

If you intend to use the Internet as a major part of your publicity efforts, you would be well-served by reading one of the many books devoted entirely to Internet publicity. Your local business or computer bookstore, or an online bookstore, like www.amazon.com, will yield a variety of resources.

A basic Internet publicity program should at least include sending press releases to print and broadcast reporters via e-mail. Many reporters now prefer receiving information this way because it is much easier to review and categorize press releases on a computer. Using e-mail is faster and cheaper for the sender as well. Photographs can be sent electronically and saved as low-resolution (100-dpi) scans. If a reporter needs an actual print, he or she will contact you.

In addition to sending press releases to reporters at magazines, newspapers, and radio and television stations, you should deliver newsworthy information to Usenet newsgroups, Internet mailing lists, online newsletters, e-zines (electronic magazines), and Web sites that are relevant to your industry.

Developing a distribution list

There are several ways to build a distribution list for non-print-and-broadcast media. One method is to use your favorite search engine, such as www.yahoo.com or www.webcrawler.com, to search for key terms like "construction newsgroups," or "construction electronic magazines." A rather unique searching tool on the Internet is the Mining Company (www.miningco.com). It specializes in links to electronic publications, newsgroups, and Web sites. A search of "construction Usenet newsgroups," for example, yields more than 2000 related sites. You are sure to be able to develop a valuable list if you start there.

Once you have built a list, you should get to know the reporters, in the case of e-zines and newsletters, and organizers, in the case of Usenet newsgroups and Web sites. Each site has its own rules governing what is appropriate or inappropriate to post. To avoid negative feedback, and ensure that your posts are of value to readers, you should get to know these rules of engagement in advance.

Software

The software needed to launch an online publicity effort is familiar to most Internet users today. The most obvious software tool is a word processing package. Once you have written and spell-checked your press release, you can import it to your e-mail program, which is probably the most important component of your online setup because you will use it to send information and read responses. There are several commercial e-mail software programs available and major Web browsers come with one built in.

Sending your message

One important thing to remember about sending information on the Internet is that you should not simply take your written press release and stuff it into an e-mail. The speed with which people move from one message to another, either e-mail or newsgroup postings, demands that you get their attention. You should therefore provide viewers with a short and catchy invitation to read your information. If they are interested, they can read the release or link to your online newsletter.

Publicizing your company on the Internet is a new and challenging way to expand awareness of it. If you utilize the Internet in these pioneering days, you will quickly differentiate yourself from your competitors.

Crisis Management

Crisis management is an aspect of public relations that is especially important for construction industry companies. Unfortunately, the accidents that can occur on a construction site are of a magnitude that often attracts media attention. Every contractor should know how to work with the media if a crisis occurs.

The basic elements of your crisis management plan should include

Establish a hierarchy of people within your organization who are authorized to speak with the media. Start at the top with people who are comfortable with and knowledgeable about providing background information about your company. Make sure you can reach these people at any time of the day or night in case a crisis manifests itself.

Establish a policy of complete honesty in dealing with the media. There is no use lying about any facts because chances are they will be verified by the media and reported. If you are found to lie, you will lose your credibility with the media and public, which will be very hard to recapture.

Gather facts as quickly as possible. Don't sacrifice thoroughness, but gather facts quickly. Particularly if the electronic media is involved, reporters need information quickly, and if they don't get it from you, they will look elsewhere, possibly gathering misleading or damaging information from people lacking knowledge about the situation.

Never say, "No comment." It is always construed as being evasive. If you do not have the information they need, say so. Explain that you are gathering facts. Tell them what you do know, and explain that you will divulge all facts as soon as they are available. If you stonewall reporters, they will dig up information in other ways, perhaps by speaking with job site workers. Chances are that misleading information retrieved in this manner will be worse than what they would have learned if you were forthright.

Respect the media.   Their job is to provide information to the public, especially in the event of an accident. If you show that you want to help them, they will respect your need for patience in gathering and reporting the facts.

The classic case study of effective crisis management in action is the cyanide tampering of Tylenol pain reliever in 1982. As you probably remember, the makers of Tylenol, McNeil Products Company, handled its situation well.

Although McNeil, and its parent, Johnson & Johnson, had no ongoing public relations program for Tylenol and no emergency public information plan at the time of the tampering, it quickly prepared a course of action. Management strategists worked together to formulate three stages of action:

1. Identify the problem and take immediate corrective action.

2. Cooperate with the authorities in the investigation to find the killer.

3. Rebuild the Tylenol name and capsule line, including the Regular Strength capsules, which had been recalled along the Extra-Strength brand, which had been tampered with.

Faced with a crisis of such magnitude, the company chose to handle inquires from the press and consumers openly. All communication between the media and the company had to be channeled through the corporate communication department. Only in this way could open, clear, consistent, legal, and credible communications be managed.

Since the Tylenol incident, many other companies have developed crisis management strategies.

Note

1. Marcia Yudkin, *6 Steps to Free Publicity,* Penguin, New York, 1994, p. 79.

11

Marketing on the Web

Introduction

The Internet is revolutionizing the way people live and do business. Communication via e-mail, information exchanges and news groups, and research of virtually any item of information is now commonplace on the Internet.

Many companies are seeing the vast potential for establishing commerce over the Internet, marketing their services or products, and developing communication channels with clients or prospects. Most are setting up in an area of the Internet known as the World Wide Web. Unlike other areas of the Internet, which are mainly text-based, the Web uses a graphic interface that offers interactive capabilities including animation, sound, and full-motion video.

The steps leading up to the development of a Web site are essentially the same as those needed to develop any other marketing material. The difference, of course, is the medium itself. The Web has advantages and disadvantages that make using it very different from using advertisements, brochures, newsletters, and direct mail.

This chapter will give you a better understanding of the advantages of a Web presence and steps needed to establish one or improve the one you already have. It will help you understand why you should have a Web site and how companies in the construction industry are using them. You will see that the same marketing rules that apply to the development of any sales material apply to developing a Web site.

Nevertheless, Web marketing is different from other forms of marketing, and you can make your Web site more successful by taking advantage of these features. You will see a step-by-step guide of what to do and will get an idea of what it will cost to get online. Finally, you will learn how you can increase the likelihood that people will visit your site once it is posted.

Should I Have a Web Site?

Are Web sites for everybody? Maybe not, but any company that has a story to tell and products to show, sells merchandise, or has the opportunity to be pre-qualified in front of a prospect would benefit in one way or another.

Communication via the Internet is clearly the trend of the future for individuals and businesses. People who use the Internet surely must see value in it for education, research, communication, or commerce, or else Internet usage would not be increasing by 60 to 85 percent every year.

Electronic commerce

Electronic commerce (e-commerce), or shopping via the Internet, is projected to grow at an astronomical rate, with annual sales of well over $10 billion in the year 2000. There are several advantages to shopping online. The most obvious is the time saving. If you want to send a gift to the accounts payable clerk who expedites payments at your largest client, visit a gift site like www.lifetimes.com (see Fig. 11.1). If you need to book a flight to Houston, you can visit www.delta-air.com, plug in your departure and arrival times and locations, and review a choice of 10 itineraries that meet your needs. At that point, you can buy your ticket. Need to find a book on roofing systems?

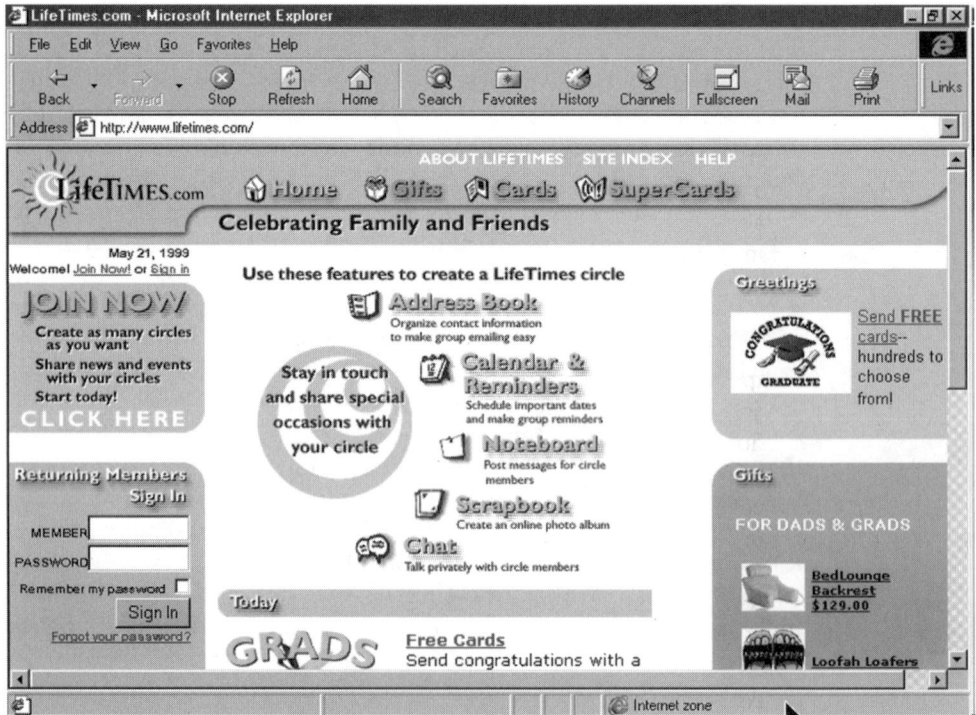

Figure 11.1   There are numerous sites where you can buy gifts for clients or friends.

Visit www.amazon.com and search its database of over 15,000 books about construction. You'll find over 100 about roofing, alone.

Want to research bridge engineering? Go to www.enr.com and access its database, powered by Northern Light, which gives access to *ENR* archives dating back to 1995 and to over 4500 other news sources. Once you find the articles you need, you can purchase them for a nominal price. These days, you can purchase just about anything on the Internet, and some economists predict that the Internet will virtually eliminate cash transactions.

The ability to do research before making a purchase is also an advantage of the Internet. If you are refinancing your home mortgage, buying a sports utility vehicle, looking into possible vacation destinations, or buying a laptop computer for your home office, you can compare the features and benefits of almost any product on Web sites and in articles on the Internet.

Promotion

Even for firms that cannot feasibly sell their products or services online, a Web site offers an opportunity to mix graphics and words in a dynamic fashion to promote their services. It allows for quick updates to allow visitors to keep abreast of new projects, awards for safety and quality, promotions, and new hires.

When a prospective client wants to learn more about several firms competing for its business, a well-organized Web site can provide a strong first impression (see Fig. 11.2). It can convey competitive advantages, show photos of similar experience, and provide contact information.

In an age when the selling points that differentiate one company from the next are few, a well-organized, inviting Web site could help make your company stand out. Web sites are great equalizers. A small firm can easily develop a Web site that rivals that of a large firm. In fact, it is usually the smaller firms that embrace new technology faster than larger ones in an effort to give them a competitive edge.

Conducting business using extranets

A rapidly growing application for the Web within the construction industry is on the job site, where members of the project team can have shared access to information via secure Web sites known as *extranets*. This type of site, which can only be accessed via a password, can allow the owner, architect, engineer, contractor, subcontractors, and key consultants to communicate and share job information. According to Abba Blum, CPA, an information technology expert with the construction/real estate industry services group of Aronson, Fetridge & Weigle of Rockville, Maryland, most job-related extranets share at least the following:

- E-mail directory of all project participants
- Progress photographs
- Change order forms and logs

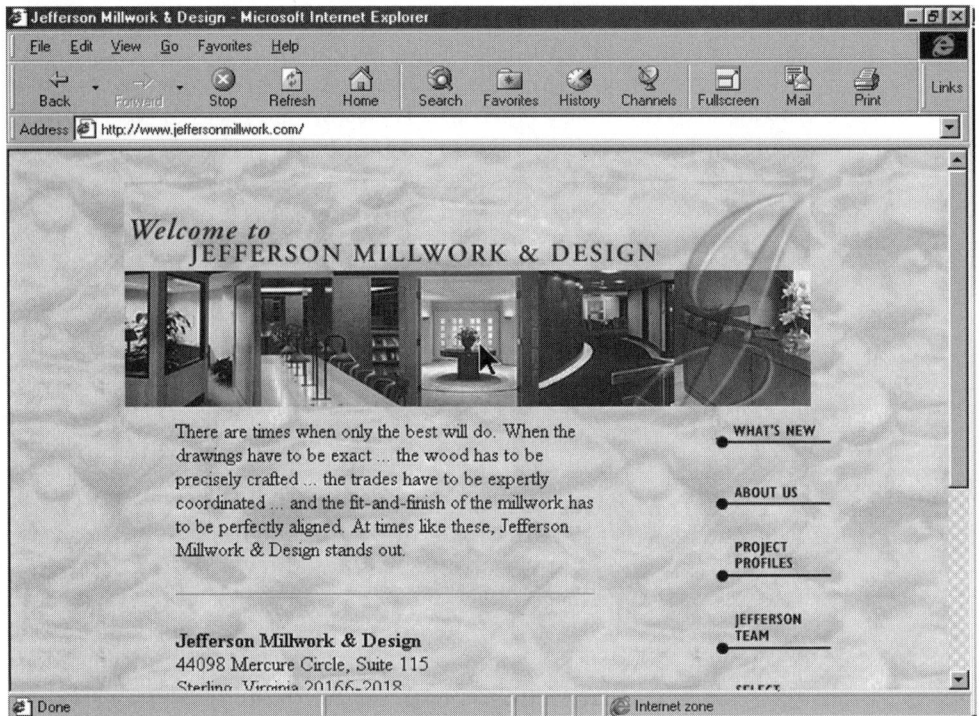

Figure 11.2 A well-organized Web site makes a strong first impression.

- RFI reports and logs
- Meeting minutes
- Project schedules and completion time frames
- CAD drawings
- Correspondence files

"Imagine how much time will be saved by having all of this information available at your fingertips," says Blum.[1]

For example, if an owner wants to make a change that affects the mechanical room, this request for a change could be submitted to the architect, who would design it on AutoCAD and upload the revised plans to the extranet Web site. The mechanical engineer could produce working drawings and post them for the general contractor and subcontractors to download. The cost of the changes could be determined using popular software like Timberline and posted onto the site with a change order form. Once reviewed, the change order would be approved and posted back on the Web site. The entire process would take a matter of days instead of weeks and would save a considerable amount of money for all the members of the project team.

Certain states are already requiring that job-related Web sites be set up, and it is becoming a requirement of many federal projects as well. One example of an extranet in action is at the National Institutes of Health (NIH) in Bethesda, Maryland. Bovis Construction is providing construction inspection, management, and consultation services for the 50-building complex.

"The NIH project site has allowed us the ability to put to the test technology that we only talked about a couple of years ago. We are using computer technology to communicate between us, the architects, general contractor, and the NIH to provide fast turnaround on questions regarding project specifications," said project manager Charlie McSweeney.[2]

Architects are embracing the use of extranets to set up project-specific Web sites, too. Many are using ProjectCenter, the project extranet available to AIA members through AIAOnline (www.aiaonline.com).

In addition to a password-accessible area, a project extranet can also have a section accessible by the general public. According to Abba Blum, these sites typically contain the following:

- Current pictures of the project, sometimes using video for real time images.

- Renderings of what the site will look like when completed. Some of the more sophisticated sites allow the visitor to conduct virtual walkthroughs of the site when it will be completed to get a sense of different perspectives and views.

- Leasing information

- Information about the project team, including hyperlinks to their Web sites.[3]

Communicating with employees via intranets

Web sites used exclusively by employees are called *intra*nets (as opposed to the *Inter*net). These sites may be attached to your external Internet Web site but are protected from outside viewing by passwords. Employees can access the site and look up things like their retirement account, payroll information, available sick and vacation days, and information that would usually be found in employee newsletters, like marriages, babies, volleyball results, and so on.

Many companies find intranets to be exceptional ways to communicate with employees. They can be updated frequently to include meeting schedules and training opportunities, short-notice events like blood drives and United Way updates, and much more. They can save companies money because updating intranets costs far less than printing numerous memos and newsletters. They can utilize many of the same design elements of an external Web site, so development costs can also be minimized.

Recruiting

Finally, and perhaps most importantly in today's competitive job market, an informative and exciting Web site can be an excellent recruiting tool (see Fig. 11.3). Clearly one of the largest groups of people using the Internet are young adults, primarily college students. When they begin searching for a job, they

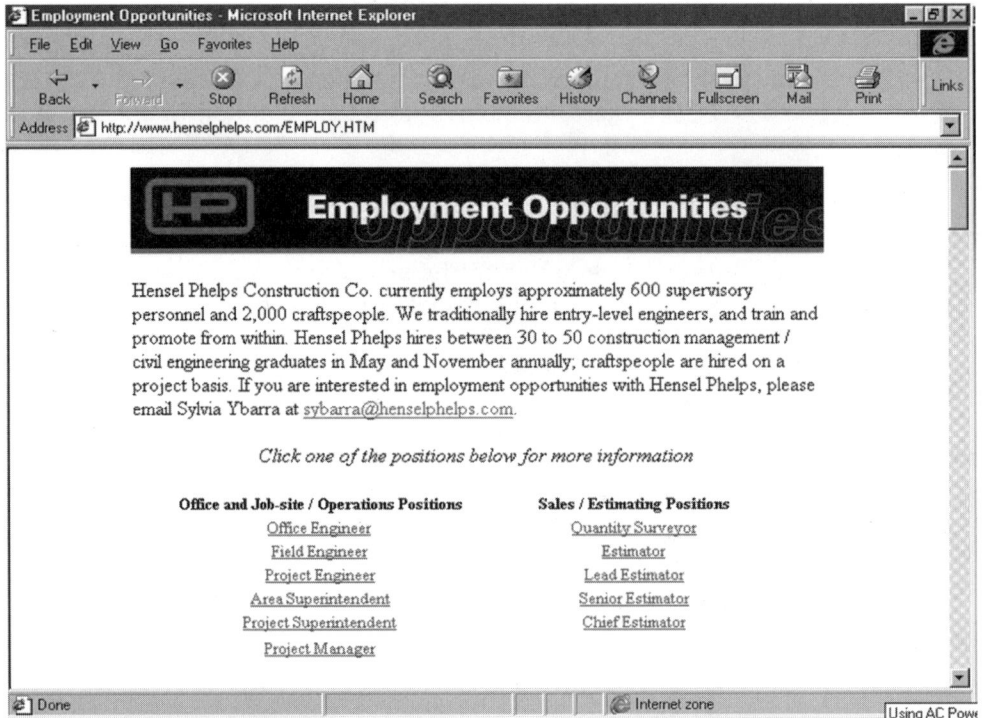

Figure 11.3 The first place many potential new hires will look is your Web site.

will research firms via the Internet and Web. If you are trying to recruit top engineers or business students, you can be assured that the first thing an interested young candidate will do is check your Web site. Candidates will want to know everything they can about you before they invest the time pursuing a job opportunity. If your competitors have a Web presence and you do not, you will be one step behind right away, and you will be sending a signal about your willingness to embrace new trends and technologies. That alone should make it worth your while to develop a Web site.

Planning for a Web Site

Planning for a Web site requires the same strategic thinking as planning for any other type of marketing materials. Step away from the medium for a moment and ask yourself the same questions you would ask if you were producing a brochure, a direct mailer, a newsletter, or an advertising campaign:

- What are my *competitive advantages?*
- What are the *benefits* of my competitive advantages to my prospective buyers?
- Who is my *target audience?*
- What are their *needs?*

- What do they need to *know* about me?
- How can I get them *interested* in what I have to say?
- How can I make them take *action?*

You probably answered most of these questions when you developed your marketing communication plan (Chap. 8).

These are the critical marketing questions anyone must ask—whether you are a retailer, an airline, an attorney, or a synagogue—before embarking on a marketing campaign. Marketing using the Web is no different. Remember that the key to effective marketing is understanding the needs of your target audience and then explaining how your goods or services will satisfy their needs.

Take Advantage of the Web's Strengths

The Web is different from other marketing materials, so when you develop a Web site, you should take advantage of its strengths. As part of the marketing communication planning process (Chap. 8), you already evaluated the advantages of other materials before deciding to use them for marketing. Advertising, for instance, has a broad reach and is good for establishing awareness. Direct mail can be targeted to well-defined groups and is good if you have a small target audience. Published press releases have greater credibility and reach a broad audience, but their placement is unpredictable.

The Web's unique advantages

The Web has many unique advantages you should take advantage of:

1. *Multimedia.*   Like television, the Web combines text with still pictures, video, and sound. Pictures of current and completed projects, product lines, and key employees are frequently posted on contractors' Web sites. Also showing up are video clips with sound that take visitors on virtual tours of job sites or show products being used. The technologies used to create the multimedia presentations are becoming less expensive by the day, and most computers can easily view them. In addition, many firms are posting Microsoft PowerPoint presentations on their Web sites so that prospective clients can view sales presentations at a time that is convenient for them.

2. *Interactive.*   Web sites allow for an immediate response—in other words, an order. A small general contractor in Maryland recently searched the Internet to compare lumber prices. He found a supplier in Canada whose prices were lower than those of any supplier he could find locally. He invested $500 in a plane ticket and a day's travel time to inspect the lumber and then made the purchase. He saved nearly a half-million dollars by doing his homework on the Internet. More and more suppliers are selling products on the Internet. Initially, it will be a convenience to existing clients who will reorder supplies without the need of a salesperson. As buyers become more confident of e-commerce, more business will be conducted this way, saving time and money for all concerned.

3. *Less expensive.* The cost to build and maintain a basic Web site can be less than a comparably sized brochure and far less than most forms of advertising. Once the Web site is designed and laid out using Web development software, it is almost ready to post. A few hours of work on the source code to make the site work across browsers are all that is needed. After a brochure is designed and laid out, you will incur production prices like printing, which typically is twice what it costs to design and lay out either the brochure or Web site. When creating an advertising campaign, placement costs are 6 times the cost of design and layout. It costs less to revise Web sites, too. To update photos or text in a brochure, you will have to reprint it completely. On a Web site, photos and text can be replaced instantly, without incurring any cost beyond the time it takes to scan the photo, paste in the text, and post them to your Web site.

4. *Dynamic.* It is easy and quick to update Web sites. How many times have you agonized over whose picture to include in your company brochure or which products or services to include, because of the rapid changes taking place in your organization? The Web eliminates that concern. You can update your Web site whenever you want, from your office. Better Web sites showcase new products or services, post current press releases, and always offer something fresh and different for frequent visitors. One of your goals when developing a Web site is to get prospects to visit it often. Like direct sales, the more times you get in front of your prospects, the more likely you are to get their business.

5. *It's one of the ways business is conducted today.* The simple fact that everyone else has a Web site is enough to motivate some companies to develop one—not to copy, but to keep up. There is no questioning the fact that marketing on the Web is the wave of the future. Although there is still some question about its value today, developing a Web site gets you established now for what will become an increasingly important way of doing business in the future.

The Web's perceived disadvantages

The Web also has some disadvantages, although generally they are not as bad as they are perceived to be:

1. *Accessibility.* Not everyone is connected to the Internet. Perhaps some of your prospective clients are not yet using e-mail and surfing the web. However, with Internet use nearly doubling each year, you can imagine that before long, nearly every business will be connected.

2. *Slow modems.* Modem speed affects how fast you can view Web pages, and this is something you should be conscious of when developing your site. Make sure your photos are not too big and that no single page is so complicated that it will take long to download. However, modem speeds keep getting faster—56K (kilobits per second) modems set the standard, and dedicated ISDN lines can move information as fast as 128K. T1 and T2 lines are even faster. These modems send information over telephone lines. Soon, this technology will be abandoned in favor of cable modems. How much faster is cable than conventional modems or ISDN? For data moving from the server to your

computer, the rate ranges from 3 to 10MB (megabits per second). An 18MB file that would take 2 hours to download on a 28.8K modem would take about 30 seconds using a cable modem. As you can see, as technology continues to make downloading quicker, the concern about speed of downloading lessens.

3. *Cost.*   Cost is a perceived disadvantage. *The Wall Street Journal* reported in 1997 that the average Web site cost over $1 million to develop. The Web sites the article referred to were developed by major corporations and filled hundreds of "pages." However, as was mentioned earlier in this chapter, the cost to create a Web site is often less than other marketing materials. Creative costs charged by Web designers and programmers are only marginally higher than those charged by graphic designers working in print media; they are higher because their craft requires computer programming skills as well.

What Makes Web Sites Succeed?

The well-known architectural axiom popularized by Frank Lloyd Wright, "form follows function," should be your guide to developing a Web site. People go to Web sites to get answers. Think of the times that you have visited Web sites. When you couldn't easily find the information you wanted, what did you do? You probably left the Web site and looked for another one.

A study by User Interface Engineering supports this. The organization assembled 50 people with varied Web experience and put them on a scavenger hunt of nine different Web sites to gather information. The results were revealing. The most expensive site, a virtual community that cost $20 million to develop, was the most difficult in which to find this specific information. Although the site was enormously entertaining, it took too much time for people to navigate to get answers. The most functional site was a $50,000 site that included a simple index on the home page, which made information easy to find.[4]

For a Web site to be effective it must

- *Be graphically appealing.*   We are a visual society. Like any marketing material, an attractive, well-designed Web site will attract and keep visitors and reflect positively on your company.

- *Be fast to download.*   Large and complicated graphics are slow to download, which often prompts people to stop and move on to another site.

- *Be easy to navigate and provide information logically.*   Your attention to detail and logic in the planning stage will pay dividends in the long run. The following section explains how to organize a Web site so it is easy to navigate.

- *Take advantage of the Web's capabilities.*   Use multimedia and interactivity, and update it frequently.

- *Be interesting.*   Do a few fun things to keep people on your site and entice them to visit often. Like any sales opportunity, the longer you can spend time with someone, the more likely you are to eventually get his or her business. Provide links to your favorite places, and include contests or anything to be a little different and a little fun. But don't forget, stress function over form.

What Should I Do If I Want a Site?

Developing a Web site is not a daunting task. After reading the previous chapters in this book, it should be no more difficult than developing a brochure, direct mailer, or publicity campaign. You may not be able to do everything yourself. However, if you take the time to plan carefully and follow the steps in this and previous chapters, you will be able to do much of the work yourself and have a good understanding of the tasks you should outsource.

The following is a step-by-step guide to developing a functional Web site.

Establish your goal

Developing a Web site is not unlike developing any other marketing material. First you must determine your goal. Is it to provide information about your company to prospective customers, to recruit prospective employees, primarily for the use of existing employees, to conduct commerce, or to exchange data with specific companies? Each of these goals suggests a different approach to developing a Web presence. You may have one goal or a combination of goals to satisfy your needs. Your goal (or goals) will be unique to your organization and your industry.

If your goal is to provide information about your company, your Web site will be developed similarly to a company brochure, placing emphasis on the benefits you offer clients. It will include photos and lists of completed projects, biographies of key employees, and contact information. If, on the other hand, your goal is to attract prospective employees, your emphasis will be on specific job opportunities, training programs, employee benefits, company picnics and softball teams, the history of your company, and so on.

If your goal is to communicate internally with employees, you might want to develop an *intra*net.

If your goal is to conduct commerce, your site will place heavy emphasis on products, with photographs, detailed descriptions, testimonials, and pricing information. It will be developed using tools that allow for exchange of money.

If your goal is to exchange data with specific companies, you will probably set up an *extra*net site (as described earlier), which is accessible only by a password. Here, you will post information that is needed by a specific group of people, like a project team. These sites will have a minimum of graphics and will emphasize easy access to information.

Create an outline

Once you have established your goal or goals, you should make a list of the different things you need to present on your Web site to satisfy them. If your goal is to provide information to prospective customers, your list might include:

- Company history
- Competitive advantages
- Completed projects, divided by category

- Awards
- District offices
- Biographies of key personnel
- Safety program
- Maintenance division
- Directory with hyperlinks to e-mail
- Client list with hyperlinks to their Web sites
- Hyperlinks to Web sites of interest

Make a similar list for whichever goal you decide upon. Spend a lot of time developing the list. Brainstorm with others in your company who are involved in whatever aspect of the company you are using the Web for, whether it is marketing, personnel, or project management.

This is like the programming stage of the development of a building. You are deciding what you want the Web site to be like. Furthering the building analogy, the next stage is where you begin the actual design of the site.

Develop an organizational chart

A Web site is developed very much like an organizational chart. It starts at the top with the most important and general interest page and then fans out into pages with specific functions.

Most Web sites start with what is called a *home* page. This page introduces your company and typically contains a very brief description (if any) and an index of the main categories contained in your Web site. Figure 11.4 shows the home page of a general contractor's Web site.

After the home page, the organization chart flows to main categories and then to subcategories, until all the information you want to present is delivered in a logical manner. Figure 11.5 shows a Web site organizational chart for a hypothetical contractor.

Create the outline and flow chart on paper and get consensus from all the people who need to be involved in the decision-making process; then proceed from there.

Obtain the necessary tools or outside resources

Once you know what will be in your Web site, you should decide what you can do yourself and what you need to outsource. There are several elements, all requiring different skills.

Writing. Writing for the Web is different from writing for other marketing material. Writing advertisements demands brief, attention-getting statements. Writing direct mail lets you guide a prospect through the decision-making process more thoroughly but still within space limits. Writing for the Web

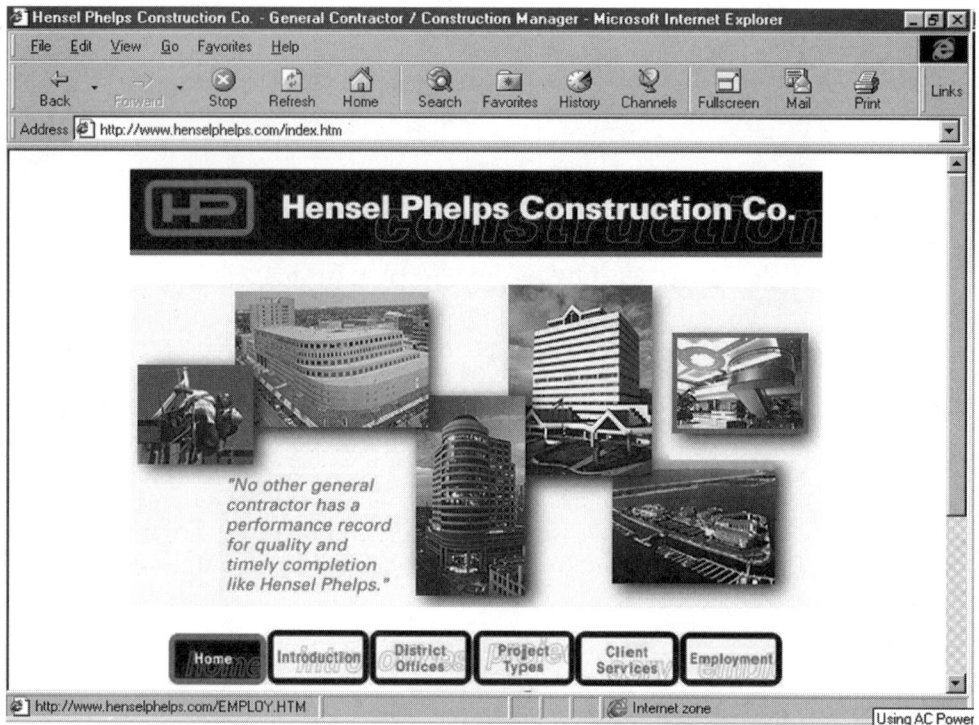

Figure 11.4 The home page typically introduces your company and contains a brief description or quote and an index of main categories in your Web site.

allows you to provide as much information as you feel is needed to get your points across. Web copy can scroll down a page or link to other pages indefinitely. So as long as it is easy for visitors to navigate to another part of the site when they have enough information, you can write as much as you feel is necessary.

Rules of good writing apply, of course, such as writing short, concise sentences and using good grammar. If you are a good writer and can edit your own work, you should be able to write your own Web site.

Design.   Designing for the Web is different also. Whereas designing for ads and direct mail requires a graphic approach that grabs attention, design elements on a Web site serve to aid visitors on their journey to get information. Design is critical because it helps convey your competitive strengths and creates an important first impression. But the design you choose should not slow down information exchange.

Off-the-shelf software is available for developing Web pages. PageMill from Adobe (www.adobe.com) and Front Page from Microsoft (www.microsoft.com) are the two most commonly used software programs for designing Web sites, and NetObjects Fusion is a basic program that adds some higher levels of

Certified Construction Corp.
www.certified.com

| Home Page |

Introduction	**Completed Projects**	**Representative Clients**	**Key Personnel**
- History - Description of services - Competitive advantages	- List of project categories that link to separate page for each. Each category page includes a list of projects and 3-4 photos.	- List of clients with links to their Web sites	- Bios of key personnel and company directory with links to e-mail

| **Awards** |

| **Safety Program** |

| **District Offices** |
- US map with link to each district office

| **Links** |

| **Health Care**
with project list and photos | **Commercial Office**
with project list and photos |

| **Northeast** |

| **Multifamily**
with project list and photos | **Government**
with project list and photos |

| **Southwest** |

| **Midwest** |

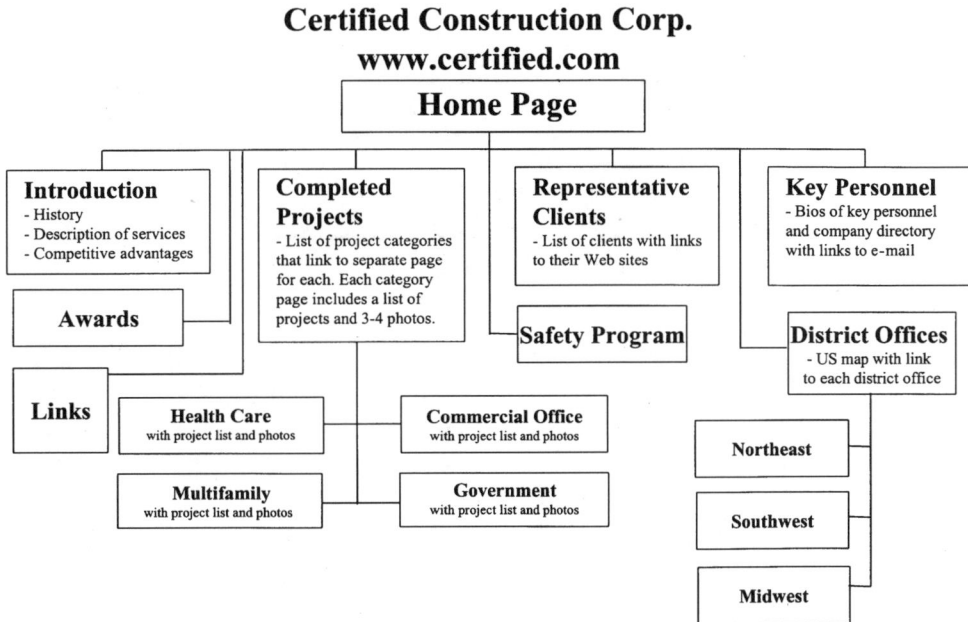

Figure 11.5 Another home page.

"functionality." These programs generate the Hypertext Markup Language (HTML) programming needed to create a functioning Web site. They provide templates in which you can plug in visuals and text, or you can import design elements created from other graphic design desktop software like QuarkXPress or Aldus PageMaker, from Adobe. You can also import photos from programs like Adobe Photoshop or illustrations from programs like Adobe Illustrator. Even services like America Online (www.aol.com) have templates that the average person can use to create simple Web sites; however, they would not create the kind of site most businesses would need to set them apart from their competitors.

Of course, these are just tools, and they don't necessarily give you the skills you need to design a Web site that has the right mix of marketing savvy and creativity. And, the level of multimedia sophistication and programming needed to develop the better Web sites have increased dramatically in recent years. For these reasons, the business of Web site design and development has grown dramatically, with the better Web development firms investing heavily in training their designers and developers.

Programming.   A certain amount of computer programming is required, even if you use the off-the-shelf Web publishing software described above. In addition, other more sophisticated programs, such as Macromedia Flash (www. macromedia.com/software/flash), The Network Director (www.nrsinc.com/nrs. 2200.asp), and Macromedia Authorware (www.macromedia.com/software/

When centering an object on the page. PageMill will insert the code as
<P ALIGN=CENTER>*stuff being centered would be here*</P>. Although some browsers will
"understand" this, others expect to see it written as <CENTER>*stuff being centered would be
here*</CENTER> (unless the "object" is a paragraph of text; then it would be written
<P><CENTER>*stuff being centered would be here*</CENTER></P>. If you can go into the
HTML coding and recognize these situations, they can easily be fixed. It will require some
time to view the page in the different Internet browsers to check for these kinds of
inconsistencies.

Figure 11.6 Basic HTML programming.

authorware), add more of the "bells and whistles" that many of today's Web
sites feature. These programs can be used to add higher-end multimedia and
interactivity functions to your Web site. For example, with Shockwave
(www.macromedia.com/software/shockwave) you can create games, animated
interfaces, interactive ads and demos, streaming CD-quality audio for music
and speech, and instructional and educational presentations.

HTML is the primary code in which Web pages are written. Internet browser software, which provides a graphical connection to the Internet, reads the
HTML code. Web publishing software will "write" this code for you through an
interface that is very similar to desktop publishing software.

Although there are many different types of Web browsers available,
Netscape Communicator (www.netscape.com) and Microsoft Internet Explorer
(www.microsoft.com) are the most commonly used. Therefore, at a minimum,
someone who is developing a Web site must have a basic understanding of
HTML and how it works. Web publishing software will write code that is not
interpreted the same by different Web browsers, and therefore someone must
edit the file at the HTML level (removing or modifying tags, etc.). See Fig. 11.6
for a brief lesson in HTML programming.

Build your Web site

Once you have acquired the tools you need and have decided which, if any, of
the steps you will outsource, you can begin to build your Web site. Web sites
should be built in steps, as discussed in the following subsections.

Home page and typical pages.   Your first concern in building the site is how it
will look. Better Web sites have a consistent look from one *page* to the next. A
page is basically described as information that is accessible on your screen
without navigating to a different screen. In reality, a page may contain so
much information that it amounts to several printed pages, if you send it to
your printer.

The home page sets the tone for how your site will look. Your home page
must accomplish certain things:

- Welcome visitors by letting them know whose site they are visiting

- Capture the visitor's attention by being graphically appealing

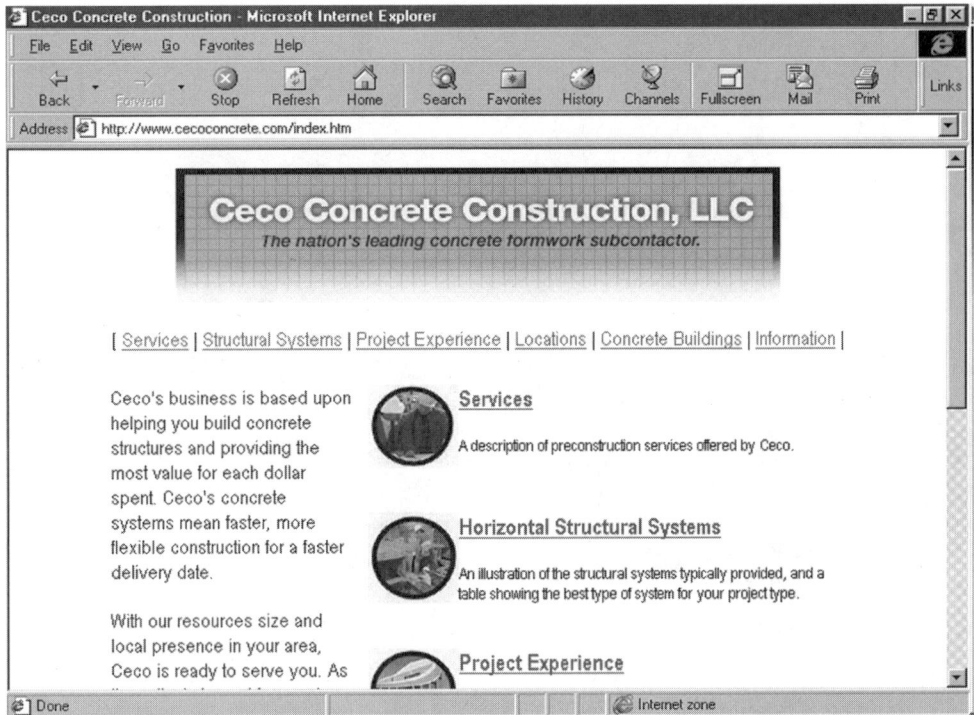

Figure 11.7 The home page sets the tone for how your Web site will look and tells visitors where to get more information.

■ Provide enough information, either through graphics (photos or illustrations) or text to let visitors know what you are promoting

■ Provide a clear index of the information contained in your site

■ Provide contact information, with hyperlinks to an e-mail address that can be used to send a note or request additional information

The home pages shown in Figs. 11.7 and 11.8 show variations of how to present the critical information mentioned above. The common elements of most home pages are

■ A header that shows the name of your company

■ A visual representation of what your company does

■ Icons or lists of main heading pages

■ Contact information

Once you have built your home page, you should build one or two main heading pages (see Figs. 11.9 to 11.11). They should tie together graphically with the home page so that it is always clear that visitors are still on your site. They should completely present the information you need to give and

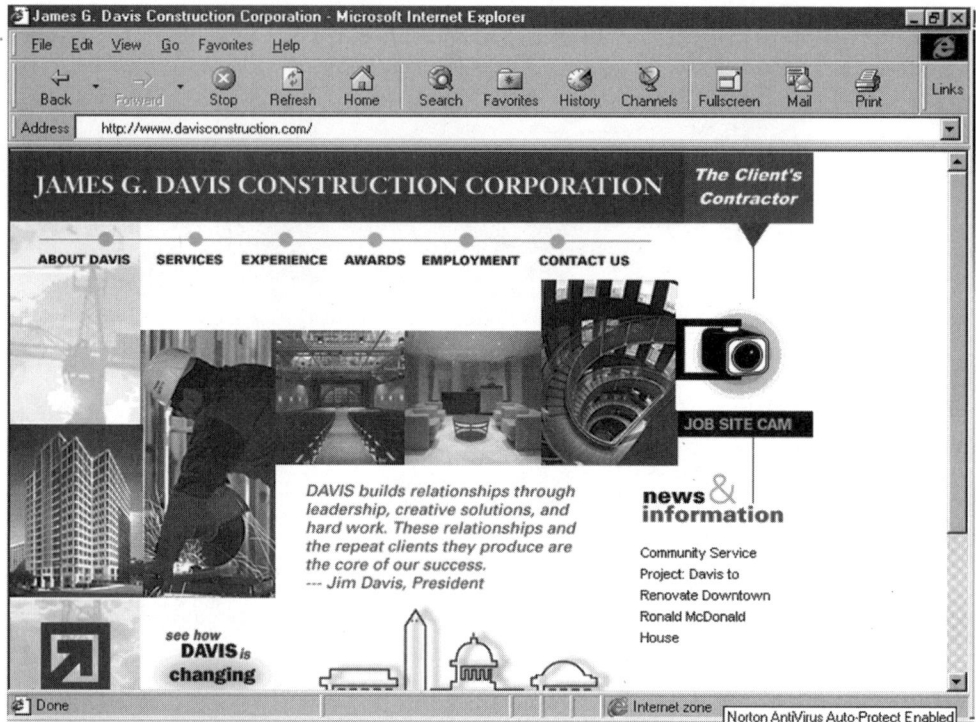

Figure 11.8   Another home page.

then offer at least one easy way to navigate to either the home page, another main heading page, or a subpage within the section they are visiting (refer to Fig. 11.5).

After you have built one of your main heading pages, you should build at least one subpage under that main heading page (Fig. 11.12). Here, too, the look should be consistent with the main heading page and the home page and should logically present all the information you need to give visitors. Also, you should make it easy to navigate either back to the subpage's main heading page, another main heading page, or the home page.

If you follow this guide, you will develop anywhere from four to six pages. Before proceeding, it is good to review the layout and typical text of these pages to decide if it has the look and functionality that you want. You can modify your design and text at this point or move on and fill in all of the pages with text and graphics.

You will develop the Web site on a computer, not on the Internet. You can upload it to the Internet if you already have a Web address and Internet service provider (ISP; see the subsection Select an ISP and Register a Web Address for details). It is easier to understand how it will look on the Internet if you review it there, although the software on which you build your site will allow you to preview it even if you cannot upload it to a Web site.

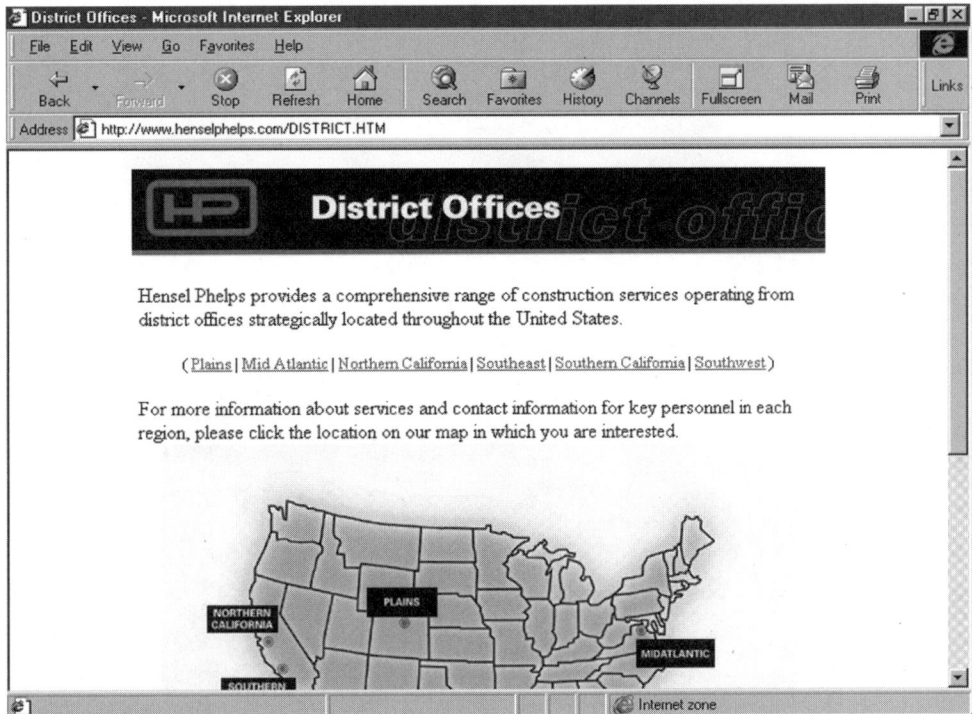

Figure 11.9 Main heading pages should tie together graphically with the home page.

Establish links

Most good Web sites enhance their functionality by providing easy access, via hyperlinks, to other Web sites that provide additional information that supports your organization's mission or competitive advantages.

Think about your own company. What kind of information would help your target audience (whether it is prospective clients or employees, existing staff, or project team members) do its job better? Make a list of all the things you can think of. It might include magazines, associations, government agencies, weather forecasts, traffic cameras, mapping services, banks, or furniture suppliers. You may even want to include nonrelated but interesting links like professional sports teams, local colleges, and concert halls. Again, one of your overriding goals is to keep people on your site for as long as possible and to keep them coming back.

How to link pages.   Linking your site to other sites is easy. All you need to know is the Web site address, or if you are linking to an e-mail, you need to know its address. The HTML code will take the following form:

<A HREF= "http://www.frostmiller.com">Frost Miller Group</A> or

<A HREF= "mailto:kmiller@frostmiller.com">Kevin Miller</A>

Figure 11.10   Main heading pages should tie together graphically with the home page.

In this example, Frost Miller Group is the site being linked and Kevin Miller's e-mail address is being linked.

Web sites that include hyperlinks to other sites should be developed using a *frame* layout. A frame is a design technique that basically creates separate, smaller pages on one page (see Fig. 11.13). The frame can contain your list of links and be designed consistently with other pages. When visitors use hyperlinks to visit other sites, your frame will remain on their computer screen, allowing them to click onto your page and navigate back to your Web site at any time.

Select an ISP and register a Web address

Selecting an ISP and registering your Web domain name, or address, can actually take place at any time in the development process. You may want to take care of these technical aspects at the very beginning to increase the likelihood you will get the Web address you want (i.e., www.yourcompany.com) and to get technical assistance from your ISP.

The ISP.   Your ISP is your link to the Internet. To have a Web site, you must have an ISP or host your own site with a dedicated server, which is costly,

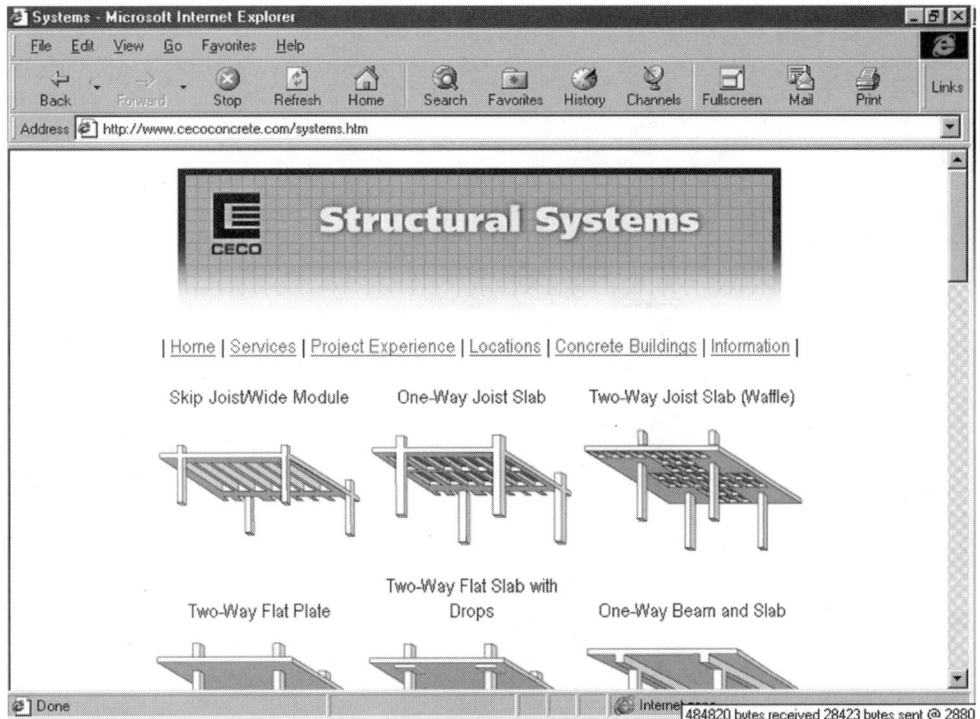

Figure 11.11 Main heading pages should tie together graphically with the home page.

maintenance-intensive, and not recommended in the first several years of having a Web site.

When you contract with an ISP, you will typically get e-mail as well as Web hosting. In addition, you will need to have some way to hook up with your ISP. This can either be a dial-up connection via a modem and phone line or a direct connection like an ISDN or T1 line. Your ISP, depending on its size, will either have a direct backbone connection to the Internet or will purchase service from a larger provider with a direct connection.

If you have a dial-up connection, each individual computer that has access will have a separate connection. You will dial into your ISP's computer (your computer does this automatically on your command once you have programmed the proper phone numbers using e-mail software, like Eudora Pro), which will provide you access to the Internet and your e-mail. If you have a direct connection, like an ISDN line, multiple computers can tie into an in-house server via your company's internal network. You can also set up an in-house e-mail server that will receive all e-mail messages for individuals in your company and disseminate messages to their proper destination.

Web sites are housed (or "hosted") on the ISP's servers. Servers are simply high-capacity hard drives. The number of Web sites hosted on each server will affect the system's performance and reliability, so the fewer sites on the same server, the better.

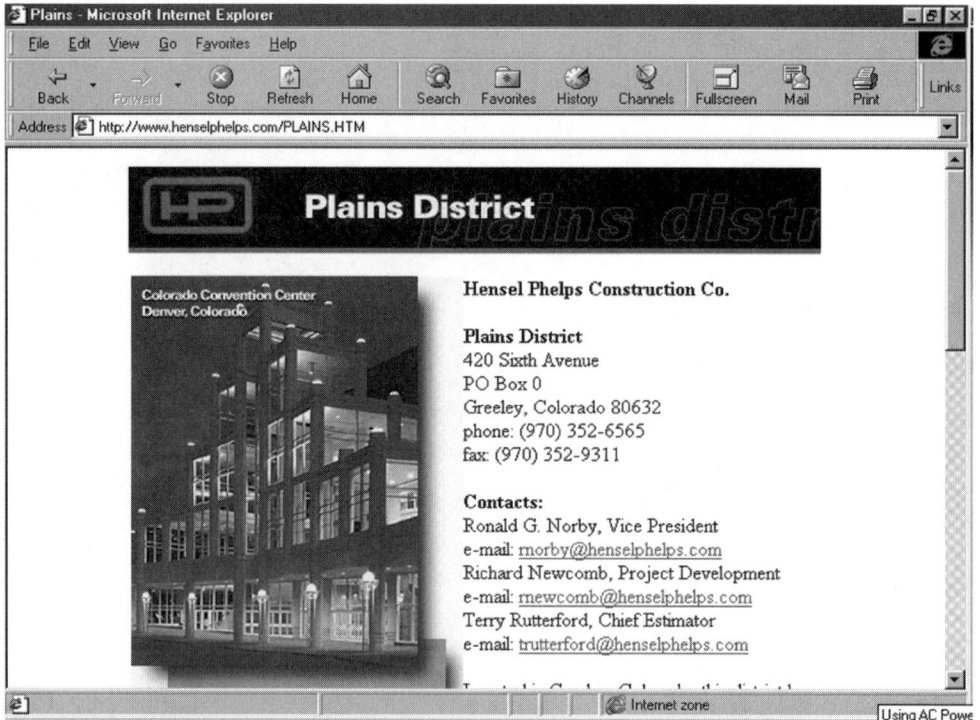

Figure 11.12   A subpage, too, should tie together graphically with the pages leading to it.

The ISP you choose depends on your needs. You should meet with representatives of a few ISPs who are located in your area to get an understanding of the services they offer and their costs. Make sure that you select one that will give you the level of service and attention you need. Because of the growth and traffic patterns of the Internet, you will frequently have problems connecting to your ISP, regardless of who it is. You will also have frustrations with e-mail and other services. Make sure you have a good rapport with people working at your ISP because you will be speaking to them often. Even the smallest ones provide you with more than enough memory to host an expansive Web site.

Register a Web address.   Your ISP can help you register your Web site address. The best domain names include your company name. For a company named Shock Electrical Supplies, the best domain name would be www.shock.com. However, if it is unavailable because the word *shock* is so common, you could probably register www.shockelectric.com, which would be less common. It is an easy process if you want to do it yourself; however, if you register, you need to have an ISP to which you can point the Web address.

To register, or even just to find out if the domain name you want is available, you can go to the InterNIC (http://www.internic.net). Network Solutions, Inc.

Figure 11.13 This Web site utilizes a frame layout. The area on the left stays on your screen so you can always navigate back to the home page.

runs this site and lets you search available domain names using WHOIS. Network Solutions is a company that pioneered the development of registering Web addresses ending in .com, .net, .org, and .edu (see Fig. 11.14). The Department of Commerce has an agreement with Network Solutions to develop and maintain some of the key administrative functions of the Internet including the domain name registration system.

Maintenance

It is important to maintain your Web site. After all, one of the Web's greatest advantages is its ability to let you present timely information. Web development firms have the ability to design, develop, and maintain your Web site. Simple maintenance, however, can be conducted in-house with just a bit of HTML training and understanding of a Web publishing software package. Even a word processor or simple text editor can be used to create or edit your Web site. HTML files are, in reality, just text files with an .htm or .html extension.

If you are going to maintain your Web site in-house, you will need to have the ability to transfer your files to and from the server where they are housed. File Transfer Protocol (FTP) software, such as WSFTP, must be used to perform this task. You simply download the original file (transfer it from the

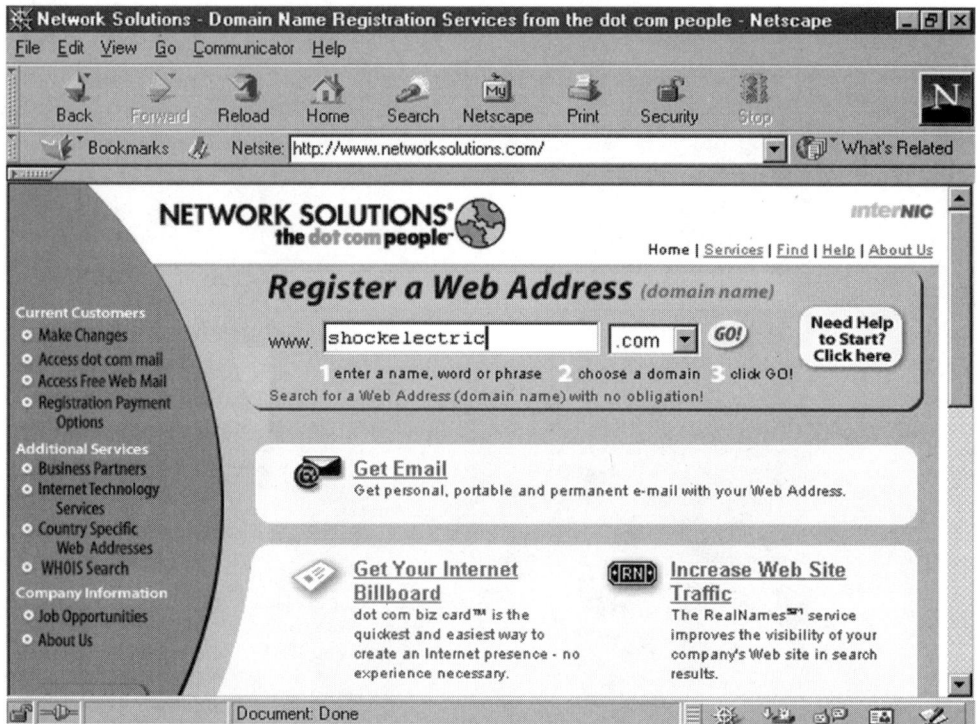

Figure 11.14 The InterNIC site is where you can go to search the availability of domain names like schockelectric.

server to your hard drive using the FTP interface), open it in your HTML editing program, make your changes, and upload it back to the server using FTP.

This way, you can frequently add press releases, replace project photos and information, add biographies of key staff, or correct a problem with the coding from your desktop.

How Much Does It Cost?

The cost to develop a Web site varies dramatically depending on your needs. Registering a domain name costs only $75 for 2 years, and $35 for each year thereafter, at the time this book was published. That represents a *decrease* in cost from a few months earlier when the costs were $100 and $50, respectively. However, some ISPs will absorb the cost of domain name registration to encourage you to sign up with them for Web hosting.

The prices charged by ISPs vary, depending on the services they offer. Most ISPs charge a small set-up fee, a monthly charge for unlimited Internet access, a fee for each business e-mail account, and a charge to host the Web site. A company with 10 to 15 e-mail addresses and a Web site should expect to pay about $200 to $300 a month for dial-up service, and 50 to 100 percent more for a direct connection to the Internet, like an ISDN, T1, or T2 connection.

Creative costs for a basic but functional, original, and engaging beginning Web site range from about $5000 to $15,000. A recent study of *Fortune* 1000 companies by Forrester Research showed average costs of $304,000 for promotional sites—on-line brochures; $1.3 million for content sites—sites that offer information or entertainment content; and $3.4 million for fully transactional sites—secure commerce sites with online ordering, online database, and multimedia graphics.[5]

Another cost to consider is updates and maintenance. Larger firms hire webmasters to manage their sites. Simple updates can be handled in-house, as discussed above. If a small or midsized firm needs to outsource updates and maintenance, it should budget anywhere from the high hundreds to low thousands monthly or quarterly, depending on the frequency of updates.

If I Build It, Will They Come?

By now, we all know that simply having a Web site does not mean anybody will see it. Getting prospects to a Web site is not easy in this self-directed medium. First and foremost, the best way to promote your Web site is to include your Web address on every printed marketing piece you have. Print it on your letterhead and business cards; display it prominently on brochures, direct mail, and newsletters; and broadcast it on your television and radio commercials, too.

You can also make it easier for people surfing the Web to find you. Most Web users start an information quest by searching one of hundreds of Web directories or search engines. A *directory,* like Yahoo! (www.yahoo.com), is a true directory. The sites it catalogs have been sorted by Yahoo!'s staff into subject categories, subcategories, and sub-subcategories.

So if you are looking for an architect and you go to Yahoo! and search the word *architect,* you will get lists of Web sites for architectural firms. To narrow a search, you could add words like office buildings or California. Their search might look something like this: architect + office buildings + California. This would force the search engine to first find Web sites that meet all three criteria, then search for sites containing the three words together, and then find all sites containing any of the words.

Search engines, on the other hand, like Excite (www.excite.com), Altavista (www.altavista.digital.com), and WebCrawler (www.webcrawler.com), index individual words or groups of words found in the pages of Web sites. Think of a catalog that would index every word found in all the books of a library's collection. If you go to a search engine and search for *architect,* you will get a list of every Web site that has the word *architect* in it, which would include every article that includes the word *architect* in it—obviously a different result than a directory would yield.

Register your Web site

If you want your Web site to be found, you must register it with the various directories and search engines, both the primary and secondary ones. Most directories and search engines will eventually find your Web site. In Yahoo!'s case it is done manually, and in most search engines' cases it is automated.

There are currently about 15 to 20 major search engines that the majority of Web surfers use. In addition, there are midrange search engines, many of which are specialized. In all, there are about 250 search engines.

There are a couple of ways to register. You can go to the Web site of each of the directories, look for an area that says something like "Add URL link" (the URL, or Uniform Resource Locator, shows the type of item and its basic address and path), and follow directions for registering. Or you can access services that register your site on multiple directories and search engines. A few of these are Submit It! (www.submit-it.com/), WebStep 100 (www.mmgo.com/top100.html), WebPedia (www.webpedia.com), and Trends2 (www.trends2.com/register). Most of the major search engines charge slightly more for premium positions (toward the top of the list) and the services that register you on multiple sites charge a small fee as well.

Design Web pages so people will find them

In addition, you should design Web page headers that encourage more hits. Most search engines scan the first several words of a Web site to make a match. Therefore, you should make sure a description of all of the services you offer, your geographic location, and your competitive strengths are listed early. These types of words might be entered when people use the search engines.

Use key words and meta codes

You should also place key words or phrases into the top of your source code where they cannot be seen. These "meta codes" should be any and all words that people conducting a search might use, even if they show up in the early part of the actual text of your site. Examples for a general contractor might include construction, construction company, Boston, general contractor, contractor, quality, office buildings, retail, interiors, builder, and so on. Search engines search for these hidden meta codes as well as for text within the Web site. By placing these keywords into your Web site's code, you will be making the search engines' work easier.

Get hyperlinked

Another way to get found is to get your site hyperlinked with as many complementary sites as possible. Write a letter to all the companies you know that have any relevance to your business, give them your Web address, and ask them to link you to their site.

Advertise

Also, banner advertising is a strategy used to get people to an organization's Web site. Many popular Web sites, including the major directories and search engines, offer paid advertising that links back to the advertiser's own home page. It is estimated that companies will spend $2.2 billion by the year 2000 for banner advertising, a two-hundred-fold increase in just 4 years. For more information on banner advertising visit Double Click (www.doubleclick.com) and Net Gravity (www.netgravity.com).

With rising Web costs, measurement of results has taken on even greater importance. Although few organizations have a clear sense of what their return on investment will be, you can begin the process of measurement by counting the number of *hits* you receive. A hit is defined as the number of times a visitor goes to your site.

This method is somewhat flawed because it doesn't qualify the visitors. The most foolproof tracking method is to ask the users to subscribe to the site before allowing them access. Users must fill out a brief survey that includes name, e-mail address, and other demographic information. Upon submitting the information, subscribers lock into a password, which allows their activity to be tracked. This reduces overall traffic, but it allows for much better qualification and tracking.

Take Advantage of the Web

The Web has proven itself to be a valuable source of information as well as an excellent method for building relationships and generating leads. As the culture of the Internet evolves, it is proving itself to be a successful means of selling products direct and will be more successful as more secure ordering systems become available. At the turn of the century, the Web is supporting billions and billions of dollars in commerce, and it is certainly a communication medium you should begin to use.

Notes

1. Abba Blum, "Plugging In: New Technologies for the Building Industries," *Bulletin,* Washington Building Congress, February 1998.
2. Trish Foxwell, "Bovis Setting the Pace with Advanced Technology," *Corridor Real Estate Journal,* June 25, 1998.
3. Blum, op. cit.
4. "Cool Doesn't Cut It!" User Interface Engineering seminar, 1997.
5. "New Media for Direct Marketers," in Bob Stone, *Successful Direct Marketing Methods,* NTC Business Books, Lincolnwood, Ill., 1996.

12

Marketing Traditional Lump-Sum Projects

Introduction

Many general contractors and subcontractors continue to emphasize obtaining only lump-sum, low-bid projects, believing incorrectly that marketing is not required to secure this work. Simply waiting for project announcements in such sources as the federal government's *Commerce Business Daily* (*CBD*) (refer to Chap. 15), other bid reports, or announcement subscriptions no longer assures a company of a sufficient number of projects on which to bid. Lump-sum work, like any other type of construction contracting, requires a substantial marketing and sales effort to maintain a sufficient backlog and associated profitability to remain in business.

Most notably, the federal government, once the largest customer of the construction industry using low-bid lump-sum bidding as the standard for selecting a general contractor, is looking at alternative delivery systems. Now the government prefers to prequalify potential bidders, allowing only those contractors that can meet particular requirements to submit a bid or proposal for the project.

The same situation is even more pronounced in the private sector. Building owners and clients of the industry no longer desire to enter a contractual relationship based merely on low price. Owners, either themselves or through the architect, engineer, or other representative, are qualifying potential bidders before taking bids on a particular project.

Qualification Requirements

Owners and clients are likely to prequalify contractors and maintain an active qualified bidders list, allowing only contractors that meet their stringent requirements to propose on their upcoming work. Although the project is usually awarded to the lowest bidder, only by first marketing to and qualifying with the client will a general contractor have the opportunity to secure the work by presenting the low bid. Many government agencies have taken this one step further, establishing programs that prequalify a small select group of contractors that bid for specific types of work for that particular agency in a geographical area over an established period of time, for example, 5 years. This contracting method is referred to as job order contracts (JOCs), service order contracts (SOCs), and other acquisition terms (refer to Chap. 15 for additional information on these contracting methods).

Some clients not only qualify prime contractors but have taken the process a step further by reviewing and qualifying potential major subcontractors for their projects. Only these subcontractors are permitted to submit bids to contractors bidding as general contractors for the project. Qualification usually requires that subcontractors have specific experience for work involved on the project (i.e., restoration of masonry walls) or those that are completing a substantial portion of the field work do (i.e., mechanical and electrical subcontractors).

Clearly the requirement of prequalifying requires contractors to establish a marketing program that aggressively targets potential clients long before any particular project is ready for bidding. This includes federal government agencies.

State and local governments are also moving toward alternative selection and delivery methods. Considering the diversity of clients over a large geographical area, any contractor wishing to maintain an adequate number of projects available for bidding must maintain an extensive marketing database that includes the individual owners' specific requirements for qualifying and the time that the qualifications remain current.

Much of the driving force behind the increase in prequalification is to identify the companies that are capable of performing to the standards expected by the owner. Some clients are willing to pay a premium to select a capable contractor rather than use price as the sole decision-making factor. Furthermore, by qualifying contractors, clients potentially limit the number of disputes that can occur on a project by identifying contractors that do not have the resources to complete a project properly.

Fixed-price and lump-sum low bid work should no longer be secured within a marketing vacuum. An organized marketing program is necessary today even if a general contractor or subcontractor chooses to compete solely on price. A marketing plan for securing low-bid work may be less complex than that required to compete for other work, especially if the latter is packaged with construction services options. For example, with design-build work the client will eventually select a contractor based on price after the firm is judged qualified to submit a bid. However, there are specific marketing plan requirements that must be implemented to target the lump-sum market successfully. This chapter presents appropriate marketing techniques targeting lump-sum bid, and fixed-price opportunities.

Getting Organized to Market Lump-Sum Bid Work

The first step in organizing a marketing program for lump-sum bids is to identify a company's internal capabilities. Once a firm's niche capabilities have been identified, based upon available resources and project experience, a strategic plan addressing customers, size and type of projects, and geographical area desired must be prepared. Any marketing plan, including one for securing lump-sum, low-bid work, must address where one expects to go and how to measure success after arrival at the destination. For a detailed review of strategic and market planning refer to Chaps. 3, 5, and 8.

Strategic planning

A strategic plan addresses and reviews areas where a company has been profitable in the past, internal resources to staff particular types and sizes of projects, geographic locations where the company wishes to expand or remain, and new types of construction work the company may wish to enter in the future. Competition, financial resources, estimating capability, and market trends are all considered in a marketing plan for lump-sum bid work.

The strategic marketing plan should clearly identify the amount and types of work the company expects to sell over a 5-year period. Budgets for instituting the marketing plan should be confirmed, including the cost of personnel and associated costs of brochures, advertisements, association dues, and other necessary expenditures.

After a strategic marketing plan is adopted, a list of potential customers is developed, including direct clients and associated customers such as architects and engineers who might design work that falls within the targeted goals. Today, in many situations it is necessary to secure work with a particular customer through a third-party representative, for example, a construction management firm. These marketing considerations should be identified and addressed in the marketing plan itself. Marketing to these organizations is covered in Chap. 13.

Standard qualification package

After the marketing plan is completed and a potential customer list is developed, basic marketing literature is needed. This literature will typically be presented in a standard qualifications brochure that outlines the company's abilities and past successes in particular types of construction projects and services.

This general marketing and qualifications brochure briefly introduces the firm to a potential customer, and although not specific in nature, it should be precise enough to allow a client to understand the firm's capabilities and types of work it can complete successfully. These qualification statements should include references on past work and photographs of example projects that grab the reader's attention. Again, even though the contract award eventually goes to the lowest bidder, extensive marketing is required today to be placed on a client's bid list just to get the opportunity to submit a bid on the project.

The qualification brochure is typically not specific enough to secure a place on a client's bid list, but it should be sufficient to ensure that the customer will request the contractor to complete a formal qualifications statement for a specific project or list of long-term qualified bidders. The qualification brochure might include representative projects involving a variety of work types to enable the package to be used for a wide variety of customers. Brochures are discussed in detail in Chap. 9.

Some companies choose to prepare several different qualification packages, each targeting a different niche market. Others often incorporate qualifications into a corporate brochure, but these tend to be very limited in the amount of information presented and often concentrate on visual presentation rather than specific information.

Electronic qualification statements

Web sites are also an effective means to introduce a company's capabilities to a prospective customer. It is not appropriate to suggest a customer visit a Web page to obtain information during a personal meeting instead of bringing a hard copy document. However, the Internet is often today used as a research tool and many clients may prefer to search for potential contractors in this manner before contacting a company representative. Maintaining a Web site that specifically details a company's experience is critical to selling low-bid work or any other type of construction service. This method of marketing is discussed in detail in Chap. 11.

Whatever method is chosen to present a company's capabilities to potential customers, the experience information must first be compiled and always kept current. Nothing presents a poorer image than to send a client information that is inadequate and out of date.

Information on past projects, current projects, and personnel capabilities must be kept current. Numerous software programs are available that can create and maintain a database of a company's historical project completions and resumes of personnel. These databases can be structured to provide a customized printout of information that is formatted to meet each individual prequalification requirement.

After a database is established, it must be kept current, particularly the references. People move, change phone numbers, and join or leave companies. To ensure that a potential customer doesn't end up calling a wrong number for a reference, these database references should be updated at least twice a year.

A searchable computerized database record of projects and resumés enables a marketing person to respond to a qualification request by providing information on the company's past work that most closely resembles the proposed project. The more accurate and responsive the information provided to a client, the better the chance of being selected as a potential contractor.

Software for customer contacts

Finally, a company marketing lump-sum work should also consider compiling a customer contact database. This database will maintain records on the number and types of contacts made to each client and provide detailed information about each personal contact at a client's office such as birthdays, favorite pastimes, and so forth. The advantages and uses of this software are presented in Chap. 7.

Responding to Qualification Requests

Requests for qualification statements will differ from client to client. However, there are basic general topics that will be addressed in each request, and a marketing plan should be able to address them automatically and with precision. Basic corporate information that will be addressed includes

- Corporate experience
- Experience similar to the proposed project
- Personnel resources and their experience
- Financials, including bonding capability
- References

Corporate experience

A request will often begin by asking for details about the contractor's past experience completing work. To limit the extent of response, clients will ask about experience in a set number of past years, typically between 5 and 10 years. The number and size of projects that can be included as references are also usually limited.

In responding to a past performance question, contractors should concentrate on the projects that best describe their overall capabilities. This can and should be broad enough to ensure that the customer realizes the contractor can complete and respond to other types of bid requests.

Customers ask about past performance so they can check references about the contractor's performance and capabilities. It is critical that each and every one of the responses included is an exceptional reference for the contractor. As mentioned in the previous section, marketing departments must continually update each individual reference to ensure the information is current.

It is important to realize that a client may become overwhelmed responding to reference requests and may begin to respond less than positively or not at all if it becomes a burden. Because even an exceptional reference can take on negative overtones in a situation such as this, it is best to alternate the use of each reference to prevent this from occurring.

The historical database program should be able to sort projects using a host of search criteria that enables the company to respond without having to edit the list manually. Sort criteria should include completion date, size, geographical area, and the project manager and superintendent who managed the work. Each entry in the database should also include specific information on the client reference and indicate clearly if it is not appropriate to use the project as a reference. In addition, software databases are available that can store project photographs to be used in the qualification statements when appropriate.

By structuring the database this way, different qualification requests can be answered automatically and with the precision necessary to meet the standards to get on the shortlist to bid the project.

Similar experience requirements

Each qualification request will also require details of past performance experience that is similar to the proposed project. This request may eliminate the inquiry about past corporate experience or may be used to supplement this experience. The experience the client requests might include parameters that are considered imperative for making the shortlist. Experience on projects similar to the proposed project's size, type of work, dollar amount, similar client, and geographical area are generally included in the outlined definition to similar work.

Although no client will expect a contractor to have a project that matches its project exactly, it will be necessary to show successful completion of work that is comparable to the proposed project. Every responding contractor will have to choose those projects that best fit the parameters of the proposed project. In certain instances the contractor may not have completed projects that meet the comparison requirements but may still be regarded as being qualified to bid the project.

Obviously, if the client is requesting that the contractor have experience completing clean room manufacturing facilities and the contractor only has commercial building experience, it probably will not qualify. However, if an owner is building a high-rise structure and the contractor only has mid-rise experience, the client may put the contractor on the shortlist if the references used confirm the contractor's capability to complete the project. Additionally, the resumés and experience of the proposed project management team will be carefully reviewed in these instances.

Personnel resources and their experience

Most clients today are placing more emphasis on the experience of the proposed project management team than on the past performance of the contractor. Clients realize that the company itself does not build projects but that its employees do. Often the qualifications may only consider the past experience on similar projects of the proposed management team and not the company's experience.

Marketing-driven contractors now keep their personnel concentrated on similar types of projects to enable the personnel to develop experience and performance history in a certain type project that will impress clients. In addition, contractors will keep project team personnel together on each assignment to show potential clients that the management team has experience working together. This is a requirement that has become increasingly important today because clients want to have work completed within the shortest time parameters and without disputes caused by a lack of team spirit on the project.

Obviously with the virtual construction goals industry clients require, as outlined in Chap. 2, it is becoming increasingly important to have a proposed management team that has worked together numerous times (even though this is difficult to put into practice). This gives the client confidence that the team will not waste precious time in the early stages of the project learning how to work together. The team should have the capacity to start a project in unison and without a learning curve necessary to manage the project.

Qualification response will generally include the individual team member's educational background, length of service with the contractor, experience within the industry including types of work completed, and any professional affiliations. The qualifications will also request references on past projects that the personnel have completed that are similar to the proposed project. Again, the marketing department should maintain reliable information about the references for each member of the project team to ensure that clients can properly investigate performance capabilities of the proposed team.

Financials including bonding capability

Qualification statements will also generally inquire about the contractor's financial capabilities to perform the proposed project. If the size of the project is clearly beyond the financial limits of previous work and balance sheet of the contractor, clients will not enter into a contractual relationship with that contractor. A contractor's financial officer should provide the necessary information to the marketing team that is responding to the qualification questionnaire. This will include banking references, balance and income statements, and financial ratings from such independent references as Dun & Bradstreet.

Clients will also require the current bonding limits of the firm directly from the contractor's bonding or insurance company. A letter from the bonding company enclosed with the qualification package often suffices, but it must be current and address the specific project. Clients may also request information about any possible pending awards to the contractor that might cause the contractor to reach bonding limits in the immediate future and prevent the contractor from bonding its project.

References

It becomes evident that successfully qualifying for a select list of bidders depends upon a company's ability to maintain an adequate list of references for potential clients to use to verify a contractor's performance capabilities. A successful marketing plan for any construction company depends upon its ability to maintain excellent relationships with present customers, which can lead to new customers.

In this regard, a marketer must make it a priority to maintain communications and contacts with each existing client to foster not only repeat work but also to have a reference portfolio for business development purposes. Even if a company decides to compete for only lump-sum bid work, it is inevitable that references will be required to qualify for prospective projects that enable the company to maintain an active bid program. References then become as important as work experience in qualifying for bid work because this is the most logical way for a new customer to determine if a contractor is worthy of their business.

Additional qualification requirements

In addition to the these typical qualification requirements, clients often ask specialized questions to satisfy their corporate strategies for selecting responsible construction companies. An effective marketing program will have the resources and cooperation to respond to these situations as they arise.

Responses required may include past and current legal claims the company has pending. Clients may have a particular interest in working with contractors that do not have a litigious history. It is important that a contractor's marketing department be aware of a company's claims and be able to answer pertinent questions as succinctly as possible.

Historical safety records are often required. Although customers may not be directly responsible for safety at their construction sites, they certainly do not want to select contractors that are unresponsive to the safety issues of their employees.

Clients often request information about the contractor's schedule and cost controls even though these responsibilities are managed by the construction company. In many qualifications, clients may request specific information concerning cost growth on past projects. Even though the project is bid as fixed price, clients may wish to determine if a contractor tends to depend on change orders and compensable time extensions for cost recovery or profit on the job.

Other corporate information should also be readily and accurately available for the marketing department; this includes states where the company holds contractors' licenses, locations of branch offices, extent of equipment owned by the company, and other resources available to support the project should it become necessary. An effective marketing program will maintain and have all this information immediately available so the team can respond promptly and accurately.

Qualification format

A typical qualification questionnaire is reproduced in Fig. 12.1. Marketers should review this form and immediately assemble the information required to respond to each question. This is a good first step for beginning a marketing program for any type of construction work.

Once assembled, the information should be placed in a computerized database that can easily be updated regularly. A sufficient database will also

Figure 12.1 Contractor or construction manager qualifications.

Contractor or Construction Manager Qualifications

Date of submittal: _____

Business name: _____
Address: _____
Phone: _____
Home office address if different from above: _____
_____

Name of parent organization of above firm or subsidiary: _____
_____

Type of firm:
    Corporation   ☐
    Partnership   ☐
    Joint venture ☐
    Sole proprietor ☐

How long has firm been operating as an architectural/engineering firm? _____

If Corporation (if joint venture, answer for each firm represented)
Date of incorporation: _____
State of incorporation: _____
President/CEO: _____
Vice president(s): _____
Secretary: _____
Treasurer: _____

If Partnership
Date of organization: _____
Type of partnership: _____
General partners: _____

If Sole Proprietorship
Date of establishment: _____
Name of owner: _____

Labor Classification
    Union   ☐
    Merit shop ☐

Does the firm operate under any job-site agreements?   Yes ☐ No ☐

List classifications of work completed by own forces other than general conditions:
_____
_____
_____
_____

List states / local jurisdictions in which the organization is legally licensed or qualified to operate as a contractor or construction manager. Include license numbers.
_____
_____
_____
_____

(Continued)

Figure 12.1 *(Continued)*

Has the firm ever failed to complete any projects awarded to it? If so, provide details.

_____

_____

Has any officer or owner of the organization previously been associated with a firm that has failed to complete a project awarded to it within the last seven years?

_____

_____

Are there any judgments, claims, arbitrations, or legal suits pending or outstanding against the organization?

_____

_____

Has the organization filed any lawsuits, arbitrations, or other alternate dispute resolution hearings in association with any contracts, including subcontracts, in the last seven years?

_____

_____

Current Workload
Annual dollar volume: _____
Current year to date: _____
Last year: 20 — _____
Previous year: 20 — _____

Experience
List as attachments all construction projects currently in progress, including name, address, phone of owner and architect. Include contract amount, percent complete, and expected completion date.

Largest single contract ever managed by the office qualifying: $ _____

Financial References
Include an audited profit-and-loss statement, including income and balance sheet. Provide name and address of firm preparing the audited statement:

_____

_____

If the provided audited statement is for any firm other than that listed as the qualifying contractor, explain the fiduciary relationship between the firms:

_____

_____

If the organization's financial statement is for any firm other than that qualifying, provide evidence that this firm will act as guarantor for the duration of this project:

_____

_____

Corporate Quality Program
Does the organization's main or home office have a structured and active total quality management program?   Yes   ☐   No   ☐   If yes:

Name: _____
Date established: _____
Name of person managing program: _____
Details: _____

_____

_____

Figure 12.1 (*Continued*)

Does the organization have a structured employee training program? If so, describe: _____

List courses provided last year: _____

Number of employee hours given in program: _____

Name of person managing program: _____

Does the organization participate in a site-managed total-quality-improvement program separate from the above? If so:_____

Names of projects currently participating: _____

Provide details of programs: _____

List the recent projects for which the organization has completed structured partnering programs: _____

Does the firm have or participate in any long-term partnering agreements? Yes ☐   No ☐   If yes, provide details. _____

Describe in detail the CAD systems, budgeting processes, and constructability review process used by the firm. _____

Safety
Provide details of main-office and site-managed safety programs, including persons responsible for the programs. At minimum, include the following information:

When and how are main office safety programs provided? _____

When and how are job-site safety programs provided? _____

Provide positions and names of personnel managing these programs: _____

Provide copies of actual written programs. _____

(*Continued*)

Figure 12.1 *(Continued)*

Provide the organization's Interstate Experience Modification Rate for the last three years:

20 : _____

20 : _____

20 : _____

From the most recently completed yearly OSHA 200 log, provide the following:

Number of hours worked: _____

Number of employees: _____

Worker's compensation claims: _____

Recordable injuries: _____

Lost-time injuries: _____

Lost workdays: _____

Lost-workday case rate: _____

Recordable case rate: _____

Related Experience

By attachment, list a minimum of five projects completed within the last five years that most closely match the scope and dollar value of the project description included in the RFI. For each project listed, provide the following information:

Name of project: _____

Location: _____

Owner, with contact person, address, and phone: _____
_____

Architect, with contact person, address, and phone: _____
_____

Detailed project description, including:

    Original contract amount: _____

    Final contract amount: _____

    Dollar amount of change orders: _____

    Percent of change orders compared to contract amount: _____

    If over 2.5%, explain: _____

    Original scheduled substantial completion date: _____

    Actual completion date: _____

    Originally scheduled final completion date: _____

    Actual final completion date: _____

    Reason for delays or improvements in project schedule: _____

    Details of any site-quality management programs and innovations used on the project:
_____
_____

Proposed Team Members

For each of the proposed team members, provide the following information:

- Complete resume, including project-specific experience related to the scope of the proposed project
- Proposed position on this project
- Name, address, and phone of owner's project representative on the employee's last two projects
- Quality, safety, and additional training the employee has received in the last two years
- Partnering programs in which the employee has participated
- Involvement in any site-managed, site-quality programs
- Letter of management's commitment guaranteeing the employee's availability for the duration of the proposed project
- Safety record of the last two projects in which the employee participated
- Have the proposed project personnel previously worked together as a team on any project?

enable the marketer to search for the specific information required for each individual qualification request received from potential clients.

Marketing Strategies for Lump-Sum Work

After a contractor has the marketing data available and organized, a responsive marketing program can be implemented to secure opportunities for the estimating department. There are basic similarities and major differences between marketing plans for delivery of lump-sum work and marketing other contractual methods for construction services. Although actually securing work depends upon a capable estimating department and being low bidder, a firm may not receive the opportunity to bid a project without an effective marketing program. The ability to reach the qualification phase of a project is just as important as the bidding phase. One cannot proceed without the other.

A marketing tactic for lump-sum bid work is going directly to the potential client or its representative and convincing it that the company has the experience and personnel to satisfactorily construct its project. Getting to this point involves several distinct marketing steps, and, once completed, instituting continuous marketing to ensure that the company is considered for future opportunities as well.

Existing Clients

Any marketing program begins by addressing present and immediate past clients. It is much easier to turn existing clients into repeat clients than to secure new ones. Even those that became clients because your company was low bidder for a publicly announced project should be considered future repeat clients.

A professional marketer should keep careful track of all existing clients throughout the construction phase of a lump-sum project. A successfully completed fixed-price project presents an opportunity to sell additional work to the customer, even the possibility of negotiating the next project.

Depending upon management organization within a firm and associated responsibilities, marketing personnel should certainly maintain contact with clients through the course of a project and become a sounding board for clients. Often, any problems that might arise during a project can be settled amicably if a communication channel is kept open, particularly virtual communications directly linked to the marketing personnel.

Although marketing personnel typically do not have the responsibility or authority to ensure that the customer's project is completed to its expectations and satisfaction, they should encourage feedback from the customer that can then be forwarded to the project management team for resolution if necessary. Corporate management should support all actions of marketing and sales personnel in making a current customer a repeat customer.

Even with federal government projects, it is important to maintain excellent customer relationships because many agencies are moving beyond low-bid selection of contractors. Agencies now have access to reports maintained on the acceptability of a contractor's performance, and should a contractor be judged unsatisfactorily, the government can refuse to accept further bids from the company or can consider its bids nonresponsive. Chapter 15 reviews marketing to the federal government in detail.

Toward the end of a project or shortly after completion, the marketing director should visit the client and review the results of the work. At this time, assuming the customer was pleased with the results, the client should be asked for a letter of recommendation and if it can be used as a reference on qualification statements. If the customer agrees, the information should be stored in the database and then regularly updated. This is also the appropriate time to ask about future projects and the applicability of negotiating this work if the owner is so inclined.

Even if there is no immediate work forthcoming from this client, proper relationships should be maintained and nurtured to ensure the company can again propose on future work when opportunities do arise. Using the customer contact software programs mentioned in Chap. 7, the customer should be contacted on a regular basis through a series of structured phone calls, personal visits, and mailings as appropriate. Direct mailings should be used to inform the client about the company's current work and pertinent news to ensure that this customer is kept aware of the company's capabilities to perform future work for it. Again, it is more advantageous to maintain a current client relationship than to start from the beginning and form a new one.

Cold Calls

Although existing customers will provide a certain amount of work, all contractors require new customers to grow and maintain an adequate backlog of project orders. The first contact with these potential new customers is known as a cold call.

As previously outlined, after completing a strategic plan, potential clients are identified, which forms the basis for cold call marketing. The marketer should understand that the potential customer often has no idea who is calling or why. The initial call should be brief, including an explanation for the call, introducing the caller, and adding a few words about the company and why the contractor should be considered for future work. The call should be closed by stating that an informational package will be sent to further introduce the company. Some marketers prefer to send an introductory package first, then follow up with the personal call. Either way should produce satisfactory results.

Any information learned in this first call should be immediately documented in the customer tracking database. The information package, consisting of the general qualifications statement addressed previously, a cover letter, and possibly a brochure, should then be promptly sent to the customer. This should be done as soon as possible so the customer remembers to connect the information to the cold call.

Follow-Up

After a reasonable period of time, the customer should be contacted again to confirm that the information was received and that it addressed the customer's questions about the company's capability. It would then be appropriate to request a personal meeting at the client's convenience; the meeting might include breakfast or lunch. The customer might direct you to another representative or firm that manages the construction for its firm. The customer's wishes about whom to contact regarding future work should be respected. Although it would be appropriate to maintain contact with this person through company mailings, he or she should not be contacted further if you have been directed to another party or firm.

If the customer does not desire to meet at this time, a specific time for future follow-up should be requested. The next call should be within 2 or 3 months, with direct mailings sent during this time. After a contact is made, the marketing plan should be placed on automatic, ensuring regular contact with the client until an opportunity is identified or the client indicates it desires no further contact by, or information from, the contractor.

The information learned from this follow-up discussion as well as any future calls should be carefully documented in the contact database. The software programs presented in Chap. 7 enable a company to institute a virtual marketing program that can manage the call process and incorporate it directly into a direct mailing program. Phone calls, personal visits, and mailings should be all coordinated to maintain regular contact with potential and current customers. Software programs available today eliminate dependency on fallible human memories. Programs can automatically remind business development personnel of when a call, and what type, is necessary and provide personal information about the person being contacted.

Getting Qualified and Staying Qualified

After customer contacts have led to the company's qualification to bid for a particular owner, for a single project or multiple projects, the marketer's goal becomes one of customer maintenance. If the company is low bidder and is awarded the project, the marketing of existing customers, as previously discussed, comes into effect.

If the company is not awarded the project, the marketing goal is to ensure that the company has the opportunity to bid this client's next project. A letter should be sent to the customer thanking it for the opportunity to bid and expressing your company's continued interest in remaining qualified to bid future work. Then regular contacts should occur with the client, monitored by your marketing software program.

It is imperative that the customer contacts are not dropped because you have not received a particular contract. This client should now be considered an existing client so you don't have to start marketing again when the customer has another project to bid. Qualifying for the next project should actually be simpler because the client now recognizes your company's capabilities to complete its work. After a company is qualified, it should ensure it remains qualified by actively marketing the customer. This can be accomplished by maintaining communications with the customer via marketing tools used by the company.

Preconstruction Services: A Path to Sales

It is difficult to gain the confidence of a client before actually completing work for it, and the customer must feel a level of confidence before it permits a contractor to bid its work. Besides trying to convey this confidence using only a published set of qualifications covering past experience, aggressive construction companies equip their marketing staff with additional sales tools to help achieve this confidence level with new clients.

One effective way to instill confidence is to offer assistance in the early planning stages of a project, long before the qualifying of contractors for bidding. These early assistance measures include budgeting, schedule reviews, constructibility reviews, and other preconstruction services.

As presented in Chap. 2, teaming and more contractor involvement in the early phases of the design and construction processes is imperative in the virtual age. Even for those clients that may still choose the lump-sum low-bid contracting method, there are major advantages in not having the design completed in a vacuum without the assistance of the contractor or subcontractors. An aggressive contractor, even though it might be marketing a low-bid project, will realize the advantages of being brought in early on the project.

All construction companies should have a defined policy about offering preconstruction services to clients. For example, some contractors will provide minimal budgeting and constructibility reviews at no cost and without obligation to the customer. Although the contractor has placed these costs at risk, it may be repaid by the knowledge learned about the project before it is released for bidding, information that other competitors may not have access to. In reality, the contractor often consolidates this preliminary budgeting into the estimate for the project, and therefore no additional costs are actually expended.

Constructibility reviews that are general in nature also are often provided to customers. Bidding contractors make a constructibility review of the project's plans and specifications prior to beginning an estimate to check for errors or problems that should be questioned and then answered in the form of an addendum. Completing this process in the preconstruction phase assists the contractor and the design team and owner as well. Certainly calling the team's attention to possible problems early in the process can save the client valuable time later and possibly additional costs if the change might have resulted in a labor or material change request.

A step beyond these two services might be an initial schedule review. Again, a contractor preparing a bid for any project must prepare a schedule to determine the time needed to bid a project to produce a cost estimate for the general conditions required over the course of a project. Completing this for the client in the preconstruction phase may not cost the contractor any additional funds when marketing and estimating are considered as total costs.

It is certainly reasonable to question why a contractor would choose to offer free, no-obligation services to the customer simply for the opportunity to compete against numerous other firms for the business based on low price. One must carefully evaluate the possible rewards to be had for extending these services to the client during the preconstruction phase.

If the client and design team recognize the value in the services offered, they may very well extend the scope required with the contractor and offer a reasonable fee for these services. Becoming involved early in the planning often will build a team spirit and can lead to the client negotiating the project with this one contractor if the cost is within budget. Finally, the contractor will have built a relationship to ensure that it is immediately qualified on the customer's next project, which is the marketer's goal: once qualified, you want to stay qualified.

A general contractor can also increase its service response to the client during the preconstruction phase by involving subcontractors when appropriate. For example, if the owner is contemplating a design that can use the technical advice of a subcontractor's experience, the contractor should suggest an appropriate subcontractor to approach for this input. Here again is a case for implementing the upside-down contracting method presented in Chap. 2.

How and when these preconstruction services are offered to a potential client, particularly a low-bid customer, should be clearly defined as part of the corporate marketing program. This will to enable the marketing and sales department to be prepared to answer questions regarding these services and make appropriate offers of service support. Extending marketing of low-bid work beyond just responding to bid announcements with a lump-sum price can lead to relationships that progress from qualifying and bidding to long-term partnering with customers. Having a potential client realize that a contractor's service level is worth more than soliciting a low price from a group of contractors should be the ultimate goal of any marketing plan, including one for low-bid work.

Select List of Bidders

Some construction industry customers will prequalify a group of bidders and allow only this select list of bidders to propose on their projects. If the customer is a large corporation and geographically dispersed across this continent or the world, it may have regional select bidders lists or lists determined by a particular niche or specialty. For example, Wal-Mart uses select bidders lists with geographical boundaries, whereas a company such as Motorola might have a certain group of contractors approved for its chip manufacturing facilities and others selected for its office building work. In addition, as previously mentioned, many federal agencies have structured similar contractual arrangements, which is discussed further in Chap. 14.

Regardless of the type of select list, the marketer's goal is to have the company secure a place on the list and remain on it. Although the marketer's job is made easier by having to qualify only once for multiple projects, it becomes crucial that the company make the list to prevent it from being locked out of bidding the client's projects in the future. Remaining on the list is as important as being added to it. The only way the opportunity arises for a new company to make the select list of bidders is if some other contractor has not responded to the level of service the customer expects. As part of an efficient marketing program, the contractor must commit to bidding every project of this select list customer. If the contractor starts to selectively not bid work, for whatever reason, the customer is likely to replace it with another, more-responsive contractor. If a contractor just cannot bid, it should tell the owner why—for example, the owner should be made aware of a problematic contract.

This corporate policy should be kept in mind when attempting to qualify for select lists. For example, if the company chooses not to bid work in a geographical area, it should immediately tell the owner not to expect a bid when work arises in this area. The contractor also should not offer global capability if it cannot realistically meet this requirement. Often the client will divide its lists geographically and may permit contractors to qualify for all areas if they are capable. Although a company may feel that it is best to respond to a global capability, if estimates are to be prepared in areas the contract cannot realistically cover, the client is likely to remove the contractor from the select list entirely.

Offering preconstruction services to clients that maintain select lists must also be realistically reviewed for cost effectiveness. If it is a corporate policy of the customer to maintain a bid list and secure a minimum number of bids to satisfy its purchasing requirements, it is unlikely that offering these services will result in negotiated work or being awarded work when not low bidder. Clients using a select list of bidders should be offered preconstruction services for a reasonable fee. If provided with no-cost obligation, the return on investment is unlikely to be more than that of assisting in preparing the actual cost estimate in the bidding phase. A contractor's marketing program should define what services can be made available to clients that include its firm on a select list of bidders.

Know Your Customer

After a cold-call marketing program has been established, facilitated by a software program to manage the database of information, it becomes necessary to learn everything possible about a potential customer, including the names of personal contacts at that company. As facts are gathered, they should be immediately put into the database so that pertinent information is available to everyone in the contractor's office and facts are not forgotten.

Information gathered should include everything learned about customer contacts, including such information as their alma mater, hobbies, sports interest, family information, and even birthdays. Many companies use the database to ensure that each potential customer receives a birthday card at the appropriate time. Of course this information should be gathered over time and not by asking the customer to fill out a fact finding or survey sheet to answer this type of question. This process takes time, sometimes years, but the process of collecting the information is also a means to develop a relationship with the customer.

Contractors that provide tickets to local sporting events can use a customer's sports interest information to make appropriate use of their tickets. In addition, whenever a marketing person hears or sees information that would interest a client, he or she can inform the client, for example, by sending a newspaper clipping or magazine article on golf to a golfing enthusiast. Information gathered should not simply be stored in the database but should be constantly updated and used to make the client feel that the contractor's personnel actually cares about their relationships with its customers. This attitude will certainly foster business opportunities, including adding a contractor to the bidders list. *Swim with the Sharks,* by Harvey MacKay (Ballantine, New York, 1996), provides a good overview of maintaining an effective database for improving client relationships that lead to business opportunities.

Information should also be gathered about companies a contractor is targeting to become customers. Initial information can be gathered on public companies from a variety of sources, the best of which is the Internet and its various search engines such as Yahoo!. The Internet is also an excellent source for information on privately owned companies, including a company's own Web page, which can be a useful source of information.

After a cold call has lead to an initial appointment, the sales representative should acquire as much background information on the company itself as possible. The more information about the company the contractor's representative can discuss in the first meeting, the better the chances of a return visit and future work potential. When a client finds that someone has taken the time to learn about its company's goals and needs, the client is more likely to be willing to share additional valuable information about the company's future plans.

No one has time to spend with a sales representative who is asking very basic questions in the initial meeting. For example, nothing can create a worse impression than asking a client what type of business it is in or other uninformed questions. Do not expect potential customers to take time to deal with unprepared marketing presentations.

It is important to make the initial impression that the contractor's office is well aware of the customer's business and is prepared to discuss how the contractor can assist the customer in achieving success in meeting its future physical office and facilities requirements. Executives do not have time to spend on needless appointments; the first appointment should be beneficial to both the client and customer. If a worthwhile relationship between client and contractor is not established immediately in this first meeting, it can be expected that the future potential of this client as a long-term customer is minimal.

Although it is important in this first meeting to gather as many facts as possible about the specific requirements of a client, it is also equally important to be able to clearly present what added value the contractor can provide. This first meeting should be brief, leaving both the client and contractor feeling that future meetings and discussions will be worthwhile. Do not expect this potential customer to agree to future meetings if the first has proven to be a waste of valuable time.

The facts gathered before, during, and after this first meeting should be included in the database for future reference. Anytime the contractor completes a project that would be of interest to this client or comes across new technology or information that might be valuable, the marketing department has the opportunity to again have a worthwhile meeting with the client. The database established for the potential client should include relevant information such as potential facilities requirements, types of products or services the customer sells, financial information including growth rates of specific divisions within the company, and specific personnel who have the decision-making authority affecting construction or physical plant requirements.

In today's information society, knowledge is king, and construction firms with marketing departments that use information to assist their clients will be successful in the virtual age.

After Qualifying, Price Is Everything

Although the information presented in this marketing strategies chapter is useful for all types of construction marketing, for lump-sum marketing it is useful only to attain a place on a select list of bidders. After a firm has been placed on a bidders list, price will typically govern the final selection of the contractor.

Although the marketing department may have been successful in getting the firm on the bidders list, if the contractor's estimating department is not successful at being low bidder, the process starts all over for the client's next project. Although price is the ultimate selection determination, the contractor's marketing department must be prepared to maintain relationships with the client regardless of the outcome of the bidding phase.

This is especially true if the contractor, for whatever reason, chooses not to bid the project or submits a bid that is out of the competitive range. The marketing department must be able to respond decisively and quickly should any of these situations occur. The customer must be comfortable with the contractor's response during the bidding phase, and it should be the contractor's assigned marketing representative that makes necessary contacts with this client. Although price is key to receiving a project after qualifying, the marketing department is never fully removed from the process. Even if the contractor is low bidder, marketing contacts should remain active to ensure that the contractor is awarded the project and not some other bidder.

Likewise, after contract award based on a low bid, the marketing continues throughout the construction of the project to maintain effective relations to ensure that the customer will permit the contractor to bid on future projects. Qualifying, even for a continual select list of bidders, is never the end to a successful marketing program; it is only the beginning.

Follow-Through Marketing

Marketing, regardless of the type of construction service sold, is a constant strategy for every construction company. A previous section presented the basic components of a marketing program for lump-sum contracting that also can be used for marketing any contractor's services. These basic functions must be carefully tailored to each individual contractor and its specific market sector goals.

The success of any marketing program incorporates the basic principles presented in previous sections and then modifies the program to meet the requirements of the contractor's long-term goals. The marketing personnel must know their customers, their management, and their goals, or the marketing program will fail. Customize the marketing program after the basic program is structured based on the contractor's mission statement (Chap. 3), management's requirements, and capabilities of the estimating department. The following ideas should be considered when tailoring a basic marketing program for specific goals:

- Selecting targets: rifle versus shotgun approach.
- Developing a better relationship with the estimating department.
- Moving from lump-sum to negotiated work after the sale.
- If concentrating on bid work, take care of the subcontractors.
- Is partnering on low-bid work possible?
- After-sale marketing.

Selecting targets: Rifle versus shotgun approach

There are two schools of thought on bidding lump-sum work. The approach used by those contractors that bid everything in sight is the shotgun method. These contractors believe that by bidding large numbers of projects continually, they will be low on a sufficient number of them to maintain a healthy backlog.

Contractors that are very selective about which projects they choose to bid use the other approach. These contractors will narrow prospective bid projects by size, number of expected competitors, and type of work. They bid only on projects on which they feel they have a 50 percent chance or better of being low bidder.

In the shotgun approach, a marketing program can be very generic in nature. Almost any project the marketing department brings in to bid would be acceptable to the estimating department. Rarely would a marketer have to explain to a customer why the company chose not to bid a particular project.

The goal of the shotgun approach is to ensure that a sufficient number of projects are continually presented for estimates. This might require a larger marketing staff or one that does not become as closely involved with the customer as might be required in marketing negotiated work. Also, it is likely that after-sale marketing by these contractors is not a high priority, preferring instead that the marketing department immediately begin chasing other clients with potential bids. The importance of appearing on continuous select lists of bidders, rather than pursuing clients with individual projects, is more important to this type contractor than it is to the one using the rifle approach. Select lists are more likely to bring a steady stream of projects, the mainstay of the shotgun approach, and thus become a major goal of the marketing program.

Marketing for the rifle approach requires the marketing program to be more selective and concise, pursuing customers seeking very limited numbers of contractors to bid on a project. Certainly, using the rifle approach requires the marketing program to target customers that provide better odds for the contractor to be the low and successful bidder. In this situation the contractor might target niche markets that have limited competition either by type or size of work or geographical location of the proposed project.

After-sale marketing also becomes more critical to these contractors, with the goal of ensuring that each and every client becomes a repeat customer. In rifle marketing, knowing the customer is as important as offering preconstruction services whenever appropriate. The marketing goal becomes one of offering any means possible to limit or eliminate competition to increase the odds of being the successful low bidder.

Obviously there are contractors that believe in a balance between shotgun and rifle approaches. In such cases the marketing program must be tailored as necessary to meet the requirements of the contractor's management. For such construction companies, targeting customers that believe in long-term relationships using a uniform select list of bidders becomes a major goal. The marketing program must allow for continuous after-sale marketing to maintain rifle targets, while permitting a continuous flow of bid opportunities. This is the same approach as shotgun marketing but with the after-sale marketing to customers the contractor believes will bring ample rifle bid opportunities in the future.

Whatever the situation, the marketing program must be consistent with or compatible with the capabilities of the estimating department. Certainly the estimating department will govern the success of a lump-sum contractor's marketing program. The estimating department must be a success before the marketing program will be considered successful. Regardless of how many select lists the contractor makes, the company cannot become successful if it is not the low bidder often enough to produce a backlog of profitable work.

Know your estimating department

If a marketing department brings in a continuous flow of project bid opportunities in work that the estimating department is uncomfortable bidding, it is likely that both departments will be judged unsuccessful by management. If the marketing department in an effort to "look good" brings in so many projects that the estimating department cannot keep up and turns down work constantly, both will again be judged unsuccessful. In this case the marketing program would begin to deteriorate due to clients removing the contractor from their select list for being unresponsive. If the marketing department is promoting a rifle approach while the estimating department advocates the shotgun approach, neither will be satisfied, and management will begin to review the situation negatively also.

Clearly it is imperative that, after a basic marketing program is structured, the department incorporates the goals of the estimating department so that both can be successful. Internal regular meetings between marketing and estimating departments are mandatory, especially for lump-sum contractors. In these meetings the marketing department must first determine the capabilities and desires of the estimators and then learn how the estimating department approaches bidding.

Regardless of what the marketer's goals are, if they are not in complete agreement with those of the estimators, there will be internal conflict that will eventually render any marketing plan ineffective. Estimators should have input into the types of customers, projects, and number of bids the marketing program will target. This customizing should then be reviewed consistently to ensure that the goals of both departments remain the same.

For example, if an estimating department preferring the rifle approach announces that it is becoming overloaded with identified opportunities, the marketing department should make appropriate adjustments. This might include advising certain customers that the firm cannot presently meet proposed bid dates due to the amount of bidding occurring in the marketplace. Advising a client of this likelihood before a bid is released to a select list of bidders is more appreciated than waiting until after the shortlist of bidders is notified. The client has the opportunity to add another bidder if appropriate, or in certain situations may actually delay the bidding if other proposed contractors are operating under the same current conditions. Any client will realize that it will receive more competitive bids when the contractors have sufficient time to prepare a proposal.

Marketing lump-sum work can only be successful if closely associated with the estimating department. This will ensure that both departments' goals are in unison and will permit both to be successful.

Moving from Lump-Sum to Negotiated Work after the Initial Sale

Although it is the intent of the marketing program for lump-sum bids to secure sufficient select bidders list positions, most contractors would not turn away an opportunity to secure negotiated work. Lump-sum marketing will rarely present a negotiated opportunity during initial marketing (although marketing for negotiated work often presents opportunities for bidders lists). However, there is no reason that once an initial project is secured by low bid with a customer that the follow-through marketing should not emphasize the contractor's ability to perform additional work on a negotiated basis.

The ability to present current customers with alternative contracting methods depends a great deal upon the success of the project management during construction. Obviously a company cannot expect to secure negotiated work or even future work with a client if the current project is not completed satisfactorily. Typically a customer would have to be very impressed with the contractor's performance to consider negotiating the next project rather than once again using the select bidders list format.

As previously discussed, it should be the responsibility of the marketing department to maintain a direct line of communications with the customer throughout the project's completion to ensure that a satisfactory relationship is maintained. Although it would be the project management's responsibility to correct any problems the owner brings to the contractor's attention, the marketing representative is often used as a sounding board by the customer when a dispute or problem arises on the project. If the marketing personnel have any expectations of maintaining a relationship with a current customer after the project is complete, they must keep in constant contact with the client during the construction phase.

Many contractors might consider that the marketing department's efforts end when a project is secured and do not begin again until the customer has another project to build. This is not an effective marketing plan that results in long-term relationships and repeat clients. Any customer would expect that its initial contact within the contractor's firm, the marketing representative, would always be available to assist the client whenever necessary. A customer should never get the impression that the marketing department is only interested in the client when a new project is being proposed because it is likely the contractor will not again receive the opportunity to bid additional work even if its last project was a complete success.

By maintaining an effective and positive relationship that was formed during the initial marketing phase, the marketing department will have a much better chance of securing future work, possibly including the opportunity for negotiated work.

If You're Marketing Lump-Sum Work, Take Care of Your Subcontractors

As previously reviewed, the purpose of a marketing program for lump-sum work is to ensure the contractor is given the opportunity to submit a bid for a project, usually through a select list of bidders. Once the initial marketing program has achieved this goal, it becomes crucial for the contractor to submit sufficient numbers of low bids to achieve a profitable level of work and a sufficient backlog.

Because most contractors subcontract a large percentage, if not the majority, of field construction on any typical project, to be low bidder the contractor is dependent on subcontractors to achieve the best overall pricing submitted to the owner. This dependency requires the contractor to maintain an excellent relationship with local subcontractors to achieve long-term success.

Many aggressive contractors' marketing programs will include specific programs designed to maintain relationships with subcontractors. The goal of subcontractor marketing is to achieve a competitive advantage over other contractors bidding work in the same area. Clearly the marketing to subcontractors becomes as important as marketing to achieve a position on a select list of bidders. One without the other will not lead to long-term success.

Subcontractors will deliver the best terms and conditions to contractors with whom they have developed win-win relationships. Obviously when a company is comfortable doing business with a particular firm, knowing what to expect during construction, including payment terms and field management, it will naturally offer the best terms to this contractor. These favorable terms will include the most advantageous pricing and schedule that will then assist the contractor in submitting consistent low bids. The low bids developed in this manner will enable everyone on the project to achieve success rather than result in situations that occur when contractors must resort to bid shopping to develop low bids.

Contractors have different opinions about who has responsibility for managing relationships with subcontractors. Some will place this responsibility with the estimating department, others with the marketing department, and sometimes both departments will become involved. In any case, it is important to have a communications line developed for subcontractors beyond the project site management team. Although it is not the intent of this communications and marketing to replace the authority necessary at the job site, it is necessary that subcontractors be aware they have the ability to discuss problems with a contractor's office directly when appropriate.

Simply having communication lines open on bid day for taking pricing will not achieve long-term success for any contractor. Subcontractors that become dissatisfied are likely to not bid or may submit higher pricing to adequately reflect the additional cost of doing business with a particular contractor. This higher pricing may be reflective of slow payments, poor job site management that delays and unnecessarily adds costs for a subcontractor, or the lack of any personal relationships within the contractor's organization. Whatever the reasons, a contractor that does not promote effective subcontractor relationships will never achieve long-term success in any market.

Marketing programs directed to subcontractors become just as important as client marketing when a contractor depends upon low-bid work. Any marketing program for lump-sum projects should include precise goals to achieve effective subcontractor relationships. These goals might be similar to client marketing, including regular visits and communications with subcontractors. Many contractors will have a separate software contact management program for subcontractors that is very similar to that of client contact marketing discussed previously. These contractors realize that a project successfully completed for a client occurred because of the cooperation of their subcontractors and are likely to share this success with the subcontractors, including acknowledgments in the contractor's marketing newsletters.

Developing relationships between subcontractors and contractors is presented in Chap. 17. Although much of the advice in Chap. 17 is targeted toward negotiated or design-build projects, the majority of the marketing suggestions can be used for any type of construction work, including low-price work. Contractors depending upon low-bid work to build a backlog must realize the importance of subcontractor relationships that can ensure they achieve effective pricing on each project for which a bid is submitted. The marketing suggestions in Chap. 17 can be effectively incorporated into any lump-sum marketing program.

Is Partnering on Lump-Sum Work Possible?

After a contractor has achieved a sufficient backlog of work through the low-bid process, marketing programs for lump-sum work then focus on maintaining existing clients for future repeat work. Many contractors will include partnering in their retention marketing as one way to achieve repeat work. Partnering is a quality management process used to improve communications flow among business partners. In the virtual world, partnering on all construction projects becomes crucial for success, even on fixed-price work.

Partnering on lump-sum work has been referred to as an oxymoron, in that partnering on what is typically an adversarial contract may not be achievable. Lump-sum contracts can lead to adversarial relationships among the project team members. Partnering is often attempted to counteract this situation, but it is a challenge to maintain. As soon as a profit-reducing situation arises, even with partnering concepts in effect, the contracting parties will typically do whatever is necessary to maintain their positions.

Clearly, partnering is designed to eliminate or drastically limit the number of disputes that arise on a project that become full-blown disputes. Most standard industry contracts contain language that specifically spells out the process by which disputes are settled, including the legal venue and parties that have ultimate authority at the project level. Although this contractual language intends to prevent unnecessary outside legal involvement, it often prevents the establishment of effective dispute resolution at job site, lower-management levels.

Postproject Completion Marketing

Marketing is a continuous process that never ends. It continues through the selection of a contractor for a list of bidders, through the bid and sale process, throughout the project's construction phase, and then starts anew after the completion of the project. In fact, the most important phase of marketing may be the retention of past or existing clients rather than marketing new clients.

It is universally recognized that the marketing costs for maintaining existing clients are considerably less than the cost and time it takes to sell to a new client. Any marketing program should be structured to emphasize existing client maintenance. All marketing plans should include a carefully structured program to market to and maintain existing and past clients.

The first step in client retention is an open and commutative program that encourages feedback from the client. The software contact program should include all existing and past clients to ensure they receive regular updates on the contractor's progress on work-related items. Personal visits should not stop because a project has been completed for a customer, even if it is clear this particular customer may not have another project for the immediate future.

It is important to recognize that this client, even though it may not have any forthcoming work, will be used for references to secure other clients, may recommend the contractor to others, or have tenant, remodeling, or other type of work to be completed in this just-completed project. All these potential opportunities make it necessary to maintain a close relationship with every client on a continuous basis.

Part of effective after-sale marketing includes a quick response to punch list or warranty work. By maintaining relationships with subcontractors as previously described, the contractor can respond as required to a client's request without unnecessary delay when a subcontractor must be involved again. Many contractors maintain a database on warranty expiration of their projects, advising the client when the warranty is about to expire and actually request a visit to inspect the project and compile a final list of warranty work to be completed. Nothing enhances a contractor's reputation for quality work better than actually calling upon a customer to determine if repair work is necessary. Such a system should be included in any construction marketing program, including lump-sum work.

Other marketing opportunities for repeat clients include offering assistance during the planning stage of future projects, other previously discussed preconstruction services, and inviting the client to inspect other work in progress that may be of interest to it. Any opportunity to maintain contact with a present customer should be available to the marketing department to use whenever it is deemed appropriate to maintain customer relationships that provide repeat work.

Marketing Directly to Purchasers of Construction Services

Introduction

The general objective of any marketing program is to develop targets and clients as high up the "food-chain" as possible. In other words, marketing should be directed close to, if not directly to, the ultimate customer or purchaser of the construction services. Targeting a marketing program in this manner purposely eliminates as many intermediaries as possible to avoid any possible conflict of interests in marketing to companies that may not have mutual interests and goals.

For example, general contractors prefer to target building developers rather than architectural firms that might have a contract with the developer. The marketing program should not exclude others in the food chain but should direct the program to the ultimate customer and include all other possible contacts as appropriate.

Marketing programs structured in this manner are more likely to succeed than those that let others determine their outcome. Early and direct marketing by a contractor to the ultimate customer may prevent an architectural or construction management (CM) firm from making an arbitrary decision that excludes the contractor from being selected. Of course, the program must be structured so as not to interfere with any preexisting relationships or intermediaries with which the client chooses to operate. The latter is particularly true when a client chooses a CM or other owner's representative to manage and direct its construction program, including the selection of a contractor and subcontractors.

In this situation the client specifically chooses not to be contacted directly, and any attempts to do so may negatively affect the relationship with the client and/or the firm representing the client. Marketing can and should be directed specifically to the client until the client selects a firm to represent it or specifies another contact. In fact, direct contact with the client initially may be necessary to find out who represents it in real estate and what its physical construction requirements are; then the marketing should be directed to the owner's representative as suggested.

Should a client have no intermediary representative, a relationship should be forged to ensure that when an occasion arises, the contractor is given the opportunity to propose or negotiate for any work released. Should the potential customer later choose to hire a representative for the construction process, the contractor would be given an appropriate opportunity to propose for the project based upon this preexisting relationship.

The marketing techniques presented in this chapter are designed to assist in developing strategic goals to market directly to the direct purchasers of construction services, mainly the owner, architect, engineer, and others, including CMs and owner representatives. Although every contact made is important in marketing construction services, only marketing to the firm and person within that firm that can make the ultimate buy decision will result in a sale.

Marketing to Owners

The owner is the ultimate consumer of the construction services required to fulfill physical space or facility needs. The clients produce a saleable product or deliver service using these physical facilities. Although construction needs may be essential for producing a product or service, the physical plant or facility is only a secondary requirement; therefore, most clients or users of construction do not maintain a construction engineering staff to manage facility construction.

The owner then chooses from a host of available options to manage its construction, including outside representatives such as the firm designing the project or a specialty firm that manages both the design and construction, usually as consultants only. In larger organizations the client may have a construction staff whose responsibilities range from selection and oversight of consultants only to more hands-on management of both the design and construction phases.

A marketing program for any contractor should begin by targeting potential clients, using information gathered during marketing to determine the appropriate means to facilitate a sale with those clients.

Pinpointing the decision maker

Marketing and business development, no matter how well planned and executed, will be unsuccessful if it is targeted to a person unable to make the ultimate buy decision. Initial marketing might be required at several levels of management and to a number of personnel within an organization to determine the proper decision maker. This initial marketing should not only determine the client's future construction requirements but also the person who has the authority to make the selection and buy decisions.

Each organization manages the purchase of construction services differently. Only personal and direct contacts with each individual organization will enable a marketer to determine the proper person to which the marketing program should be directed. Although this initial investigation may only require a phone inquiry to determine the appropriate contact, often the process is more difficult, particularly in larger organizations that have different divisions and multiple branches or locations.

Through a series of planned contacts with the organization, the contact should become confident that the contractor can assist his or her organization achieve its physical construction goals. The contact will then share the information necessary to finalize the sale. In the case of a larger firm, the contact should describe the organization and note if there are multiple contacts required within the firm, such as separate operating divisions or profit centers that manage their construction needs independently.

As information is gleaned from initial marketing contacts, it should be entered into the client contact database presented in Chap. 7. This allows the contractor to establish a defined marketing program that targets the appropriate decision makers within the organization. The marketer should not be afraid to ask questions during this phase of marketing. Do not assume that the initial contact is the person with the authority or responsibility to make buying decisions. Initial contacts are often directed to lower-level managers simply because of protocol and the time availability of the decision maker. The marketer should politely inquire how the company operates and makes decisions regarding construction services. From this information a plan can be established to market directly to the decision maker(s).

In the case of larger or global organizations, the contractor may have to market to several different people and even several different departments at multiple locations. Depending upon the contractor's strategic plan and scope of work capabilities, including geographical goals, the marketing may be directed at all or only select groups within the organization. Marketing to the entire facilities group is advisable because transfers within organizations occur frequently, and the person handling work outside the contractor's capability might be transferred to a position managing projects within the strategic planning.

In the same manner, although it is mandatory to market to the right decision maker, it is not improper to target additional facilities personnel within an organization. Over a period of time, changes are likely to occur (i.e., promotions resulting in a decision maker being replaced by someone who has already been educated about the contractor by this multiple-level marketing). Further, when people change firms, they may actually present the contractor with sales opportunities with another customer it was not previously targeting.

To reemphasize, targeting the proper decision maker is key to any successful marketing program. Once this decision maker has been identified and included in the contractor's marketing program, expansion of the program to multilevel marketing within the same organization can increase the benefits of the program over a longer period. This multilevel marketing provides insurance to cover possible personnel changes that include promotions, retirements, and other changes in the key decision maker's position.

The owner's perspective

A marketing plan that targets individual clients, including corporations and developers, should encompass what can be expected to be the customer's needs and chief concerns in choosing construction services. A marketing program that incorporates the owner's perspective will be considerably more successful than a generic program aimed at all types of clients.

Although it is neither feasible nor necessary to implement a different marketing program for each targeted client, the contractor's marketing program should be capable of isolating groups of customers. For example, marketing targeted to manufacturing or process facilities work serves no purpose when directed toward developers of commercial property. Some contractors send a generic newsletter featuring all their projects to all potential clients. Such an approach could defeat its purpose if a customer thinks the newsletter emphasizes projects that have no relation to the work it needs completed, even if the contractor is capable of performing the work required.

Contractors that market and complete a wide range of construction services and types of work should similarly be willing to differentiate their marketing program among these various group of services. Software and hardware available today are capable of providing a cost effective means to customize marketing literature to highlight specific types of work. Every company newsletter or brochure can be easily and quickly changed to reflect different projects, using the basic information and changing only specific items to reflect an individual client's interest.

When preparing a marketing piece, consider the value of a client's time and provide only information that is pertinent to that customer for maximum impact. General-interest pieces are likely to end up in the trash. In the information age, the excessive amount of information that every manager and executive receives prohibits a review of every piece of literature received.

Virtual construction requires the same customization of marketing literature that the construction product itself does. Clients no longer have the time to spend reading superfluous information; therefore, any contractor's program must be specifically targeted to the client's needs.

This customization of marketing products requires the marketer to use the information collected from personal visits and from sources such as the Internet to develop a concise list of items of interest for each potential and present customer. Once customized marketing tools are prepared, the contact software database can produce specialized mailing lists from all the clients in the database.

For example, a newsletter containing general information can be customized for several groups of clients. Place a lead article on a specific type work or project in a prominent place in the newsletter. The type of project featured can be changed for each client group. This can be done easily through software programs, permitting a newsletter to be changed several times without major effort. Then the contact software is used to correlate the individual newsletters with specific mailing lists that are compiled based on information in the database. For example, a newsletter highlighting a land development project will be sent to all database clients marked as land or commercial developers.

Clients receive a marketing piece that is specialized to their specific interest in construction services. This is a useful way to maintain contact with clients using the owner's perspective. Similarly, individual meetings held with the client should include discussion of information that is customized for the client.

Once a contractor has established contact with the decision maker, as described in the previous section, visits and meetings other than socializing events should respect the contact's time. The virtual society leaves little, if any, time during a business day for unproductive work. When meeting with a client to market or sell construction services, the meeting should have a specific purpose and time limit established. The purpose should clearly be made from the owner's perspective.

For example, the marketer should come prepared to discuss how the contractor can assist the client achieve its goals on a particular project. This might be a detailed explanation of how the contractor can manage a particular project to save considerable overall time, thereby shortening the construction schedule. The information provided should be specific and customized, not generic like the tired and worn expression, XYZ Construction can build your project on time and under budget.

This information is worthless to the client. Today contractors must present specifics on how they can reduce schedule parameters to build quickly. Otherwise, you can expect the client to lose interest immediately, seeing no reason to conduct business with the contractor except for the low-bid commodity issues so prevalent in the industry for most construction companies today.

From the client's perspective, time issues, cost, and innovation are essential to maintaining competitiveness in its own industry. Construction marketing should concentrate on how a contractor's capabilities can be employed by the client to improve competitiveness and profitability.

Too many contractors in the construction industry rely on outmoded marketing and sales efforts that provide old information on completion of projects on time and under budget. Today, aggressive contractors have to have their operations departments participate in marketing by providing suggestions on innovative project approaches that can save the client time and money. These suggestions should be offered in the project design or planning stages, when they can be readily adapted into the program.

Virtual age marketing incorporates the expert advice a contractor can provide in early project phases rather than relying on outmoded marketing techniques of advising potential clients that it has the experience to build the project and then waiting for the client to call for proposals. Clients are realizing the importance of contractor participation earlier and earlier in planning to maximize their effectiveness in reducing costs and schedule parameters to improve their profitability. Contractors whose marketing programs provide experience-only information and then wait for requests for proposals will find a shrinking marketplace and will be competing with progressive contractors who are selling their services and expertise in the planning stages of a project instead of in the postdesign phase.

Marketing in the information age requires contractors to move beyond selling their past performance and to market their abilities to assist clients in reaching their own goals. Contractors that emphasize the customer's perspective and requirements will in turn reach their goals, providing a win-win situation for all. A successful marketing plan must initially provide information on the contractor's experience to satisfy the client that the contractor has the ability to provide required services. Then the plan must incorporate proactive information that is useful to the client. This approach should continue after the initial sale to maintain a long-term alliance with the customer.

Select list and other prequalification requirements

The initial marketing contact with many clients, particularly large corporations, will involve the submission of qualifications that provide a means for the customer to qualify the contractor's ability to perform work for them. Due to the large number of construction companies vying for the client's business, customers institute the qualification process as a means to ensure that only the most capable contractors are selected to complete work.

This initial qualification process is typically a mandatory process, but it by no means restricts further marketing to the firm, as discussed previously. However, the contractor is obligated to follow the formatted processes within the client's organization to become a qualified supplier. The marketing plan should then become a proactive information-providing system based upon the owner's perspective.

Client qualification statements may vary, but they typically request the same basic information needed to judge a contractor's capability and performance record. The standard qualification form shown in Fig. 12.1 is typical of those used by many corporations. Information requested includes previous experience, personnel resources, quality and safety programs, cost and schedule performance, and financial resources.

Each qualification questionnaire is customized to fit the individual client's interests and needs. A customer interested in contractors that are capable of completing microchip-processing facilities will inquire into a contractor's capabilities in this area, including past performance on similar work.

In a 1998 survey conducted by the Construction Marketing Committee of the Associated General Contractors (AGC), industry clients ranked prequalification criteria in order of importance; the results of that survey are as follows (listed in order of importance):

1. Financial issues
2. General reputation
3. Size versus project
4. Past performance with the specific owner
5. Safety record
6. Experience/technology
7. Related experience
8. Past litigation
9. Labor relations

Any construction contractor's marketing program should include a database of corporate information to compile these qualification statements, emphasizing these selection criteria. A contractor's marketing department should have a database of personnel resumes, project experience including references, project photographs, quality and safety performance, and financial data; this database should be able to compile a qualification statement in minutes. Project photographs are often very useful in showing a client details on previous projects, and all photographs should be digitized and stored on CDs so that the marketing department can produce a customized qualification statement incorporating these visual images.

References are critical when attempting to market to a new client, and the marketing department must ensure that the contractor's references are constantly updated. If a client is unable to reach a reference because of inaccurate information provided or a contact that can no longer provide a good reference, the contractor may not be qualified or added to the shortlist.

Qualification statements are typically the first step in introducing a contractor's capabilities to a prospective client. Even though a contractor may have a standard qualification statement or brochure, the client will require the information in its own format so that each contractor is compared on an equal basis. Additional information on qualification statement preparation is presented in Chap. 12.

Final selection criteria

After the initial qualification and shortlisting of contractors, clients, particularly the federal government, will require further and more detailed information that is submitted with the bid or proposal. The amount and type of information varies, but marketers should be prepared to use their databases to support the responses necessary during the sales effort.

In the AGC client survey mentioned previously, owners ranked their final selection criteria in the following order of importance:

1. Price
2. Project team
3. Safety record
4. Preconstruction services
5. Control systems
6. Current workload
7. Past performance with specific owner
8. Local experience
9. Communications
10. Quality improvement programs

Note that the project team is significantly more important than it was during the prequalification stage. In the initial shortlisting, clients seem more concerned with the general experience and capabilities of the contractor. In the final stages of competition, the client recognizes that it is the contractor's personnel who are directly responsible for the success or failure of a project. Therefore, the client becomes more interested in the experience of the proposed project management team and in knowing if the members have experience in working together as a cohesive team.

Marketing the required expertise

When marketing directly to clients, make sure that the client requires construction services that your company can perform. If a contractor does not have the expertise required by a customer, there is no reason to spend the time or effort on this client unless it is perceived that the client might require other services in the future. Once the initial marketing has been completed and a relationship formed with the right decision maker(s), the marketing program should become a relationship-oriented program based on services required by each specific client.

Marketing programs that continue to rely on providing information about a contractor's current and recently completed projects do not benefit the client. Such marketing may keep a company's name in front of customers, but it does not provide useful information to the client. By considering the owner's perspective, contractors should realize that the client has no use or need for this simplistic information.

This creates a necessity for advanced marketing programs to be customized and made as useful as possible to each individual customer. To meet this goal, a marketing program may supply detailed information about how other clients in similar industries are cutting physical facilities maintenance costs, provide updated construction cost data, or advise the client of new advances in construction products. Any creative ideas of a similar nature provide a customer with information it can utilize in its own planning. Information-based and customized marketing is a requirement in the twenty-first century.

Today, marketers must realize that people are no longer as apt to read through all their mail as they might have been just a few years ago. Unsolicited bulk marketing mailings no longer have the impact they might have had in the past. Few employees have the time to review these unsolicited mailings, and if they appear to serve no purpose, they will be immediately added to the trash. It is essential that the marketing program provide useful information and that it be specific to the nature of the client's industry or immediate needs.

A prospective customer in the textile industry certainly will not be interested in receiving information about a contractor's construction applications in the automotive industry. As discussed previously, software can create customized mailings that can be addressed to customized mailing lists prepared from the company's marketing database. Once these customized mailings have begun, the next step in marketing to clients, particularly present customers, is to move toward marketing based on service applications that create win-win situations for both parties.

Marketing using preconstruction services, planning, and budgeting

An advanced marketing program, used by aggressive construction companies, offers their customers planning, budgeting, and other preconstruction services. These services are not provided without some type of reimbursement. Such service-based marketing should be a win-win situation in which both the client and contractor are rewarded.

In the virtual age, time is as important as cost issues. Therefore, the ability of a contractor to positively affect the overall outcome of a project is more likely to occur in the planning and design stages than in the actual construction phase, the latter often dependent on pricing rather than knowledge issues. Highly successful construction companies are those that emphasize meeting the client's needs.

Contractors are more likely to suggest options that save time and money during the planning stage than after the design is already completed or construction has commenced. Industry clients have begun to realize the importance of a contractor's direct input during these early stages and have begun hiring contractors or consultant firms who can offer these preconstruction services. Contractors selected to provide these preconstruction services typically have the opportunity to complete the construction phase. This new industry paradigm requires contractors to reevaluate their services offered and reflect these changes in their marketing program. Contractors who do not may find themselves locked out of an increasing market share of opportunities for construction projects.

Increasingly, contractors are using their ability to provide preconstruction services in their marketing to potential clients during project planning stages rather than only emphasizing their construction phase capabilities. Some construction firms have offered limited preconstruction services without cost to clients, typically initial budgeting and schedule reviews. These initial services are offered free in expectations that a potential customer is likely to request more detailed services for a fee or award the contractor the construction phase of the project on a negotiated basis.

Unfortunately, providing the more valuable information age services without charge can defeat the contractor's ability to become successful by providing this highly valuable knowledge. The construction phase is now often regarded as a commodity service, with contractor selection based solely on lowest price. This is similar to grocery stores that compete basically on price; the firms supplying information to the grocery business, such as software suppliers for the cashier scanners, are more profitable than the grocery chain. Likewise, virtual construction firms are those that are highly compensated for providing knowledge about the industry versus those contractors that provide only the commodity service, the physical construction phase.

The purpose of a sophisticated marketing and sales effort is defeated when the most valuable services a contractor can provide are given away free in the hope of negotiating the low-margin construction phase.

Marketing programs that incorporate and offer preconstruction services to clients should reflect the market value of these services by including an appropriate fee for these services. Offering an owner a complimentary initial budget review may be appropriate to gain a client's confidence, but marketing advanced services to the client should become the goal rather than just attempting to be added to the bidders list of contractors once the design is completed. Successful contractors are those that will take this initial entry with a client and expand upon it by selling an extended range of preconstruction services. They will convince the client of their ability to save time and costs over the entire project cycle, from planning to construction.

Preconstruction services generally include constructibility reviews, budgeting at several phases of design completion, schedule reviews, permit assistance, and other planning phase services that a client might require. Many contractors have already begun to address these capabilities in their marketing programs and are building a performance history of preconstruction services equal to their construction phase services.

Marketing potential clients with preconstruction services is becoming standard within the industry as a direct means to facilitate sales of the construction phase. However, progressive contractors are already realizing the capability of these early phase services to produce fees as great or greater than the physical construction phase, and are structuring their marketing programs to take advantage of this. In other words, marketing programs can be used to sell preconstruction services as the end product rather than using these services to gain access to the construction phase.

Even if the marketing of these preconstruction services is not successful, the contractor has provided the client with the knowledge that it is available to complete the construction phase. At minimum, the contractor should also be able to propose on the construction phase after design completion.

If a contractor is successful in marketing and selling preconstruction phase services, the opportunity to expand this scope of work to include the construction phase should also be a goal of the marketing program. Often contractors develop teaming relationships during the preconstruction phase and are therefore able to negotiate a contract for the field construction activities.

If you are successful in selling preconstruction phase services, marketing must continue to ensure your company has an opportunity to complete, or compete for, the construction phase. If your marketing of preconstruction services is not successful, you must continue marketing for an opportunity to participate in the construction phase, even if another contractor is chosen to complete preconstruction services.

It can be expected that as construction firms become proficient and profitable in supplying preconstruction services, they may choose not to participate in the construction phase if it does not exhibit sufficient profit potential. This would leave available the commodity portion (field construction) to those firms that choose not to supply preconstruction services.

Marketing for repeat work

With any potential customer a contractor decides to target, marketing becomes a continuous and never-ending process unless a determination is made that this customer no longer fits the contractor's strategic plan for growth. Marketing should not be structured only for those projects that are already in the planning or design stages, but for future projects as well. In effect, the marketing program should establish a long-term or strategic alliance with the client.

This long-term relationship should never be dissolved simply because a project is awarded to another contractor. Marketing programs should continue with each client regardless of short-term setbacks. Owners may choose to develop relationships with several contractors and construction service suppliers, deciding which company best fits their needs for a current project. In particular, large global companies have sufficient construction service requirements to utilize numerous contractors, even though they may establish strategic alliances with one or a few contractors based on geographical areas or types of work required.

Marketing programs that target clients and owners directly must be designed to run continuously and without regard to any one particular project. Marketing plans target clients, not projects.

During client marketing, when a specific project opportunity becomes available, the contractor's sales effort must become involved. This sales effort might include the estimating department, project management, and possibly the marketing department. However, the sales effort should be separated from marketing, and whatever the outcome on this one project, the original marketing plan remains in effect.

Sales efforts target a specific project; marketing efforts target the client. In fact, marketing programs should be blind to individual projects, concentrating on a client's long-term global requirements. Structuring the marketing program in this manner will lead to long-term relationships that produce repeat work from the same client rather than an individual hit or miss target for each project.

A client marketing program then becomes structured to be a continuous information-providing service that incorporates the owner's perspective as discussed in the previous sections. If the marketing program is structured to only present experience that fits a client's present requirement, it might develop one-time project opportunities but not a long-term relationship that presents continuous repeat opportunities. The marketing plan should target the development of a long-term alliance, with each customer targeted by a contractor.

Forming alliances for long-term growth

Whenever the marketing and sales effort have combined successfully and lead to repeat work with a selected customer, the ultimate goal should be that of forming a strategic alliance with the customer. Alliances are being used by all types of corporations when selecting suppliers for a variety of services.

An alliance is a business relationship formed to benefit two or more parties in their routine business needs. A department store might form an alliance with only one supplier of paper goods to permit timely deliveries of inventory and the best overall pricing. A franchise corporation might form an alliance with a contractor to construct all its store locations within a selected geographical area. Internal alliances between companies in the same industry are also created to better position the alliance team to secure work they might individually not be able to sell. In any case, an alliance is structured to be a win-win situation for all involved team members.

Forming an alliance with a client for its construction needs that leads to multiple repeat projects is probably the most desirable position for a contractor. Once an alliance is formed, the teaming relationship permits the members to concentrate on their respective businesses, eliminating the need for unnecessary marketing and sales efforts between the members. An alliance allows a corporation to outsource its physical facilities requirements to construction and engineering firms and concentrate on active management of its own professional services. The corporation can then increase its productivity rather than having to manage construction requirements that do not necessarily add to its profitability.

The formation of an alliance is a result of long-term relationships built over time and not the direct result of a marketing program. Occasionally a corporate client may announce its intention to select a team of alliance partners for a variety of services that might include construction, but alliances are not formed overnight. A company choosing to form an alliance will select a contractor that has satisfactorily completed multiple projects for it. The continual marketing and sales efforts described in the previous sections are the ones most likely to lead a customer to choose to form an alliance with a particular contractor.

Once a relationship with a client has resulted in numerous repeat projects, the marketing plan remains in place to continue the relationship. At this time, however, the contractor may decide to forge upper-level management contacts that can lead to discussions about the benefits of an alliance partnership. Typically the contacts to develop an alliance are outside the marketing department, although this department's support will be necessary to support the alliance initiative. In fact, the formation of an alliance requires involvement of all departments in a contractor's office, including estimating, sales, marketing, and management.

An alliance partnership is not a legal partnership or relationship, but it does require the development of a specific document outlining the goals and operating standards for the alliance agreement. For a contractor the agreement might include fees associated with specific types of services, types of work and geographical coverage to be included, and management relationships including contracting terms and conditions. Likewise the client will spell out its expectations about the contractor's service, including resources of personnel and equipment, quality standards, schedules and response time, and the contractor's personnel to be assigned to managing the alliance.

Once a strategic alliance is operating, the alliance management team should oversee additional marketing to this client. Although it may do no harm to continue direct marketing to the client's personnel outside the alliance team, it might cause unnecessary confusion about the contractor's strategy when such a long-term relationship has been forged.

Alliances are discussed further in Chap. 16. In situations where alliances are not likely to be created or the client has not the desire or capability to institute such a measure, marketing to a repeat client as part of the continual marketing program is in order.

Owners associations: Marketing gold mines

Once a client marketing program has been established, prospective clients must be identified. Additionally, exposing the contractor's capabilities to a potential customer will maintain a steady source of leads that can be sdeveloped. One of the best sources of this information is associations and groups whose members may require the services a contractor has to offer.

There are hundreds of national associations that a contractor might identify as potential marketing opportunities. Each association has individual membership requirements, but most have service or supplier membership levels that allow contractors to join even though they might not be directly involved in the association's targeted industry. For example, the Natural Rural Electric Cooperative Association (NRECA) is made up of electrical utilities from all parts of the country. Contractors that provide services to this sector, such as power generation construction, are able to join and be involved with a large number of potential clients. The best means to find associations that might be appropriate for marketing is to contact the National Association of Association Executives in Washington, D.C., at (202) 626-ASAE. It produces an annual membership directory that lists associations by name and cross referenced by category (e.g., real estate-related associations, etc.).

Associations include groups of shopping center and mall developers, warehouse developers, facility and plant managers, colleges and universities, and so on. Once a group of associations is identified for membership, there are numerous effective means to promote a contractor's capabilities to the association.

Upon joining the association, the contractor should carefully review membership rules so as not to offend the association with its marketing efforts. Often the association's membership roster is not to be used for direct mail promotions; therefore the contractor should not add every member firm to its mailing list.

Membership is usually on a local or regional chapter basis, although the membership includes access to national meetings and conventions. The first step in marketing to the group is to attend the local meetings and use personal introductions to members to begin relationships. Once a member has given you a business card, it is appropriate to follow up with a letter introducing your company's capabilities and then you can add this individual contact to the firm's marketing database for continual marketing. You should then request a meeting in the contact's office to personally introduce your company.

A contractor can extend the marketing program to this association to include supporting lunches, dinners or speaking engagements. Aggressive marketers will often make presentations to the group, presenting useful information, not just an infomercial on the contractor's capability. An audience that respects the speaker's knowledge is likely to inquire further about the person and his or her employer, leading to sales opportunities. Speech topics should address the association's membership interest, which for a contractor might include presenting innovations in contracting methods used by electric utilities to reduce costs and schedules.

If you want to be asked to present to a group or participate in a discussion panel, you must make the committee that handles speaker and panel topics aware of your capabilities and your proposed topic. The audience and association's regulations should be honored, and you should never use a speaking engagement to present an extended advertisement for your company; this would only alienate the membership. Save the commercial for the introduction or printed handouts. Often an association will permit a speaker to put up a display booth during the meeting or to otherwise distribute brochures or other marketing information in return for giving the speech, thus preventing the necessity of working your capabilities into this presentation. Speaking is an excellent means of exposure, involving little, if any, direct costs for this marketing.

Local or regional chapters of associations also have numerous other marketing opportunities available, including sponsorship of golf and tennis outings, advertisements in newsletters, and general corporate underwriting for the group's goals. Each association typically has a national meeting that can provide national exposure if it fits within a contractor's marketing program.

At the national level, the same marketing opportunities are available, as is the ability to sponsor a display booth. The contacts made from members visiting a contractor's booth can be very rewarding and produce numerous marketing leads and client contacts. Attendance at these national meetings can also become a means to renew acquaintances with clients you have met at previous conventions.

Overall, joining and participating fully in various client trade associations can be a valuable resource for a contractor's marketing program and should be an integral part of any aggressive marketing plan.

Virtual marketing to clients

The virtual age of instantaneous communications has created a whole new category of available marketing tools, including the Internet. Today when over 70 percent of all companies have e-mail available for their employees, there is no doubt that an e-mail message is much more likely to be read than a piece of direct mail. With the amount of information every employee is exposed to daily, certain information is discarded immediately to leave time for what is considered more important. E-mail messages are typically regarded as important and even those sent from unknown sources are opened and read.

This does not mean that a contractor should send mass e-mail to potential clients. Continual mailings from unwanted sources and unnecessary information are likely to cause the e-mail recipient to become disenchanted with the sender. Marketing by e-mail should be selective, informative, and short. In most cases it should refer the recipient to a contractor's Web page if more information is desired.

For example, an e-mail message might be sent to selective clients announcing a new type of construction process, with the full detailed story posted on the contractor's Web site rather than attempting to e-mail an attached file. Clients do not appreciate having to wait for a file to download unless they specifically requested the information contained therein. E-mail should be kept short, with additional information available to those interested by cross referencing to another Internet address.

Voice mail is also being used as a tool for marketing. Many companies now will use computers to dial up clients' phone numbers and leave voice mail messages after work hours. When the client arrives at work the next day, the message is waiting. This type of marketing is distasteful to most people and should not be used by professional construction companies.

Employees today continually refer to the Internet for information gathering on a variety of topics, including searches for suppliers for construction services. Today almost all contractors regardless of size or type maintain an active Web site or home page for client searches. The Web page is often the first contact a client may have with a potential contractor. A client's personnel might use the Internet to search for contractors capable of completing a particular project.

This requires that all construction firms maintain an active Web page, updated regularly with project and personnel experience. The Web page becomes a virtual qualification statement and newsletter detailing the contractor's capabilities. Additional information on using cyberspace capabilities for marketing in presented in Chap. 7.

Debriefs: A path to more effective marketing and sales

Whatever the outcome on any proposal or bid to a client, the marketer should ask for a complete debrief of the competition for the project. Marketing never should end with the client because of a missed opportunity; this client will have other work opportunities, and all marketers should continue to market to all customers.

As soon as practical after being notified of not being successful for a particular competition (or even those projects on which a contractor is successful), the marketer should ask the customer for a debrief. Debriefs should cover specific reasons why the contractor was not selected and reasons for selecting the successful contractor. The amount of information the customer is willing to release varies greatly and often depends on the relationship the marketer has with this client.

The information learned in these debriefs should be used to better prepare the next proposal or bid for this or any other client. Aggressive marketers use these valuable learning tools to the maximum extent possible to improve their success.

Marketing to Architects and Engineers

Many contractors, especially those that emphasize low-bid lump-sum work, develop marketing programs that concentrate on architectural, engineering, and other design firms. If a contractor does not target clients and owners directly, design firms are the best source for becoming aware of potential projects. Even if a contractor has developed a client marketing program, it should supplement this marketing with a program targeting architectural firms to ensure adequate coverage of the entire marketplace.

The first choice after direct client marketing

Using traditional methods of construction, clients will select an architectural firm to complete the initial planning and then design their construction project. Because architectural and engineering firms generally depend upon clients directly for their workload, they have aggressive marketing programs that target clients and owners directly.

Unless a client has hired a consultant to choose a design team and contractor, the design firms will often be the first to become aware of potential projects and clients that have design and construction requirements. Aggressive contractors' marketing representatives should develop networking relationships with the major design firms in their geographical area to ensure they are made aware of these potential projects regardless of who is selected for the design and planning stages.

In many cases, owners will involve the design team in the selection of potential general contractors. This further necessitates the development of a marketing program that includes architectural firms. Most architectural and engineering firms' marketers will inform contractors' representatives of the projects they are currently designing, unless the client has specifically requested that this work be kept confidential. This information enables the contractor to begin marketing the potential customer, if it is not already doing so.

Maintaining a marketing program for design firms also ensures that the architect knows about the contractor's capabilities and can make appropriate recommendations, if asked by the client for suggestions of contractors to propose on the project or for references if the client is investigating a contractor's capabilities.

Decidedly, marketing to architectural and engineering firms is as important as marketing directly to clients themselves. Building marketing relationships with key design firms that are in niches that are similar to the contractor's can result in a cooperative marketing effort that saturates the market. This joint effort ensures that the business development department does not miss any good opportunities.

The requisite two-way street of sharing information

Any marketing program directed to architects and engineers should have two purposes, one for ensuring that the contractor has the opportunity to propose on any of the design firm's work and one that shares information about marketing leads between the firms. The latter is done by maintaining an extensive networking base, including architectural and engineering marketers, that emphasizes relationships with persons who themselves target high on the food chain of customers.

Marketing networking is based on sharing information, not on just being the recipient of information. Any marketer must realize that fellow marketers and business development personnel are working to gather as much intelligence on potential projects and customers as possible. This situation results in the more aggressive and successful marketers maintaining relationships and networks with similarly successful marketers, namely, those who provide as much information as they receive.

It does not take long for a person to decide if a network contact is able to provide information that will foster a long-term networking relationship. Today, marketers, especially those representing high food chain organizations, do not have time to simply be information providers for a contractor's marketing team. The information learned from an individual's contacts must be shared as appropriate to foster contacts that will provide even more valuable information in the future.

There are ethical considerations involved in networking successfully. One cannot take the information learned from one architectural firm and immediately pass it on to other architectural marketers. This is especially true for information that is not yet public knowledge. For example, an architectural firm's existing client may have given indications to this firm that it is about to start another project. This information is then provided to a contractor's marketer either on a confidential basis or, at minimum, the network contact understands that the information is not to be shared with others.

Obviously the contractor's marketing representative must use the gained knowledge on an exclusive basis and under no circumstances share this information with others. Becoming a successful networker depends on the ability to gain the confidence of others by ensuring that confidential information will never be shared unless specifically approved by the person providing the information.

The majority of information shared with others should be learned from potential clients, not gathered from other marketers. Marketers who depend solely on the information learned from other marketers are doomed to failure. A successful network requires each person to directly investigate clients and glean sufficient market intelligence to share with others. Networking should be a process of "covering all the bases" without having to contact every possible client.

Mature and successful marketing networks are often developed by competent marketers who realize who has the best contacts with each potential customer. A network participant then maintains his or her own selective relationships, sharing information learned with the group's other network members. In return, each receives information about other clients he or she may not be marketing directly. The network greatly increases the ability to cover far more clients than any one individual is capable of alone.

Many networks actually form into structured groups that meet at established times to share information. Often these groups consist of marketing representatives from noncompetitive organizations. For example, a network group might meet the third Tuesday of every month and include representatives from an architect, engineer, contractor, various subcontractors, and real estate, tenant, and banking firms. Each member is expected to come to the meeting prepared to share information about prospective projects and customers in return for receiving information on other possible project leads.

Networking is an art, and every successful marketing representative must perfect its capabilities to develop sufficient multiple networks to gain the maximum market knowledge in any field, including construction and engineering. For construction marketing these network contacts should be as high up the food chain as possible.

Alliances to extend a firm's capabilities

Creating a strategic alliance between two or more companies is becoming increasingly common; many of these alliances act as virtual companies. As stated in Chap. 2, a virtual company exists to perform one or more specific tasks and then disbands and reforms only as necessary to complete additional tasks. In construction, alliances between design firms and contractors are common, particularly in design-build work. The architect, engineer, and contractor form a virtual organization to complete the design and construction of a specific project; then they disband and operate individually until another design-build opportunity is available.

Alliances can also be created for marketing to specific niche markets in which the contractor and design firms both excel. The firms' marketing groups agree to formally share information about this niche and might even create joint marketing materials to reduce costs. Alliances can greatly increase effectiveness of any individual marketing program and save costs at the same time.

Alliances should be formal agreements, although not necessarily legal in nature. Specific goals and targets, the responsibilities of each firm's marketing group, and other pertinent terms should be clearly identified in a document. This permits each company's management to monitor the progress and success of the venture.

For a detailed review of alliances and their vital role in the virtual age, refer to Chap. 16.

Budgeting and constructibility reviews for design firms

Offering potential clients assistance in the planning and preconstruction stages of a project was discussed in this chapter and in Chaps. 6 and 12. Similar marketing techniques can be applied when marketing to design firms. Offering architectural and engineering firms preconstruction services that include budgeting and constructibility review can be an effective means to build relationships.

Although most architectural and engineering firms can perform budgeting, contractors can often provide a more detailed response, especially when constructibility issues are involved. In most cases the assistance the contractor can provide in these early project stages to improve the project's construction cost and schedule are more effective than assistance provided in the construction phase. Aggressive design firms recognize the importance of this input and often solicit the assistance from recognized contractors.

As discussed in Chaps. 6, 11, and 12 and earlier in this chapter, these valuable services should never be given away, nor should any other professional organization expect to receive these services without obligation. A contractor must establish policies relating to offering these services to architects and engineers. In alliance situations, and especially for design-build work, the contractor may be responsible for providing these services during the design phase.

The marketing department should become thoroughly familiar with its firm's policy on preconstruction services and be able to address the issue if it arises when it is marketing to architects or engineers. Often contractors will offer these services on a professional basis for a design firm to include with its proposal to a client. However, the contractor's marketing team should fully review the possibility of being excluded from bidding or proposing on a project if it becomes involved in the design or preconstruction phase. Some clients, particularly the federal government, will preclude contractors from bidding a project if they are actually involved in the design phase.

Offering preconstruction services can be an effective part of any contractor's marketing program, but a corporate policy must be established to govern these services. Included in this policy should be professional fees, resolution of conflicts of interest, and a means to determine if participating in the design phase will preclude the submittal of a proposal for the construction phase.

Joint ventures and joint marketing for niche markets

When a contractor and architect or engineer have completed a significant number of successful projects together, management might determine that securing additional work of a similar nature might be better accomplished by working together. In this situation the organizations may decide to form a joint venture for specific types of work or niche markets.

Although these ventures are usually for design-build projects (marketing design-build work, including joint ventures, is presented in Chap. 14), often the firms will have unique capabilities in a particular type of design and construction that leads to a joint marketing effort regardless of the expected contract format. For example, a contractor and engineer that both have extensive experience in refrigerated warehouses would prepare a joint marketing program. The program would solicit business from clients and the partners would be able to handle any type of contract the owner chooses. While joint marketing efforts have been historically more prevalent between architects and engineers that complete similar services, teaming between architects or engineers and contractors is becoming much more common.

The team agrees before marketing commences that its goal is to service the client in whatever capacity required, even if it should mean that only the architect or contractor would actually be given the work. The goal of joint marketing efforts is to saturate a particular niche market together and sell all work available, regardless of which team member ultimately secures the work. Obviously if such a joint marketing plan over a period of time clearly favors only one of the team members, an analysis should be completed to determine if the marketing plan requires revisions or if the parties are better served by marketing separately.

Joint marketing efforts should be a formal program that complements each organization's own plan. Conflicts of interest should be fully disclosed before commencing the program, including any customers that have negative opinions of their firms. Joint marketing efforts should never be used to disguise an attempt to build a failed relationship with a particular client. In addition, each team member should recognize that certain clients will have already forged long-term relationships with a competitor that the joint marketing cannot overrule. Each team member should also provide information on any other alliances or teaming agreements it is involved in to prevent any internal conflicts of interests. Certainly ethical and professional standards must be maintained at all times, and no team member should involve a potential client that it is pursuing already through another alliance.

Joint ventures and the advantages in marketing using this method are covered fully in Chap. 16. The methods presented in Chap. 12 can be applied to many types of ventures, including those between a contractor and design firm. Joint marketing efforts can be a valuable tool, and any contractor's marketing efforts should analyze the viability of implementing such programs.

Virtual geographical expansion for expanded sales

In the virtual world, where anybody can work from anywhere at anytime, the ability to market to expanded geographical areas today is almost limitless. Numerous design firms are completing work well beyond their traditional geographical market, including overseas projects. Virtual capabilities eliminate the requirement for a design firm to have a local physical office to complete work in a specific geographic area, thus eliminating any geographical barriers to performing or selling work.

Design firms with these virtual capabilities can create new opportunities and geographical markets for contractors as well. Often the design firm participates in the selection of, or recommends a contractor for, the construction phase. Design firms completing virtual designs in a new area may not be aware of local contractors' experience and therefore choose to recommend contractors they have worked with previously in other areas.

This situation makes it imperative that a contractor's marketing department recognize that it must target architectural and engineering firms outside its immediate geographical area. Clients today realize they can select the best design firm regardless of its physical office location, and it is becoming increasingly more common for foreign firms to be selected to complete domestic work. A contractor must investigate all design firms, regardless of location, that are likely to complete work in its own niche market. Once a list of appropriate design organizations has been identified, these firms should be included as marketing targets and appropriate marketing should be implemented.

Likewise, local architectural and engineering firms that are targeting projects beyond the local marketplace should be made aware of a contractor's capabilities to follow these organizations into new geographical areas. The virtual world requires that contractors' marketing extend well beyond their local geographical area even if they choose to complete work only in the local area.

Targeting work opportunities, even in a local area, will become increasingly difficult in the immediate future as world boundaries become obsolete. Marketing programs must address these issues to ensure that project opportunities are not lost because of insufficient marketing to organizations outside a specific geographical area.

A required membership

There are numerous organizations that can benefit a contractor's marketing program as well as a marketer's networking ability. The Society of Marketing Professional Services (SMPS) is a group that contractors targeting architects and engineers should join and support. SMPS is a professional group representing marketers of professional services within the design and building industry. The majority of the members represent architectural or engineering firms, but membership is also offered to contractors, suppliers, and subcontractors.

SMPS has a national chapter and numerous local chapters throughout the country. Meetings usually involve an informational lecture followed by an opportunity to network with fellow industry marketers. A national meeting each year includes numerous lectures on improving marketing skills at beginner and advanced levels.

SMPS provides significant opportunities for construction marketers to meet fellow marketing professionals and expand their base of networking contacts. This organization should be included in every contractor's program that targets architects or engineers, including design-build teaming.

Marketing to Developers, Construction Managers, and Owners' Representatives

After establishing programs that target owners, architects, and engineers, contractors will often find that their programs require expansion to incorporate various owner consultant organizations that specialize in representing owners during the building process. As corporations concentrate more on their income-producing business sectors, they increasingly outsource numerous responsibilities, including management of the entire construction process and design. This mandates that aggressive marketing programs include consultants that have oversight control over a large portion of the building industry.

Although the emphasis for marketing should be on direct contact with owners and clients, anytime a client tells you that a consultant manages its construction program, your marketing must shift directions to include the marketing to this consulting firm. Owner representative firms manage the construction programs of all types and sizes of companies and corporations.

Large international firms often outsource their construction programs as a means of concentrating on their core business. Small firms often outsource due to their limited internal resources. Whatever the situation, a contractor's marketing department must be able to immediately identify the client's structure for construction management and incorporate the necessary activities to market appropriately for this work.

Many types of organizations provide owner representative services, including architects, engineers, construction management firms, and commercial real estate companies. Today an increasing number of firms also specialize exclusively in owner representation.

Many architectural firms aggressively market their construction management services to a client immediately after being awarded a design contract. Construction management by an architect might involve construction phase responsibilities including selection of the contractor and continuing oversight of the field construction activities. These services generally go beyond the design adherence management that an architectural firm will typically include with their design contractual responsibilities. Most design contracts, including industry standard formats such as AIA documents, contain language making the architect responsible for numerous construction phase oversight measures, including aesthetics and conformance to the plans and specifications.

Often architects use marketing plans and a construction management organization to move beyond this basic concept and incorporate active construction phase management as part of their industry service capabilities. A contractor marketing to an architectural firm for work opportunities must recognize that the two firms may become competitors for the same project at times. This should not change the contractor's marketing program, even for architectural organizations that aggressively market full-service construction management capabilities. A contractor's marketing representative must recognize the possibility of competing with architects for construction management services, and proprietary information should be kept confidential in situations where the firms might be competitors. At the same time, the marketer should realize that more often than not, the architect will not become involved as a CM and, therefore, will provide work opportunities for the contractor.

Contractors will also become involved indirectly or directly with real estate developers. Owners may choose a real estate development company to manage their physical building requirements, including complete development of a property and building program. In this situation, the contractor would be working directly for the developer under a possible build-to-suit lease program the client has initiated. In other business relationships, the customer might choose a developer's property as a location for its building needs and the developer might become involved in managing the construction process, though indirectly, for the client.

The contractor must recognize the importance of marketing to real estate developers for project opportunities but also recognize that this developer might compete with the contractor occasionally. In addition, developers are likely to have instituted an alliance with other construction companies for their building needs. The marketer must be aware of the sensitivity of certain information, using only applicable information to network with the developer.

It is entirely possible that contractor's marketer might become involved in a situation that requires involvement with a developer that already has an alliance structured with a competitor contractor. For example, a client with a building project asks for assistance to find an appropriate site or a build-to-suit opportunity involving land the developer controls. In this situation there is no reason not to expect the developer to enter into an exclusivity agreement with the contractor despite having the competitor alliance.

A written confidentiality agreement should be signed between the developer and contractor as a safeguard that the information exchanged will not be discussed externally. Then the team can proceed to submit an appropriate joint proposal to the client. Alliances are always structured to permit team members from forming other teaming agreements, if appropriate, to protect business opportunities when they arise. Marketers should understand that alliances have become increasingly common today, and they must educate themselves about competitor alliances in their local market before inappropriately divulging proprietary information to a possible team member. Additional information concerning alliances is presented in Chap. 15. CMs often complicate marketing programs due to the number of firms that offer construction management services, including general contractors, architects, engineers, real estate firms, owner representatives, and construction management firms. Not only do a wide range of organizations offer construction management services, the type and level of services they offer also varies widely. For example, a real estate firm might offer interior fit-out CM services only, an architect might offer general oversight during the construction phase, and construction management firms might handle complete project management and eliminate the need for a general contractor.

Clients might select a CM firm to manage multiple contracts by subcontracting the entire project either at risk or for fee only. Other clients hire CMs for one or multiple projects to provide oversight of the general contractor. Although the first situation eliminates any work opportunities for a general contractor, the second type of contract, oversight, presents it with sales opportunities.

Because construction management firms typically provide both types of services, a general contractor must ensure that construction opportunities are not missed by not marketing construction managers. However, it may be best to consider a CM firm a competitor and be very selective about the information provided to and shared with the firm.

Real estate firms also offer owner representation services, with some firms offering these services on a national basis. Real estate firms capitalized on this niche market by becoming involved early in the process of project planning when being hired by a client to complete space or land searches for planned projects. This early involvement enables aggressive firms to extend their scope of services by offering clients management oversight of the design and construction phases after assisting in selecting a site for the project.

Construction marketers must be aware of local real estate organizations that offer client representation services that could provide sales leads for the contractor. Numerous real estate firms offer interior fit-out representation, with some even providing contracting services for this type work. This is another situation where a firm that might provide leads may also be a competitor and must be treated with caution. Situations such as this create difficulties for marketers but also present possible project opportunities. The marketer must gather sufficient data about the real estate firm and make a decision on the extent of marketing and networking that is appropriate on an individual basis.

Real estate firms, CMs, and other professionals have lead to the creation of a specialty niche—companies that exclusively offer owner representation for all physical office and building requirements. Owner representation firms offer services that can include real estate services, architect and contractor selection and oversight, financing, and permitting assistance. They typically do not offer direct construction management or contracting services and, therefore, rarely compete with general contractors.

Consulting firms are excellent sources for work leads and networking opportunities. Although the marketer must collect the background information necessary to begin a marketing program for the typical owner representative firm, these firms offer professional services and will rarely form alliances with contractors that could create conflict of interest situations when representing a client.

The organizations that provide project services to owners are as varied as the type of services each offers. These firms can present multiple sales opportunities for a contractor, but the marketer must proceed with caution and structure marketing programs on an individual basis.

The increasing use of consultants and how they affect marketing

It is common today to find clients involving consultants in their design and building programs for a multitude of reasons, and each client structures the use of outside support differently. Marketing programs must be flexible enough to support the various business relationships that are created between a client and their consultants. The numerous project organizational structures that are created by clients can create difficulties in determining the ultimate decision maker for construction services. Also, alliances that might exist with each of the client's project management team members make sorting out the appropriate contacts difficult. If a marketer does not take adequate precautions by seeking sufficient information about the team and its existing alliances, he or she may divulge information to competitors.

The first step, as with marketing to clients directly, is to identify the organization and the person within that company qualified to make the buy decisions regarding the contractor's services. Even though an owner has selected a consultant to provide support services for a construction program, the client might retain the right to select the contractor, jointly select the contractor, or transfer this responsibility completely to the consultant. To determine this pertinent information the marketer should start with the client directly and ask detailed questions about the management of its building program, authority figures, and expected services of the consultant.

The client might suggest that the marketer contact the consultant directly, making further direct contact with the potential customer unnecessary. If so requested, the marketer should contact the consultant, but the marketing program should continue to target the client directly unless the marketer is specifically told that further direct contact is not acceptable.

Once a consultant is identified as playing a key role in the selection of the contractor, the marketer must determine if the consultant has a preestablished contractor alliance or a select list of bidders that is automatically recommended to its clients. If it appears that the consultant might prevent the contractor from submitting qualifications for the project, reevaluating a direct client contact again might improve the contractor's chances of being selected to participate but could also cause problems with the consultant. The marketer might find that the only possible remedy in the present situation is to market the consultant for the opportunity to become approved for its select list of bidders on future projects.

Such situations can be expected to occur occasionally and to complicate marketing programs. However, aggressive marketers will soon become aware of the business relationships created by various consultants and begin to market these consultants directly to ensure they have the opportunity to propose on all the consultant's projects.

In fact, by successfully marketing consultants, marketers will actually increase the number of leads they get by the consultant recommending the contractor directly or through a select list of bidders. So although consultants may complicate a marketing program, when properly targeted, they can substantially increase the number of project leads for a contractor.

Qualifications are imperative

A consultant's internal staffing organization can present complications for any marketing program. Often a consultant will maintain multiple discipline management capabilities that a contractor will have to market directly. Consultants are typically more technically competent in construction engineering than a client's internal staff; therefore, the marketing to a consultant is often more technical than to a client.

Consultants managing a building program are responsible for providing the technical support that a client is not able to complete with its own internal resources. As such, consultants move beyond the personal and networking relationships that marketers use to provide sales leads. Consultants, in the client's best interest, are responsible for ensuring the project is completed with the most technically capable resources at the best value available.

Professional consultants concentrate on a contractor's technical capabilities and will typically qualify contractors for each project they manage, regardless of any preexisting relationships that might have already been established with the client. The marketing program must sufficiently introduce the contractor's capabilities to the entire technical and project management staff of the consultant. This ensures that the contractor will be given the opportunity to submit qualifications for projects managed by the consultant that are deemed within the contractor's capability.

Proper completion of the qualification is imperative to become a finalist for a prospective project. Qualifications statements are discussed in Chap. 12 and earlier in this chapter. In addition, most consultants will extend the qualification process beyond the experience and resource information presented in the qualification package. These additional requirements can include oral presentations by the proposed project management team and preliminary budgeting and schedules that are compared to the consultant's to determine if the contractor is proposing a reasonable plan of project management.

Marketing to consultant representatives poses more technical-intensive requirements than other marketing to other potential customers. This situation requires appropriate advanced planning by the marketing department to enlist management's support to involve other internal departments, including estimating and construction supervision, to meet the prequalification process requirements.

Recognizing the owner's reasons for involving consultants

Although any marketer must determine the appropriate decision maker when consultants are involved, it is also important to identify the reasons why the client has chosen to involve a consultant in the construction process. Marketers that do not investigate these circumstances may be missing an opportunity to provide additional services to the customer and create a long-term alliance.

Clients that outsource building program management often do so because of a lack of internal resources and a limited knowledge of the local construction market. In either situation a contractor should be able to present a package of services directly to the customer and become involved in the early planning stages of a project rather than merely waiting for the later construction phase opportunities.

A marketing program should include steps to tell the customer how the contractor can supplement the customer's internal staffing in the early planning processes, even if a consultant is managing the present process. Obviously, if the client has a mandated corporate policy to outsource all construction management to independent consultants, it would not be appropriate for the contractor to interfere with this policy. However, many corporations hire consultants because contractors have not informed them that they can provide early planning services, including preconstruction tasks, to eliminate the necessity for a third-party consultant to be hired.

As stated previously, marketers must aggressively gather sufficient knowledge about all potential clients' needs and operating requirements to be sure the contractor's marketing and sales plan includes a complete program of services to those clients requiring their services. Too often, construction marketers overlook opportunities outside the traditional marketing of construction phase services and fail to include marketing professional support services to increase their company's overall profitability and success.

Maintaining a successful contact database

The numerous contacts a marketer becomes involved with while marketing and networking require the establishment of an electronic database to manage the program effectively. Several commercial software programs are available that structure databases to retrieve required information on customers, including their use of consultants on past projects. This effective sales tool allows marketers to offer preconstruction services when they learn that a particular client is planning a project rather than reacting after a consultant becomes involved.

The database should be cross-referenced to include historical information about consulting firms and their customers. Eventually the database should provide a detailed map of the local marketplace that identifies organizations that provide consulting services, the level of these services, and their past and current customers. This information can then be used to select the best course of action in marketing to potential customers directly and to prevent wasted time on situations that will not provide potential sales opportunities for the contractor. For example, the database can be used to pinpoint existing alliances between a client and the real estate firm that performs construction management services.

In the virtual world there is no time to spend on clients and consultants that do not offer a reasonable expectation of work opportunities. A well-structured and regularly updated database can substantially increase a marketer's overall effectiveness. Software used to assist the marketer is presented in detail in Chap. 7.

Marketing the right contact

A database can also be used to instantaneously identify the right contact person and/or firm when a new project opportunity lead arises. Any marketing database should contain information on all past projects the contractor attempted to sell as well as those it decided not to market. It should include information on the professional relationships between the client and selected consultants and the appropriate decision maker involved in these situations.

This readily retrievable information can immediately identify the contact(s) that must be marketed directly and eliminates wasted time trying to determine the relationships and proper contacts to sell the contractor's services. Lost sales opportunities can be prevented by immediately locating the proper contact before it becomes too late to make a sales presentation for the project.

Inappropriate marketing alliances or teaming

In many situations it will be inappropriate to attempt to establish an alliance or teaming relationship because of the professional services offered by a consultant. Marketers must recognize that many consultants provide services with the understanding that selection of the design and construction team members is done on a project-by-project basis based on qualifications.

Although nothing prevents a contractor's marketer from networking with a consultant, it should not be expected that the consultant would automatically recommend the contractor to every client it is under contract to.

Establishing a networking relationship to develop leads on new projects and ensuring the consultant is well aware of a contractor's capabilities should be the key components of marketing programs directed to professional owner representatives.

Summary

Marketing programs directly targeting private owners, architects, and consultants can produce exceptional results if properly implemented. The key component of marketing to any of these potential clients is to be respectful of their time and directions.

Employees today have no time to waste listening to or reading irrelevant information. After an initial short meeting of introduction, a construction marketer must ensure that all follow-on meetings are beneficial learning experiences for the client. Marketing that only emphasizes how successful a construction company was in completing past projects is for the most part useless information to a client.

Customers require information that can improve their individual performance or their overall organization's success. This typically requires that the marketer be prepared to discuss the potential customer's business and how the contractor can contribute to the success of that business.

At the same time, it is important to keep a company's name in front of the customer at all times, including those times the customer may not have any active projects. Although marketers continue sending literature such as press releases, announcements of project starts or completions, and similar notations of a contractor's achievements, in the electronic age this information is becoming less valuable and most often completely disregarded before it is even read. No corporate executive is interested in reading a contractor's newsletter highlighting a contractor's superintendent's receiving a safety award on somebody else's project. Customers only have time to deal with their work requirements and problems and the people and organizations willing to help them with these situations.

Today, marketers must completely customize the information sent to each current and prospective client, sending only meaningful information that can be helpful to them on their work activities. This can include articles on how their competitors and industry members are satisfying their physical facility requirements. It does not have to be necessarily business related but can highlight personal interest topics such as sending a fall planting guide to a garden enthusiast.

For actual meeting requirements, marketers should be mindful of each particular contact's preference for meeting times. Some clients prefer to meet away from the office during breakfast or lunch so as not to take time away from their normal work requirements. Others might prefer short office meetings, keeping their meal times to themselves. In any event, the marketers must learn the customer's preferences, customize their marketing to these preferences, and be ever mindful of a customer's time when doing so.

In all situations, marketing must be directed as high in the customer's organizational structure as possible and, as importantly, must be to the decision maker who has the authority to make selection decisions. In larger corporations be mindful that this will typically require marketing to several individuals who will have decision authority for selected areas of responsibility.

Marketers must also be aware that customer contacts change jobs frequently. This requires regularly updating client contacts and also learning where former contacts within a company have moved, situations that often create additional new clients for the marketer.

Finally, every marketer must be prepared before meeting with a client, knowing not only the contact person individually, but the contact's industry, goals, and requirements. Potential customers will appreciate any marketer who comes prepared to discuss how he or she can assist them without the client first having to explain the details of the client's business to the marketer. There is no excuse for any marketer not to spend the minimal time necessary to take advantage of information available on the Internet about a potential customer and its industry before first meeting with this client. Marketers who differentiate themselves from their competitors by taking advantage of cyberspace age capabilities will clearly outdistance their peers.

Chapter

14

Design-Build:
A Whole New Era

Introduction

Some 4500 years ago, incredible monumental structures were built by the design and construction industry that still stand today as testimony to the capabilities of ancient professional builders. The industry professional, then known as a master builder, is today referred to as a design-builder. Master builders completed the design and all construction activities because subcontractors did not yet exist. Through the ages as buildings became more complex, subcontractors began completing specialized portions of field construction activities. Even so, general contractors continued to complete the majority of construction by self-performing the work with their own field crews. Today, many general contractors subcontract much of their work.

A typical organizational chart for managing a construction project ages ago looked like Fig. 14.1. Today, due to the increasing influence of legal issues and extensive subcontracting, the organization has evolved to the typical chart shown in Fig. 14.2. The layers of organizational authority used in these project management techniques are now obsolete. It is becoming increasingly more evident to both the construction industry and its customers that the project organization used originally in the industry (Fig. 14.1) is even more applicable today to make more effective use of virtual management capabilities.

Design-Build versus Engineer-Procure-Construct

Marketers should know the slight difference between design-build and engineer-procure-construct (EPC). Design-build is most often referred to in commercial construction, whereas in process (or manufacturing) and energy construction the term *EPC* is commonly used. The difference is only a recognition of the fact

333

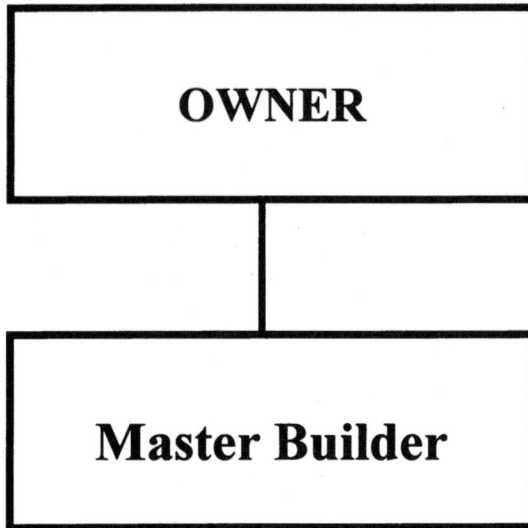

Figure 14.1   The original concept of design-build management.

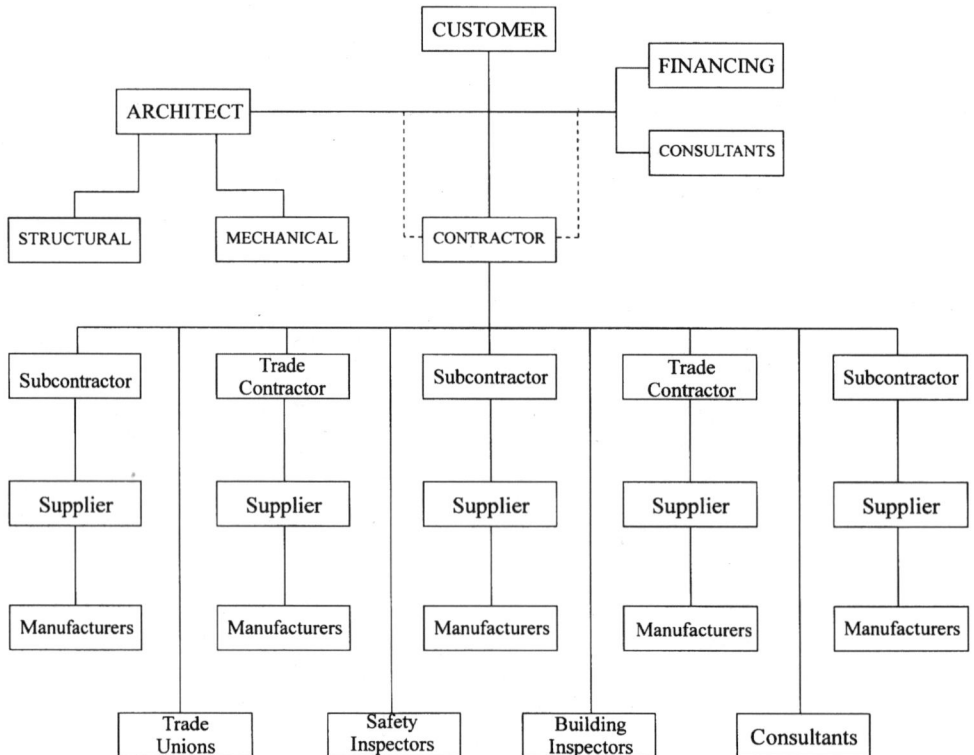

Figure 14.2   Typical project management organizational chart used today.

that in process and energy projects, the design-build contractor is usually also responsible for the procurement of equipment used in the manufacturing and energy-producing processes by the customer after completion of the project.

In commercial construction, there is usually no equipment installation involved that the customer uses for its own profit-making services. In manufacturing and energy projects, the contractor, however, often installs these profit-making processes, thus the reference to EPC rather than design-build.

For purposes of discussion throughout this book, design-build and EPC can be used interchangeably, but only design-build is referred to.

This return to our formative roots after thousands of years of industry experience indicates that design-build is once again becoming one of the dominant contracting methods. Marketers of design-build capabilities must be aware of the four main reasons causing this change in preferred contracting methods:

1. Legal and contractual simplification

2. Time compression to meet virtual standards

3. Resolution of CAD file ownership

4. Necessary early contractor and subcontractor involvement in the planning, design, and construction phases

Legal simplification

In the traditional design-bid-build and other related contractual methods, the owner contracts for the design completely separate from the construction. Once the architect or engineering firm completes the design, the documents are then presented to general contractors or construction managers for bidding purposes. The owner, by contracting in this manner, must then warrant directly to the contractors that the design documents are accurate and complete.

Industry customers are becoming aware that this contractual method requires them to accept professional risk that should remain the responsibility of industry professionals. Today more and more clients are using the design-build method of contracting to reduce this assumed risk and the resulting disputes and legal claims.

Using the design-build contracting format, the contractor and designer assume the responsibility for document accuracy and completeness. The owner transfers design and construction risk to the industry professionals. Figure 14.3 shows a simplified document warranty flow for both the traditional design-bid-build and design-build contracting methods.

The design-build method requires the industry professional to maintain the responsibility of document warranty and eliminates "finger-pointing" during the construction phase. Design-build results in a simple and effective single-point of responsibility.

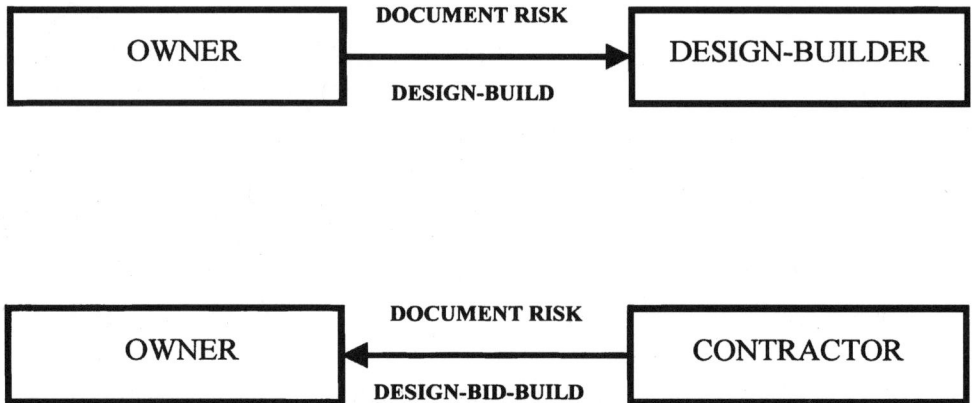

Figure 14.3 Warranty flow comparison between design-build and design-bid-build.

Time compression

Industries today measure time not by hours or days but by dollars. Customers are requiring the construction industry to compress building schedules to meet their demands to remain competitive in their respective industry. For example, a microchip manufacturer is expecting to produce in its new facility enough chips to return a profit of $500,000 per day. Then just one month's reduction in a building schedule can represents a profit of over $10 million for a client. If this one-month's income represents one-tenth of the overall construction costs, the client will obviously place great emphasis on the schedule and expect the construction industry to respond accordingly. Considering also that the client's product might become obsolete prior to the construction completion adds more emphasis to the virtual age paradigm of time being measured in monetary units.

Typically the industry's response to such a situation would be to offer fast-track construction methods. Fast-track construction is generally defined as starting the field phase of construction before the plans and specifications are completed. The stage of document completion at which construction actually commences can vary greatly. Using fast-track construction usually requires the client to assume a great deal of risk because the contractor is beginning work without a complete set of documents.

Although the general fast-track concept is good, it is understandable that industry clients have not used this method more frequently. However, the design-build contracting method can incorporate a fast-track schedule, with the designer and contractor assuming the risks for early starts without completed documents. This method requires a solid working relationship between the designer and contractor (unless the contractor is capable of providing the design in-house). Even with in-house capabilities the contractor takes risks by assuming its design and construction units can create a sufficient level of documentation prior to the start of construction to prevent unnecessary rework in the field.

Clearly, the design-build contracting method is an option available to compress overall design and construction schedules to accommodate clients as they compete in their own industries. Placing the responsibility for the design and construction with one firm enables the industry to incorporate time-saving techniques, including fast-track construction, to respond to current time standards.

CAD file ownership

The design-build contracting method also answers the pressing industry legal question of today: Who owns, is responsible for, and controls the files of completed computer-aided drawings (CAD) for a specific project? Under traditional methods of construction the client receives a set of documents, which is used to obtain contractor's proposals and build the project. Even though the owner becomes responsible for the documents as described previously, the project drawing files, usually provided today in electronic format (CAD), are not being used on the available electronic database linkage between team members.

The computerization of the construction process demands that CAD files become the nucleus of project communications and data files. This includes the linking of estimates and schedules with CAD files to improve the overall quality and timeliness of project completion.

Design-build ensures the client that the CAD drawings and files will be available for use throughout the entire design and construction processes, something that has yet to become standard with design-bid-build contracting methods. Usually, even though the client has paid a professional fee for the design and document completion, the documents cannot be added to or changed during the construction phase without the architect's input. Design-build contracting eliminates this ownership issue, creating a contracting method that can utilize network linkage via CAD files as a common denominator to improve the overall construction processes.

Early involvement by contractors

To further implement practices that improve the overall quality of the construction process and compress project schedules, general contractor and subcontractor involvement early in the project planning stage is imperative. Contractors' input in the planning and design stage can contribute to improvements in the later construction phases to a higher degree than if the contractor is not involved until after the design is completed.

Using the design-build format, the contractor becomes involved immediately in the planning and design and can provide suggestions to improve the overall constructibility and quality of the completed project. This early involvement also increases the opportunity to improve the scheduling process, including, if appropriate, the implementation of fast-track starts, as discussed previously. The industry will likely continue to move toward further implementation of design-build contracting methods, with contractors reviewing their strategic plans to determine if design-build can become a financial growth opportunity.

Design-Build Capability Strategies

For any contractor deciding to begin the marketing of design-build services, a series of strategic decisions must be made to structure and implement an effective marketing program on an individual basis. The numerous differences between design-bid-build and design-build require a more precise and selective marketing program.

Once a decision is made to proceed and sell design-build services, the first and foremost step is determining how to provide design capabilities to clients. Most general contractors and construction managers do not have "in-house" architecture and engineering capabilities. This requires the contractor to purchase a firm with these capabilities, hire a complete staff of architects and engineers, or form one or more alliances with architectural and engineering firms to jointly market design-build. The latter is typically the most common and realistic option to provide design-build services for most general contractors.

The first two choices, buying a firm or hiring a staff, typically become feasible only after a contractor has a sufficient backlog of design-build work to make the investment required in these options cost effective. Finding an existing design firm that specializes in the range of building types a contractor markets is often not feasible. Most in-house staff would not be able to design a wide range of project types or have sufficient knowledge of all the geographical areas in which a contractor might market. So for most contractors the alliance method of marketing design-build services is the most practicable.

There are recognizable differences between in-house capabilities and outsourcing designs, and marketers must prepare themselves to emphasize the strengths of the method chosen to complete design-build projects. Both methods have specific advantages and disadvantages that must be addressed in marketing programs.

Outsourcing design work

For most contractors, outsourcing architectural and engineering work is the only option available to complete design-build work. Without an internal capability to perform design work, contractors can execute the design process either through an alliance or by simply subcontracting the design to an architectural firm.

Alliances

For most contractors, teaming to gain access to design resources is the most effective way to maintain design-build capabilities. Even if the contractor limits design teaming to subcontracting, marketers will find that architectural and engineering firms desire to market their capabilities jointly through some informal teaming arrangement.

The first option is to form an alliance with a selected architecture or engineering firm to market design-build projects. Alliances are usually structured to require that the team will market work exclusively in a particular geographical area, for a specific niche market, or a combination of both. For example, a contractor can form an alliance with a design firm to compete for all types of commercial building design-build projects in a specified geographic area.

Often a contractor will team with several design firms based on expertise and geographical conditions, ensuring that none of the teaming relationships presents a conflict of interest with other teams. In the same manner, architects and engineers can be expected to maintain associations with multiple contractors based on similar criteria. In choosing appropriate team members for a design-build project, certain items should be carefully evaluated.

Typically alliances between architects and contractors are created based on past working relationships and are usually promoted by the respective firms' marketing departments. By specializing in a market segment, contractors and design firms will have often worked together and have developed a relationship that includes their respective marketing departments networking together and sharing information. This relationship might lead to the creation of a marketing alliance to target specific market niches and include the marketing of design-build work to clients desiring this type of contracting method.

Alliances are not usually created without a past working relationship because it is difficult to convince a client of the alliance's capability without past experience. Those teams with an extensive history of completing projects together, even though they may have not been design-build, are in a better position to explain the creation of the alliance to prospective customers. This explanation includes the development of marketing tools that describe in detail projects previously completed by the alliance.

Obviously it is difficult, but not impossible, to explain to a prospective design-build customer the reason for contracting with a team that has not previously worked together. There can be specific reasons for an architect to team with a contractor without having a previous work history. For example, a unique or unusual project may be proposed by a client that brings together these two firms because of their past experience of having completed similar work, although not together. This might be a project that involves highly technical installations such as medical or computer chip processes, a patented construction process, or selecting an architect with a design expertise not previously used in the contractor's geographical area. These and other similar situations create conditions where a client can recognize that a previous working relationship is not as important as each firm's expertise in the type of work required.

Subcontracting

The second option is for the contractor to subcontract the design package to an architectural or engineering firm in much the same manner as the actual field construction is subcontracted. In this method the contractor markets design-build work but will typically make a selection of the design firm before the actual proposal is submitted to the client. This enables the contractor to be free to choose what it feels is the most appropriate design firm for a particular project without having to form an alliance or jointly market with only one architect.

This is often the choice of large contractors that construct a wide variety of work types. They prefer to market their services exclusively and only choose a design team when necessary to finalize a proposal for a specific client. Of course an architectural firm can market design-build work in the same manner and choose a general contractor only for a specific proposal.

This subcontracting of design is probably the method most frequently used by most construction and design firms. Subcontracting design allows the contractor to market independently, using its preferred methods without having to commit to a specific architect. It permits the contractor to identify and choose the most appropriate design firm for a specific project after the sales opportunity has been identified. This can prevent the contractor from having to submit a proposal with a prechosen architect that might not have the expertise required for a particular construction project. Certainly subcontracting design work has greater flexibility than alliances.

Subcontracting is prevalent in federal design-build contracting. The federal government proposal process allows a contractor to decide on the most appropriate teaming arrangement after a specific project proposal is released. However, most contractors establish teaming relationships before a project is officially released for proposals. This teaming arrangement is based on intelligence gathered during marketing before a proposal is released, as discussed in Chap. 15. From the information gathered directly from the potential customer, the contractor can decide on the most advantageous teaming relationship and reach agreement with a design team before the proposal process begins.

Although the team can be created after the request for proposal is officially released, the contractor may find that the most experienced design teams for the required work type have already been committed to other contractors. Although subcontracting is perfectly acceptable for federal design-build work, the contractor's marketing staff should identify potential projects before they are advertised and establish required teaming relationships as early as possible to ensure the most competitive proposal on a technical basis.

Internal design resources

Contractors with internal design capabilities typically aggressively market design-build work unless the organization is structured so that the design and construction units are independent operating units that market their work separately. Even then it is likely that design-build work would be completed with internal resources rather than outsourcing either the design or construction.

Unless the firm is a multinational large corporation, the internal design and construction capability is for production of specific niche markets rather than an attempt to maintain the capability for a wide range of design-build capabilities. Even firms that maintain internal design resources will typically not maintain a full complement of resources that would include architects, structural engineers, mechanical and electrical engineers, landscaping designers, interior designers, and so forth. Contractors that claim to be true design-build firms often have to outsource major portions of the project's design to other firms or team as would any contractor to provide the design expertise required.

Few, if any, contractors that have internal design capabilities or architectural and engineering firms that have construction capabilities will complete design-build work exclusively. Organizations with both resources will market their design and construction resources separately as well as jointly. Finding a true design-build firm, or master builder, is uncommon, and few firms can truly market as being solely master builders. The economic reality of the industry requires that firms be able to respond to all types of service requirements from individual clients.

Although the industry has seen a sharp rise in the use of design-build delivery methods in response to increasing time pressures and legal issues, it still has other dominant contracting methods, including construction management and design-bid-build. Most firms understand that to remain competitive and grow they must have the capabilities to complete a range of service requests rather than operate exclusively using a single contracting method.

So although firms may aggressively market their capabilities as having internal design resources, few can provide a complete design package and even fewer operate exclusively as master builders. This situation must be considered when addressing the marketing of design-build capabilities and whether or not the contractor has internal design resources.

Marketing Using the Various Teaming Arrangements

Marketing design-build using the outsourcing method should emphasize the contractor's ability to select the most proficient design team as well as the appropriate construction personnel for the specific project requirements. Rather than relying only on internal resources that might not have the expertise or experience required, the contractor must select the most capable architectural, engineering, and construction team available. This means that even firms with internal design capabilities must solicit outside resources, if necessary, to ensure that the client receives the best overall value.

The absence of internal design resources should never be viewed as a disadvantage. A client may choose a design-build firm that has internal design resources, assuming that dealing with one firm is a benefit over a firm that outsources its design. However, contractual arrangements can be developed to ensure that the client single-points the responsibility for both the construction and design, regardless of how many firms are involved. The client can effectively depend solely on the contractor to complete the work regardless of any internal teaming arrangements.

In many situations, by eliminating internal design resources, the contractor can claim to lower costs for the client by not having the higher overhead necessary to maintain a full design staff. The contractor should be able to convince the client of its ability to bring in the necessary design resources in a "just-in-time" manner through subcontracting or alliances as required.

With respect to both internal capabilities and formal alliances, the marketer must emphasize the resources on hand but still be prepared to respond to requests that exceed the in-house capabilities. Marketing materials prepared for internal design work and alliance ventures will concentrate on the capabilities and experience of the present resources. At the same time, if the contractor chooses to solicit work that exceeds the capabilities of the internal design resources, the marketing material should reflect its ability to do so. This might include describing work history of design-build projects using resources outside the internal or alliance resources.

Alliances are often formed for specific niche markets and marketing materials are carefully prepared to limit the capability descriptions to work specifically targeted by the alliance agreement. This situation requires that the alliance partners have additional marketing tools, exclusive of the alliance materials, if work is targeted separately that falls outside the alliance agreement. Many contractors participate in multiple alliances and market design-build work independently, requiring several separate marketing programs and related materials to cover the extent of work types and geographical areas the contractor intends to strategically target.

If a contractor has prepared multiple marketing tools, the marketing staff must exercise diligence in marketing and not interfere with the marketing strategies of the various alliances formed. This includes producing marketing materials that do not overlap the intentions of individual alliances. Multiple-level marketing is best structured so that the marketing representative makes an initial sales call with generic sales tools, and specific niche or alliance materials are not introduced to the customer until the customer's specific requirements are determined.

A contractor's marketing staff may find the best arrangement is to produce a generic marketing campaign that introduces the general scope of services offered by the contractor and its alliances. This enables the marketer to introduce more specific marketing tools, including qualification statements, when appropriate and without confusing the potential customer. Many corporate clients searching for the most experienced design-build contractor for their particular requirements might regard a contractor as too broad based and general in format if the marketer addresses all the contractor's alliances during the introductory meetings.

Marketing materials for design-build are structured to emphasize the contractor's capabilities, whether they are in subcontracting, alliances, or in-house design resources. Each type of design completion method has its advantages and disadvantages and the marketer should be prepared to address these issues:

Subcontracting advantages
- Permits the best design team to be chosen for the specific project without having to depend on preselected or in-house capabilities.
- The marketing staff can promote design-build without regard to any preestablished alliances or internal capabilities.

Subcontracting disadvantages
- Clients might perceive that the contractor is not a true design-builder if it doesn't have proven design capabilities.
- The most appropriate design team may have already committed to another contractor or alliance.

Alliance advantages
- Marketing can emphasize the ability to bring in just-in-time-talent for the design and construction phases to increase cost effectiveness for the owner.
- Marketing capability is usually increased for all alliance members by the promotion of the alliance design-build capabilities.

Alliance disadvantages
- Client might perceive that the alliance is limited to the niche work targeted by the alliance marketing.
- Formal alliance agreements can create conflicts of interest when team members attempt to form other alliances or market work individually.

Internal advantages
- Clients are able to deal with one firm for design and construction.
- Marketing approach is consistent to all clients.

Internal disadvantages
- Internal resources might be limited in work type and outside resources needed to support internal resources.
- Clients might perceive that contractor's design team can be pressured by the construction team to reduce costs by reducing quality of design.

A proven capability to complete design-build projects is more important than having the internal resources or an alliance structure necessary to complete design work. A contractor with a proven track record of teaming successfully with a variety of architectural and engineering firms to complete a wide range of building projects has the advantage in marketing to a variety of clients. Because every construction project is unique, a contractor able to team successfully with numerous design firms may actually be more advantageous than a company with internal resources that operate with a limited number of niche capabilities.

Quality Issues

Questions concerning the contractual method that produces the best-quality product for the client often arise when marketing the various types of design resources. Contractors with internal capabilities will debate that a single source of responsibility ensures the client that the design and construction team are managed under uniform circumstances, which yield the highest-quality product. A single entity usually means that the design and construction personnel have worked together previously and are better able to produce a tested and quality product.

However, the argument can be made that outsourcing the design incorporates a check and balance system into the design and construction process between the companies involved. The internal resource method has no independent oversight, but teaming puts responsibility on the individual organizations to carefully watch over each other's output to protect their reputation.

It is unlikely that either method produces a better or higher-quality product. No reputable firms would put their reputations at risk by slighting a customer over quality or related cost issues. Any design-build contractor, regardless of how the design is completed, should produce the highest-quality product as contracted. Lowering standards will only result in loss of reputation and future contracts, as it would in any industry.

Initial Design-Build Marketing

The marketing of design-build services by a contractor that has not yet completed a design-build project is difficult but possible when approached correctly. Entering the design-build market should be based upon a proven performance of past project completions in a particular type of work. Under these circumstances a contractor should obviously team with an architectural firm that has done design work the contractor has completed under a different contracting method. A contractor might choose marketing design-build to an existing repeat customer, using a design team that has also completed services for this same client.

The marketer must then carefully explain the benefits this client would receive by choosing to complete its next project under a design-build contract, including those discussed in the introduction of this chapter. Once a contractor has completed its first design-build contract, the marketing team continues to build upon this experience, either by forming alliances for niche markets or by continuing to subcontract the design or a combination of both types.

In many situations, it is the client that actually takes the first step in changing its preferred contracting method to design-build from methods previously used. For example, a corporation might determine that a project's completion requirement mandates a fast-track start, resulting in the use of a design-build contract. This company then solicits proposals for the work and will typically encourage contractors and designers that have completed work for it previously to team and compete for this proposal. In this way or through similar situations, many contractors begin their design-build experience.

Design-Build Is an Option Available to All Contractors

Most industry clients understand that design-build is not necessarily a reinvention of how construction projects are completed, but a change in the contracting method and resulting reporting procedures for the contracting parties. The architect still completes a design and the contractor still builds from this design, but now the two are teamed together rather than contracting separately with the client.

Some will argue that having a contractor in charge of a design-build project rather than the architect is like the tail wagging the dog. This could not be farther from the truth when, in fact, the design-build process requires that the architect and contractor form a team, with both responsible for each other's work.

The contractor becomes involved in the design phase, offering constructibility reviews and detailed budgets, and the architect participates in the construction phase, offering oversight and quality and code reviews. The teaming results in a product with better overall quality simply because it is created in an atmosphere of partnering and cooperation rather than using contracts that establish adversarial relationships between the parties.

Design-build is one of the industry's approaches for compressing project time schedules, improving quality, and reducing legal claims; therefore, many construction firms may consider offering or competing for design-build contracts. The lack of actual design-build experience should not be a reason for ignoring this growth market segment. The ideas presented in this and the following sections can help contractors establish a marketing program for securing design-build projects.

Marketing Design-Build by Teaming or Alliances

Contractors attempting to market design-build without previous experience will certainly, at minimum, want to show the client that they have created a successful contractual relationship with their chosen architect and engineer on previous projects. The more previous joint experience, the stronger the joint qualifications statement will appear. In addition, if the alliance is marketing a special niche category, it will want to describe the similar work completed previously in this niche, preferably working as a team.

Any separate experience the contractor or architect has with previous design-build projects should also be included, even if it was not completed jointly by the alliance. Marketing materials should sufficiently describe the individual organizations' experience to assure the prospective client that the team has the ability to complete a project using the design-build contracting method.

BEST

BE CONTRACTORS		DESIGN-BUILDERS		ST ARCHITECTS
55 Years Experience		➢ Over 100 Years Experience		42 Years Experience
350 Professional Staff		➢ 500-Plus Professional Staff		160 Professional Staff
26 Design-Build Projects		➢ 50 Successful Design-Build Projects		24 Design-Build Projects

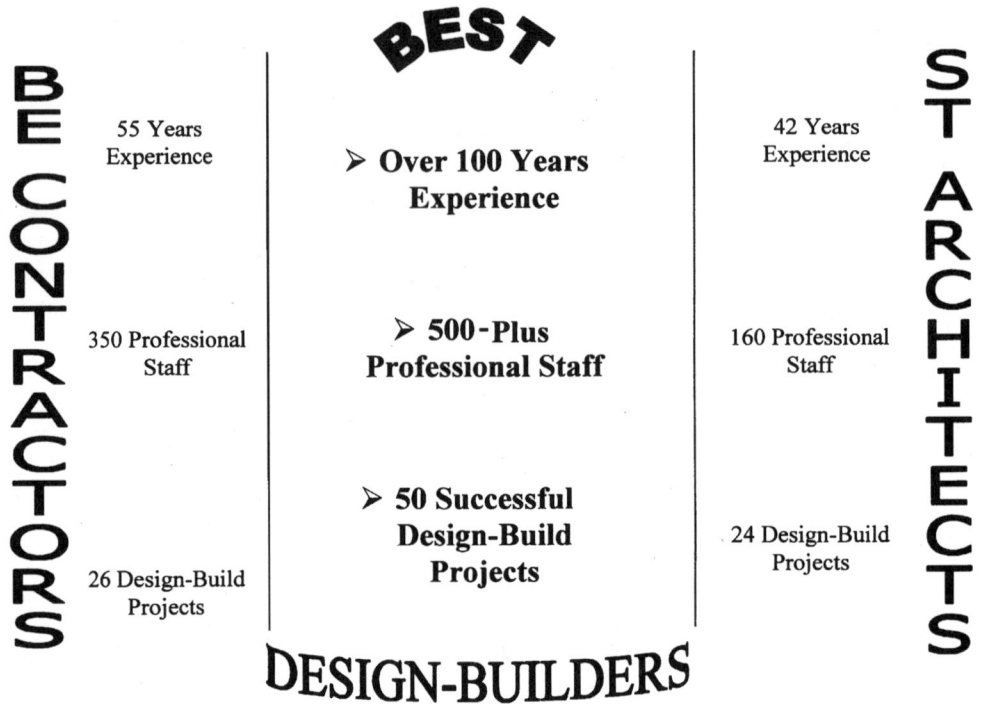

Figure 14.4 Combining experience to market as a team.

This experience can and should be supplemented by each company's similar project experience to further emphasize overall capabilities. Many marketing qualification statements and brochures combine the experience of several individual companies to portray a combined "virtual company" experience history. Refer to Fig. 14.4 for an example of marketing team experience as a virtual company.

An initial teaming arrangement might include the actual sharing of personnel between the member companies beyond the marketing personnel. A contractor might include a staff member from the architect at the job site to review submittals and shop drawings, whereas the architectural firm might appoint the contractor's project manager to provide input during the design phase. This sharing of personnel enables the team to highlight in marketing materials how the firms are currently working together.

The alliance should first target existing customers of both organizations as potential clients for the design-build strategy. These clients are among the most likely to accept the venture relationship based upon past working relationships with at least one of the firms. These past customers should also be used as references for the new design-build teaming arrangement.

In most situations, alliances will create an independent marketing program to market their joint design-build capabilities. To provide the perception that the alliance is based on long-term commitments, alliances should prepare an entire package of marketing materials, including business cards, brochures, letterhead, and associated materials. A formal agreement should be made between the members that sales representatives from one or more of the companies are assigned full-time to marketing the alliance.

The alliance then begins to operate as a virtual company, making use of available resources, offices, financial capabilities, and combined experience of the associated companies. In effect, it is acting as a company that in reality does not exist, the definition of a virtual company.

An alliance should create a separate name for the teaming relationship. For example, BE Construction might team with ST Architects and form an alliance called BEST Design-Builders. Although the BEST organization does not actually exist as a stand-alone organization, it operates as a virtual company. Contracts awarded to BEST would typically be executed by either BE Construction or ST Architects doing business as BEST Design Builders.

Although many companies prefer to create a joint venture, an alliance does not necessarily require the creation of a separate legal entity to operate. However, the client must be told how the team will execute contracts and any pertinent internal business relationships, including financial guarantees and joint venture agreements, between the team members. In most cases the client will prefer to deal with a single point of responsibility, usually the construction company, for both the design and construction, because this is one of the main benefits of design-build contracting. Therefore, the contract and marketing literature should clearly provide details on the roles and responsibilities of the parties involved. Chapter 16 explains further the structuring of alliances.

Exclusive Contractor Marketing for Design-Build

Many contractors with in-house design capabilities will choose to market design-build projects, featuring their capabilities on an exclusive basis. This is also true for most contractors that prefer to subcontract design work rather than form an alliance. In the first instance, the contractor will generally have the in-house resources to prepare a design-build proposal, including the costing of design fees.

When a contractor subcontracts the design, it usually will not have the internal resources to finalize a proposal without first committing to a design team to assist in preparing the proposal. So although a contractor may choose to market design-build projects on an exclusive basis, once an opportunity has been identified, it will have to commit to a design team partner. This enables the contractor to complete the proposal and include design pricing and any appropriate design response required.

Any exclusive marketing program still requires the networking and sharing of information leads, as discussed in previous chapters. However, because a contractor's marketing staff must concentrate on actual building owners and developers for design-build contracts, they cannot use the traditional marketing methods they employ for construction-only contracts.

Leads can be developed from traditional sources, such as networking with architects, but very often these same firms might be competitors on design-build projects. This requires that the contractor establish an independent marketing program to ensure that sufficient opportunities are developed that creates sales opportunities. The methods suggested in Chap. 13 should become the focal point for any design-build marketing program.

The initial marketing program might be limited to a few select niche markets the contractor claims as specialties. These areas of expertise, with the appropriate experience and references, should form the basis of the beginning marketing program for a contractor. In addition, although the contractor intends to market independently, sufficient relationships should be developed with architectural and engineering firms to provide design services and proposal preparation assistance when a specific project opportunity is identified.

Joint Marketing with Architects and Engineers

Another option available to contractors is to jointly market design-build with several architectural and engineering firms without forming an alliance. This method is more general and passive in nature, with none of the companies committing or dedicating any employees to the exclusive marketing for any company involved except its own.

Although generally the results of joint marketing are not as effective as those of an alliance, it is a means to generate leads that might not otherwise become available without associating with several design firms. Often joint marketing relationships develop from a contractor's independent marketing efforts; it eventually forms an alliance if sufficient leads or projects are generated with one specific design organization.

It can be expected that joint-effort marketing would require that the design firm from whose assistance the lead was generated would have the exclusive rights to team with the contractor and prepare the design-build proposal. This situation could cause difficulties if the contractor believes the architect or engineer involved is not the appropriate or best choice for the specific project. The contractor must recognize that the most appropriate response is to decline to propose on a project if the expectations of success due to the teaming situation involved do not provide a sufficient chance of being chosen by the customer.

In such a situation, the contractor might suggest that the architect team with another design firm that offers the services or experience required for the project. The contractor should advise the architect that the chances of success without such a teaming arrangement are slim. In most cases, the design firm will recognize the need to team to produce a satisfactory proposal.

The strategies to successfully market jointly with architects and engineers are also presented in Chap. 13 and can be employed to target design-build in conjunction with other opportunities or on an exclusive basis with the design firm. Although no formal agreement for joint marketing is necessary, the contractor and design firm should address their expectations of the effort even if only through informal discussion between their marketing representatives. At this time, any additional relationships any of the firms have entered into are explained to prevent any misunderstandings or conflicts of interest when marketing together.

Marketing Design-Build Using Preconstruction Services

As explained in Chaps. 6, 12, and 13, preconstruction services can be an effective marketing tool for developing sales opportunities in all types of construction projects, but even more so with design-build projects. When a client hires a contractor to provide preconstruction services, it is often to work in association with an architect previously chosen to complete the design. After a contractor becomes part of the project team to supply these services, it can be suggested that the client consider the existing team for completing the project using the design-build format.

Even though the architect already has a contract to complete the design separately, there are specific reasons that it would support the change to a design-build format. First, the architect would have a larger role in the construction phase of a design-build contract than the usual contract administration role. Second, the architect is likely to want the opportunity to increase its work experience and references for design-build contracts.

From the client's viewpoint, the advantages of design-build as presented in this chapter, including the transfer of design risks and fast-track starts to reduce completion times, can be sufficient to implement a change in the contract format. The marketer should completely explain these advantages to the client with the design team's concurrence to prevent unnecessary disruption in the preconstruction phase if the architect does not want to change the contract to a design-build format.

Should changing to a design-build not be an option for whatever reason, the contractor can still use the occasion to negotiate the construction phase of the project using alternative contracting methods. In any event, marketing through preconstruction services offers numerous advantages for a contractor and should be used aggressively whenever possible.

Getting Subcontractors Involved Early

Design-build is the ideal contracting format to encourage the participation of subcontractors early in the design process. The contractor, through relationships formed on previous projects, can invite appropriate subcontractors to participate in the preconstruction process. Subcontractors are experts in their particular construction specialty and can provide valuable insight into constructibility issues that can increase a project's quality, reduce overall construction costs, and compress the project's schedule.

A contractor can add value to a design-build proposal by including subcontracting participation in the design phase in its proposed management plan.

The discussion on upside-down contracting included in Chap. 2 can be used by a marketer to explain to a client the advantages of subcontractor involvement and the value-added benefits created by using this method.

Subcontractor involvement can also increase the effectiveness of the contractor's capabilities in niche markets when a subcontractor has a major role in the project. For example, on a historic renovation project involving the restoration of a gold leaf metal dome, a specialty subcontractor can provide expertise on product selection and the appropriate techniques to complete the restoration. The specialty subcontractor's participation in the preconstruction phase in design and specification preparation increases the success of the construction phase more than if the design phase is competed by personnel not as experienced in the restoration techniques.

Subcontractor involvement can increase the competitiveness and profitability of design-build proposals in niche markets such as interior fit-outs, restoration, and renovation projects. Contractors also benefit by sharing risk through joint venturing with a subcontractor that has the majority of work to complete in the construction phase.

Whenever a subcontractor plays a major role in the construction phase, its effectiveness can be increased substantially by its inclusion in the design phase or earlier in the proposal preparation stage. The contractor can then include the specialty subcontractor's experience and references in the proposal to the client, increasing the overall value added and possibly being more competitive than proposers without this direct expertise.

In particular, design-build proposals that must include overall cost submittals for the construction phases require the proposal involvement of key subcontractors such as the mechanical contractor. A general contractor submitting a fixed-price design-build proposal must include the participation of these key subcontractors to ensure its competitiveness and the reliability of the cost proposal submitted. Unless a contractor has an estimating staff that can estimate for all trades involved in a project, the contractor must rely on subcontractors for cost estimates that are reliable. Many major subcontractors will not participate in nor expend the costs associated with preparing an estimate on the proposed design-build project without an agreement with the contractor to commit to using the subcontractor if the contractor is awarded the project.

Because the design portion of a proposal is usually proprietary and kept secret until the project is awarded, most design-build teams will not release their proposed designs to numerous subcontractors for pricing, to maintain the confidentiality of the design. This requires the contractor to either complete the overall cost estimate of the project in-house or use a select number of subcontractors to participate in the proposal phase. When a contractor invites major subcontractors to the proposal team, appropriate confidentiality agreements are used to ensure that the design is kept from competitors.

Even in this early phase of proposal preparation, key subcontractors can offer guidance on selection of building components and systems, make design suggestions, and schedule reviews, substantially increasing the team's chances of being awarded the project.

Subcontractors can also be used by design-build contractors to design portions of the project to reduce costs and increase quality. For example, on an interior fit-out project, a mechanical subcontractor that has internal engineering capabilities can actually design the mechanical phase of the project. The design-build contractor increases its competitiveness on the project by inviting the mechanical subcontractor to become a team member that then prepares the initial proposal design requirements for the mechanical work and completes a cost estimate for this section of the work. The mechanical subcontractor, in return for these proposal phase submittals, is assured by the contractor that it will be awarded the mechanical work. If the specialty phase of the project is large enough on a particular project (e.g., mechanical work on a water treatment facility), the contractor might choose to form a joint venture with this subcontractor to increase its overall competitiveness on the project.

Subcontractor involvement, including joint ventures, is becoming increasingly common on specialty design-build work. Many clients, including the federal government, are increasing the use of design-build on maintenance, repair, and renovation projects that are likely to include a large portion of one key subcontractor's work scope. Such projects are making it increasingly likely that design-build contractors will implement upside-down contracting methods to increase the participation of subcontractors on their projects, including the participation by joint venturing when appropriate.

Virtual Alliances with Clients

Niche specialty work is one of the major design-build applications in which contractors are forming virtual alliances with customers. Corporate clients are increasingly outsourcing numerous types of work to contractors rather than maintaining internal resources to manage these projects. Design-build contracting is often the most advantageous way to complete this work for the client, managing both the design and construction processes, without the client having to maintain internal resources to complete the work or maintain oversight management for outside contractors.

Corporations now outsource a variety of work, particularly maintenance projects, that aggressive design-build contractors now specialize in. For example, with the deregulation of electricity utilities, many utility clients are not only outsourcing maintenance work such as outage repairs but are also outsourcing the design and management of the repair work. For clients such as these, design-build niche capabilities permit the clients to contract with one source for their maintenance requirements, allowing the client to concentrate on the profit-making portion of its business.

Contractors are now marketing these alliances based on a past or present relationship with a client for a reoccurring service requirement or to a potential new client that has a need similar to a contractor's capability in the design-build niche market.

Often contractors will form alliances with subcontractors to market a particular niche service to clients or maintain a team of subcontractors to market a range of maintenance services for clients using design-build contracting. A subcontractor might complete a service that can be expanded using the design capabilities, financial resources, and marketing abilities of a design-build contractor. The team arranges an alliance to market the services exclusively to clients, with the subcontractor participating as a major team member, usually under a joint-venture relationship.

When marketing design-build projects, marketers should not overlook the opportunities available in these niche virtual alliances with clients. Marketers should realize that the intelligence gathered on a client should not be used only when the client releases a specific project, but also to capitalize on unsolicited opportunities that can substantially increase a contractor's growth, profitability, and success.

Contractual Relationships

Design-build is completed using a number of teaming relationships as described previously, including internal design resources, subcontracting, and alliances. Although using internal resources does not require a separate contractual relationship between the design team and contractor, the other methods of completing design-build do.

Although subcontracting most often involves the contractor simply issuing a subcontract to an architect for completing the required design, it can involve numerous other structured relationships, as can an alliance. Typically, a contractor does not carry or have access to errors and omission insurance that is required of designers to complete their professional services. Therefore, the contractor must contract work to a design firm not only for the professional services but also to obtain the insurance to protect the team in case of design deficiencies.

Although marketers will typically not become involved in the structuring of contractual relationships between the team members, their input into the appropriateness of the structuring of the teaming relationship may be required so that the team is acceptable to an individual client.

Subcontracting normally places some of the risk of design deficiencies on the architect or engineer completing the work. Although the contractor assumes the design risk from the client in a general contract, the contractor passes on assumed design risks directly to the architect through the subcontract. A subcontract is the least effective means to establish a true teaming relationship between the design team and contractor because the subcontract usually does not allow for the architect's active participation in the construction phase other than design administration.

A joint venture is the most effective means to establish a true teaming relationship between the design team and contractor (also subcontractors as appropriate). The joint venture can involve a specific subcontract for the design requirements and necessary errors and omission insurance, but it also expands the design team's participation in the entire contract, including the construction phase. Most importantly, a joint venture will generally require the sharing of any profit or losses incurred during the project, both in the design and construction phases.

This means that the contractor will share in the success of the design phase, including profits, and the design team will share in the profitability of the construction phase. This mandatory participation ensures that the design-build contract is completed as a team, with all parties responsible for all phases of the project. This assures the client that the design-build contractor will have a full team to participate in all phases of the contract, ensuring contractor resources in the preconstruction phase and design oversight in the construction phase.

The marketer should use the structuring of the contractual relationship between the parties to emphasize the capabilities of the design-build team in the marketing, proposal, and sales approach with potential clients. A contractual relationship that promotes a teaming relationship for the contractor's team is likely to be the most successful in fulfilling the client's needs with the highest-quality, fastest field construction schedule, and best value to the client.

The ultimate goal of the contracting relationship between the team members should be to eliminate problems for the clients while distributing the project's risks equally among all team members. Establishing a win-win situation for everyone involved is the most effective contractual means for long-term success in completing design-build projects.

Various Degrees of Design-Build Contracting

Design-build is not a new concept or process for completing construction projects. It is a contractual method that permits the client to transfer the entire professional obligation and risk to the contractor for designing and constructing the project. This contracting method is used to improve the processes involved in completing construction projects by forming a team that can more effectively share resources, capabilities, and ideas to improve costs, quality, and legal time to alleviate issues involved in construction.

Advantageous to clients is the ability to transfer the risk associated with design warranties, enjoy fast-track starts, and benefit from the participation of all team members in the preconstruction process. At the same time, however, the contractor must assume risks not normally involved with typical design-bid-build projects even though it is able to pass this risk on to its design partners contractually.

Many industry customers have recognized design-build more as a means to transfer risk contractually than as a method of improving the construction process. The federal government, in particular, has pushed the envelope of design-build to the limit by identifying it as a design-build contract but still maintaining complete control of the design.

Risk transference while maintaining design control is accomplished by the owner having an architectural firm complete a portion of the project's design before awarding the design-build contract. It is common for a client to hire an independent design firm to prepare schematics and design parameters on which design-build proposals are based. However, some clients have taken this initial planning stage to the extreme, in some cases actually completing the entire architectural design. Then they require the "design"-build contractor to prepare only the associated engineering design, such as the structural design, and assume all risks for the design work already completed.

In this situation the contract cannot be considered design-build but rather a design-bid-build with assumption of design risk. Naturally, clients often desire to maintain a large share of design input, but by completing the design to the extent that the contractor has no ability to provide process improvements defeats a major advantage of design-build contracting. Contractors must realize that certain customers will utilize design-build contracting as a means to transfer complete design risk to the contractor for an already completed design package.

Design-Build Proposals

Most clients will have an independent firm, often a design organization, complete the proposal package for their contract requirements. These proposal packages vary greatly, particularly in their percentage of design completion. The extent of preparation of the scope packages ranges from a minimal written scope of work and space needs requirements to 10, 25, or 33 percent design completion packages. These proposal packages are prepared to ensure that the customer receives a building that meets its requirements, including space needs, costs, and esthetics.

Some clients will prefer to have only a written scope package prepared to permit proposing design-build firms to submit proposed conceptual designs. This design proposal is often the first phase of a multiple-phase selection process. The client will review the proposed designs and shortlist firms that provide acceptable conceptual designs. The second phase involves further detailing of the design and cost proposals.

The extent to which the proposal design must be completed also varies greatly. To limit costs of the proposal, some customers will minimize the submittals to design outlines and a few schematic drawings. Others offer stipends to each of the proposing contractors to offset the costs of preparing the design components of the design-build proposal. Sometimes potential clients will abuse the intent of design-build proposals by requiring extensive design submittal packages from all proposers with no regard to the costs borne by the proposers. And some clients will not attempt to limit or shortlist the competition to those that have a reasonable chance of being successful, but rather encourage any organization willing to submit a design package do so.

Responding to Design-Build Proposals

Design-build teams and contractors must carefully investigate the requirements of the proposal and expected competition and evaluate their reasonable chances of being successful before committing the time and corporate funds necessary to compete. Considering the situations mentioned previously, contractors must exercise diligence in choosing appropriate design-build opportunities to which to submit proposals. Most contractors formulate strategies to determine the optimal number of proposal submittals based upon past success ratios and corporate budgets allocated for marketing.

Due to the high submittal costs of design-build proposals, most contractors will elect not to submit on potential projects unless they have more than a reasonable expectation of preparing the winning proposal. This expectation ratio ranges from a low of 33 percent to a high of 50 percent or above. Anything lower, depending upon the contractor's current workload and other opportunities, will typically be rejected.

Any marketing strategic plan should include management's requirements concerning go/no-go decisions. The high cost of proposal preparation, particularly involving design-build proposals, requires an intelligent and consistent approach to selectivity. Most contractors will have an established policy to bid selectivity, and a similar policy should be developed for proposal preparation.

The number of responses that a contractor intends to target should be delineated based upon the resources of the marketing department, budgets, types of work desired, and customer and contract boilerplate issues. In addition, anytime the contractor becomes involved in an alliance for marketing design-build work, these policies should be formally established for the team so that no misunderstandings occur when an opportunity is turned down.

Proposal Responses

Once a decision has been made to respond to a design-build proposal, there are several steps required to properly complete a competitive proposal:

- Determining who does what and when
- Establishing a response budget
- Weaving together a story line
- Inclusion of models, renderings, and other proposal enhancements
- Forming the virtual company
- Preparing for the shortlist, orals, and negotiations

Determining who does what and when

Managing a proper proposal response requires that an employee be assigned as proposal manager and be responsible for both the design and construction responses. This person should have full authority to manage the entire process and call upon all company resources to complete the proposal as necessary.

The proposal team should have specific response tasks based on response items appearing in the proposal request. A design-build proposal will usually require assignments to marketing and sales personnel, estimating, design teams including architect and structural and mechanical engineers, and construction operations personnel. Major or specialty subcontractors might become part of the proposal team when deemed appropriate to meet the requirements of the proposal or assist in preparing costs estimates for their work.

Even if the contractor is outsourcing the design portion of the project, its proposal team should respond to the contractor's proposal manager to ensure uniformity and completeness of the proposal response. The team members and assignments should be detailed in writing and distributed so there is no confusion over specific responsibilities.

In addition, a schedule for the proposal response should be created for any design-build proposal. The preliminary design documents usually must be completed before the contractor can estimate the cost of the project, so sufficient time must be established in the schedule to permit both of these requirements to be completed in a timely manner.

Establishing a response budget

Costs of proposal preparation must be managed in the same manner as the hard costs of a construction project. A budget should be immediately established by management to permit the proposal manager and team to determine the parameters they have to complete the response. For example, if the budget permits, a contractor might include funding for a rendering of the project even though the proposal request does not specifically require it.

Weaving together a story line

The most important input of a marketer in the proposal response is developing a story line that differentiates the contractor from all its competitors. The marketer should be able to outline to the proposal team a complete knowledge of the customer and its specific requirements and expectations for the response. The marketer, by knowing the customer through previous meetings and contacts, should be able to provide information on value-added items that can differentiate the contractor's proposal from all others. For example, if the marketer knows about the client's desire to be environmentally sensitive, he or she would suggest that the contractor include recycled building materials in the design specifications.

Other proposal responses that can differentiate a design-build proposal include ensuring political compliance with any community requirements and sensitive issues. This might include the contractor offering to host town-hall meetings to describe the project and processes involved during the construction. Finally, ensuring that the response is the best overall value to the client and not necessarily the lowest price can be the best differentiation. For example, highlighting the savings in maintenance costs of the proposed design can be a better value to the customer than a lower initial construction cost design.

Inclusion of models, renderings, and other proposal enhancements

Design-build proposals might require that proposers include renderings or schematic elevation drawings of the proposed design. However, these often are not sufficient to express the full details and aesthetics of a contractor's proposed design.

Aggressive contractors will usually include allowances in the proposal budget to include models, renderings, or computer-enhanced drawings. These methods can better reveal the intended design parameters than simple floor plan or elevation drawings. Being able to show the customer what to expect can differentiate the proposal from all others.

Forming the virtual company

Design-build proposals are frequently responded to as a team—a virtual company—put together specifically for this one project. Regardless of how many individual companies are involved in a proposal response, it still should appear to the customer that one company is responding to the request.

This requires that the entire proposal be uniform in appearance as well as content. For example, no proposal should be responded to with different resumé formats from each team member company. Written responses should not change from active voice to passive voice from section to section because different writers were involved. Nor should the written response refer to individual companies within the team after the initial description of the members comprising the team in the introductory paragraphs. For example, companies A, B, and C proposing together could be referred to as the ABC Association in the proposal. All referrals to individual company capabilities should also appear as one company, such as ABC Association has a total of 3000 employees.

The proposal manager should be responsible for ensuring that the final product is completely uniform and it appears that the team is one virtual company and not a product of multiple companies. The client will realize that the proposal is from a team, but a virtual company response indicates that the companies can create a uniform team that can successfully work together to complete the project.

Preparing for the shortlist, orals, and negotiations

The completion of the proposal response to a design-build project does not end with the written submission. In most cases, the client will require shortlisting of the proposers, followed by oral presentations, and possibly best and final presentations. Immediately after the proposal is submitted, the proposal manager should be responsible for making arrangements to prepare for orals and other expected follow-on requirements, even before the contractor is advised it has made the shortlist.

Rarely is there adequate time between notification of being shortlisted and the date of oral presentations to properly prepare an exceptional presentation.

Contractors should budget for these preparations in the initial proposal budget. This permits the proposal manager to begin preparations, including the use of visual aids and practice presentations.

Summary

Design-build is an industry response to the requirements of operating in the electronic age of instantaneous products and services. It has been used for thousands of years and is once again becoming a significant contracting method chosen by industry customers, if not necessarily by industry members.

Contractors expecting to increase sales, market share, and profitability should seriously consider entering the design-build market. Likewise, all construction marketers should have a reasonable understanding of the differences between marketing design-build and other types of construction services as presented in this chapter.

15

Marketing to the Federal Government

Introduction

This chapter is provided to assist firms intending to target the federal government as a client. Opportunities for work with the federal government extend well beyond Washington, D.C., with all agencies having operations throughout the United States. Aggressive marketing to federal agencies no longer requires establishing an office in Washington because the government has become a major proponent of the use of virtual capabilities. Every federal agency hosts a Web site for information that identifies contracting opportunities. In addition, the personnel managing the acquisition of construction-related services are connected to the world via e-mail, allowing any marketer to connect with these potential clients from anywhere.

The federal government is the largest single purchaser of construction and related services in the world and contractors should therefore consider adding it to their potential client list. The federal government has also restructured the ways in which it acquires construction services, shifting away from low-bid procurements. This paradigm shift has created opportunities for all types and sizes of construction companies, including those that depend entirely on negotiated work.

Today the federal government should be considered a commercial customer just like any other potential client. However, some adjustments to marketing programs must be made before targeting toward federal agencies. These adjustments, however, should not prevent marketers from concentrating on the largest available customer of construction services.

Adjusting Marketing Programs for Federal Government Customers

Federal government marketing is unlike any other. Potential projects are listed in the *Commerce Business Daily* (*CBD*). However, the ability to create a repeat business relationship with any branch of government is difficult.

Federal government acquisition occurs under strict ethics regulations that prevent any company from having an unfair competitive advantage. These regulations are mandated by Congress and are published in the Federal Acquisition Regulation (FAR). The purpose of FAR is to provide "uniform policies and procedures for acquisitions by executive agencies of the federal government." FAR is the bible for selling and marketing to the federal government. Professional marketers who intend to market to any federal agency or branch of government should have a current copy of FAR for reference.

FAR covers both the conduct of the federal government in acquiring any commercial product or service, including design and construction services, and the required conduct of these suppliers of products or services in their dealing with federal employees. While pushing the envelope of ethical behavior in dealing with commercial companies might result in the loss of a customer, unethical or illegal behavior with a federal agency may well end in civil and/or criminal penalties.

For example, submitting an inflated claim for a change order with a private sector client might be a vendor-accepted strategy for negotiating extras. Submit the same inflated claim to any contracting officer and the vendor's officers may be open to being charged with submitting a false claim and may be susceptible to fines, imprisonment, and disbarment.

Marketers must also be aware of the regulations dealing with specific guidelines about how federal employees can associate with their commercial suppliers. Part 3 of FAR, "Improper Business Practices and Personal Conflicts of Interest," reviews the required standard of conduct for anyone conducting business with any federal agency. These specific guidelines are also very important to a marketer, covering everything from buying lunches to offering marketing giveaways such as advertising pens or other gifts. Although some corporate clients may have policies governing the acceptance of such items, none are as detailed and limiting as those of the federal government.

This chapter will only present the basic operating guidelines for dealing with the federal government. Although the specific details might change, from year to year and agency to agency, the general procedures presented should form the foundation for preparing a marketing program for the federal government. The most recent issue of FAR should be used to ensure that its strict guidelines are adhered to completely and without exception.

Moving toward Commercial Practices

Although the federal government requires adherence to the acquisition rules set forth in FAR, federal agencies have considerable latitude within the FAR framework in the structuring of how their construction and design requirements are procured. Most agencies are moving away from a dependence on low-bid, lump-sum contractor selection.

Recently, FAR regulations have been restructured to permit agencies to select contractors based on their qualifications, resources, and past performance rather than low price. Although most projects are still advertised in the *CBD,* this trend away from low-bid selection requires construction company marketing to begin long before the project announcement appears.

Among the contract types included in FAR are cost reimbursement, cost sharing, cost plus incentive fee, cost plus award fee, and cost plus fixed fee. All extend federal agencies' authority to contract beyond lump-sum and low-bid contracts. Also more easily available and more frequently used by most federal agencies is design-build contracting. This movement toward commercial practices is designed to promote a greater flexibility in how construction services are secured in the federal government.

Budgetary pressures have caused the government to seek ways to maximize limited construction funding by using alternative contracting methods. Anyone marketing for federal opportunities must be prepared to respond to these new paradigms to be successful in securing federal work.

Doing More with Less

Into the foreseeable future, all federal agencies are managing substantially smaller budgets for design, construction, and maintenance. These agencies have responded by demonstrating that less can be more if properly implemented. Budgetary pressures require agencies to take a creative management approach to dealing with smaller budgets, actually creating more opportunities for private industry on a national basis. Three common themes are evident in the new federal procurement system:

- Privatization
- Alternative project delivery methods
- Outsourcing

Privatization

To leverage limited available funds, some government agencies are turning to privatization. Expect more and more federal projects to be privatized in the future, with government funding being used as the seed money to develop the project. For example, the Department of Defense (DOD), including the Air Force, Navy, and Army, has begun to privatize its family housing, both new and existing developments.

Projects such as DOD privatization housing provide new kinds of opportunities for the construction industry, but the industry must realize that marketing and selling now goes well beyond submitting a low bid. These privatization efforts require the contractor to facilitate financing for the project in addition to providing design and maintenance services.

Many of the privatization requests for proposal require contractors to provide their own solutions to the required needs of the agency rather than addressing the expected responses. Most of these privatization projects may necessitate teaming not only for the design services but also to supply the complete development package, including long-term financing and maintenance of the completed facilities.

For example, DOD projects have provided the contractor substantial financial assistance or seed money, or have been underwritten with other project guarantees. The government may provide land at no cost or guarantee an adequate level of rentals from enlisted personnel to encourage private teams to offer turnkey development. Such financing not only stretches available funds but also allows the agencies to operate with a substantially leaner in-house staff by outsourcing planning, development, design, and construction services.

Privatization requires aggressive contractors to become part of a complete development team.

The Navy has already outsourced family housing developments in this manner. Under current FAR regulations, the DOD is able to negotiate directly with developers and contractors without competition based on shortlisting the team on an initial technical and finance proposal. Because the government is, in effect, only leasing the project, selection is made based on best-value proposals rather than the low-bid procurements of the past.

Alternative project delivery methods

Budget cutting has produced much leaner physical asset holdings throughout all federal agencies. When a specific need arises, the agency seeks to fulfill requirements in the shortest time possible. Agencies can no longer take the 4 to 7 years typically required in the past to bring a project into service using traditional methods including design-bid-build.

With new congressional direction, all federal agencies are operating under an "only when absolutely necessary to serve their customers" management style for physical real estate holdings. Similar to just-in-time inventory management, new construction and renovations projects now must come into service within time parameters that are less than half of traditional delivery methods.

There is clearly an interest in design-build throughout the entire federal government. Outsourcing facilities in a complete design-build package permits the agencies to complete more projects using less in-house staffing. The procurement process is expedited, with the design and construction team selected in one step, shortening the normal procurement process by as much as a year. Design and construction schedules can be compressed, including using fast-track starts that bring both new and renovation projects on-line considerably faster than traditional methods.

Agencies are now expanding and improving the delivery methods, using such practices as job order contracting (JOC), service order contracting (SOC), and similar contracting procedures to improve the process while simultaneously reducing overall project delivery times. With these new methods, the agency preselects a contractor or group of contractors for a predetermined period of time to complete all the construction requirements at a specific project site or geographical area. The contract is based on indefinite delivery and indefinite quantities (IDIQ), with the government establishing a minimum and maximum amount of work to be awarded over the length of the contract.

If one contractor is selected, work is released through individual task orders issued either with predetermined pricing or by negotiations on a cost plus (predetermined fee) basis. If multiple contractors have been awarded contracts, the work might be issued to one contractor or competed for by the preselected contractors on price at this stage.

The number of IDIQ programs is growing continually, with each agency establishing its own program characteristics, but each program is based on the selection of the contractor or contractors by qualifications and technical capabilities rather than price. Understanding the move toward use of alternative delivery and selection procedures for construction services is a key element in marketing to the federal government into the next century. Marketing and selling to the federal government is becoming more commercial in nature as the government turns to commercial practices in its contracting methods.

Outsourcing

The "reinventing government" initiative and the general downsizing of the federal government has forced federal agencies to place more emphasis on outsourcing of services previously completed by their own employees. No agency better typifies the shift to outsourcing than the General Services Administration (GSA), which now outsources 95 percent of its services to private enterprises.

GSA outsourcing covers all project development requirements, including space planning, site selection, design, construction, and maintenance. By depending upon commercial companies to provide these services at the expected best value, government employees can increase the level of services provided to their customers. In the case of GSA, these customers are other federal agencies that lease or own office space using the guidance and oversight offered by GSA.

All federal agencies are now outsourcing the majority of their real estate requirements, including design and construction capabilities. This, in turn, lends itself to an increased emphasis on privatization and turnkey and other innovative project delivery systems that require less in-house staff oversight by the client.

Although the budget cutting process may have initially created the movement to these three initiatives, their well-documented success is now self-perpetuating their use. Certainly any company marketing to the federal government must be aware of these initiatives and ensure that its services can be adjusted and, in most cases, expanded to meet the government's requirements. This is often accomplished by teaming or structuring alliances, as presented in Chap. 16, to provide the complete scope of services required to respond to a specific request when the contractor does not have the full complement of resources available in-house.

Federal Government's Rapid Movement toward Virtual Construction

The government has aggressively adopted information age technology to increase efficiency at all staff levels and offset the effect of the downsizing of government agencies. Like their commercial counterparts, most federal employees now have access to e-mail and Internet resources. Not only does e-mail facilitate information flow between agencies and other federal employees, but it also gives construction marketers quick access to these employees from anywhere in the world.

The government's virtual age capabilities permit marketers to initiate contact with appropriate federal employees including contracting officers who manage specific construction projects without having to travel to Washington or wherever the employee is located. Everyone marketing to these federal employees should capitalize on this capability.

Federal Web Sites

A good place to begin researching information to form a marketing program for a federal agency is the agency's Web site. Every major federal agency, including Congress, has now established a Web site to disseminate information regarding the agency's activities, assistance, and of particular interest to construction marketers, current and future physical facility requirements including planned construction projects. The federal government clearly has structured these Web pages to facilitate working with the agencies, allowing everyone access to all pertinent information about doing business with the government.

There are over 800 federal government agency Web sites and thousands of supporting sites available to marketers and salespeople. For example, the Air Force has a main Web site (www.af.mil), and there are approximately 150 additional sites for individual Air Force bases around the world.

Any of these sites might become a favorite for marketers depending on their strategic plan and targeted customers. The quantity of sites available and the amount of information contained within them requires that marketers limit the number of sites used regularly to those that are most appropriate to their customer base.

A Gold-Mine Site for Construction Marketers

The number of sites available relating just to U.S. government agencies is almost overwhelming. Fortunately there are webmasters within these federal agencies that have created Web sites linked to other closely related Web sites that can become invaluable to a construction marketer.

The one Web site that all construction industry marketers targeting the federal government must have "bookmarked" on their computer is the U.S. Army's Procurement Network (PROCNET), specifically the *Worm Hole* page (pronet.pica.army.mil/misc/content.htm). It has a wealth of links (over 100) to sites dealing specifically with procurement and contracting with all types of federal and state agencies. Among the links provided are

Defense sites
- Air Force Contracting
- Army Acquisition Network
- Defense Acquisition Reform Initiatives
- Defense Link
- Defense Logistics Agency
- Director of Defense Procurement
- DOD Small and Disadvantage Business Web
- Marines
- Navy
- Navy—Acquisition Reform

Other federal sites
- Acquisition Reform Network
- FedCenter
- FedCenter Contract Resource Center
- Federal Acquisition Jumpstation
- Federal Procurement Documents
- Federal Web Locator
- Federal Government WWW Servers
- General Services Administration
- Government Information Exchange
- General Accounting Office
- Library of Congress
- NASA Procurement Home Pages
- National Institutes of Health
- Small Business Administration—Outside Resources and Procurement Hotlist
- U.S. House of Representatives
- U.S. Postal Service
- White House

Cabinet agencies
- Agriculture
- Commerce
- Defense
- Education
- Energy
- Health and Human Services
- Interior
- Justice
- Labor
- State
- Transportation
- Treasury
- Veterans Affairs

State agencies
- All States

Internet resource sites
- Electronic Commerce Resource Center
- GovCon
- PointCast Network Government Insider
- U.S. Federal Government Web Servers

Search engines
- Infospace Government Data Search Engine
- Procurement Search Engine

The links provided at the Worm Hole page take a marketer to the various Web sites, allowing access to a wealth of valuable and current information regarding acquisitions, including construction-related projects, throughout the federal government.

Virtual Distribution of Plans and Specifications

Almost all federal agencies are now distributing proposal documents directly over the Internet. Some agencies no longer make available printed hard copies of proposal documents, plans, specifications, and addendum, preferring instead to handle and distribute these documents to potential bidders in a variety of paperless transactions. In fact, it is now possible to handle the entire process of marketing, soliciting documents, proposing, and bidding in paperless form via the Internet.

Through initial marketing contacts or searches on the electronic format of the *CBD,* a firm can learn about potential opportunities with the federal government. Once a proposal has been announced by an agency, it can usually be investigated by direct linkage to the agency's business opportunities Web page. This opportunities site enables interested parties to either directly download the proposal request including plans and specifications when available or to e-mail the contracting officer to obtain a disk or CD-ROM containing the proposal information.

In some cases, agencies no longer accept written or facsimile requests for the proposal documents or issue hard copies of this information. This requires all construction firms to have the resources, including Internet access, to receive and propose on a prospective federal project. Furthermore, once a Web link has been established for a project, the agency will only issue addendum electronically and post the information pertaining to changes only via e-mail to firms that have registered at this Web site for a specific project.

Internet resources are available that will electronically monitor Web sites that are personally selected and will then notify you by e-mail message when the Web site has been updated or changed. These services, which are available free, can alert a marketer to important changes, including the actual release of a request for proposal, automatically without any necessary monitoring required by the marketer. This virtual capability allows marketers to effortlessly monitor numerous pending projects once they have registered at the Web minder site. This literally makes the computer a virtual marketer's assistant, completing chores for the marketer while making the marketer more productive. One of the sites that monitors Web sites for changes and announcements is Netmind Minder (www.minder.netmind.com).

The government's move to paperless RFPs and IFBs mandates that construction firms targeting federal agencies as potential clients must have the technical resources necessary to communicate electronically with the agency. These resources include those typically necessary for anyone communicating on the Internet, such as a modem that can communicate at speeds sufficient to download the plan and specification files, which often take considerable time to retrieve due to their size.

The Army Corps of Engineers has implemented the "paperless acquisition vision." This is the Army's concept for acquiring supplies, equipment, and services. "The goal is to harness current technology to create an electronic infrastructure requiring NO paper documentation."

The Virtual *Commerce Business Daily*

The government's move into cyberspace communications has also led to the establishment of an electronic format of the *CBD*. The *CBD* is a posting of all federal government business opportunities including design, construction, and maintenance projects. The *CBD* also posts contact award information. The *CBD* is published on every government business day, and although previously only available in a printed format, it is now available electronically over the Internet.

In addition, to save the marketer time reading through the *CBD* for appropriate opportunities, Web services are available that automatically search the daily issues of the *CBD* on specific search criteria established by the marketer. The service then e-mails search results to the marketer on a daily basis. With this virtual capability the marketer does not have to read the *CBD* each day;

instead the computer automatically conducts the searches and provides only the appropriate search information.

Included in each issue of the *CBD* are the following sections that marketers can search for work that meets their targeted sales program (the sections are referenced according to the letter identifying that section within the *CBD*):

A. Research and Development

B. Special Studies

C. Architect and Engineering Services

D. Information Technology Services

F. Natural Resources and Conservation Studies

G. Social Services

H. Quality Control

J. Maintenance Repair and Rebuilding of Equipment

K. Modification of Equipment

M. Operations of Government Owned Facilities

N. Installation of Equipment

P. Salvage Services

Q. Medical Services

R. Professional, Administrative and Management Support

S. Utilities and Housing Services

T. Photographic, Mapping and Publication Services

U. Education and Training Services

V. Transportation, Travel and Relocation Services

W. Lease or Rental of Equipment

X. Lease or Rental of Facilities

Y. Construction of Structures and Facilities

Z. Maintenance, Repair or Alteration of Real Property

As with many Internet services, searches of the *CBD* can be conducted free without even having to subscribe to the *CBD*. Another gold mine Web site for federal marketers is the Department of Commerce's CBD Net, which allows searches of the *CBD,* including past issues, at no cost. The address of this site is www.cbdnet.gpo.gov.

As the one source for projects and services that are being issued for proposals or bidding, the *CBD* is an absolute necessity for any firm targeting federal opportunities. Although the *CBD* is more a sales than a marketing tool because the projects are being announced to the general public, it still remains an excellent resource for any contractor tracking federal government opportunities.

Design-Build Contracting

Like commercial clients, the federal government is increasingly making use of design-build as a viable contracting method to reduce overall project schedules from conception to completion. The streamlining of contracting requirements through FAR now permits agencies to bring projects on-line in 1 to 3 years versus the 4 to 7 years required under the traditional design-bid-build method.

Because federal agencies operate under as-needed programming, a project is not funded until it is actually required. Because of congressional budgeting requirements, rarely does an agency have the ability to fund a project based on future projection requirements. So once a project is funded, it is needed immediately, and agencies such as GSA have responded by using design-build to compress the overall project delivery schedule as much as possible.

In addition, the reduction of in-house staff has increased the need for contracting methods that require less client oversight staff in the proposal, planning, and construction stages. Design-build will continue as an answer to federal agencies requirements, and marketers targeting the federal government must be aware of this situation and make appropriate adjustments to their marketing strategies, including the use of teaming and strategic alliances as discussed in this and other chapters.

Electronic Documentation of Projects

In addition to the electronic commerce during the proposal or bidding stage, the federal government is also operating most projects using virtual age technology. Most documentation on a project is now transferred electronically, with even the as-built drawings required to be submitted on disk or CD-ROM format.

Federal agencies are also using electronic pay requisitions. This requires a contractor to have the ability to transmit all invoices to the government using electronic formatting. Should hard copy invoices or support information be part of the requisition, they must now be scanned and transferred in a paperless format with the entire requisition.

After processing the requisition, the federal government is now issuing payments electronically. These payments are usually wired directly to the contractor's bank. The entire system requires less staffing and fits the need for doing more with less.

Project Web Sites

When a federal agency awards a significantly large, long-term project, it expects and requires the contractor to establish a separate Web site for the project. The site might be linked to the agency's and contractor's Web sites, but it will be operated independently of either.

The government uses these sites to promote community relationships, small and disadvantaged business opportunities on the project, and general communications with anyone having an interest in how taxpayers' dollars are being spent. Again, the movement toward virtual capabilities within the federal government requires that their contractors have similar capabilities, such as construction Web sites, to respond appropriately to proposal requests and associated requirements.

Making Use of Virtual Resources

Among the adaptations needed for marketing to federal government agencies, some contractors will find that they must greatly improve their electronic communications capabilities. Federal government marketing requires the ability to make use of the cyberspace resources the agencies have made available to their suppliers. At the same time, responding to government opportunities and performing work for the government requires the contractor to have these same virtual capabilities.

Not having the ability to compete equally using these resources places a contractor at a real disadvantage and will likely result in it being uncompetitive in the federal market.

The New Federal Contracting Environment

No longer can a contractor expect to sell services to the federal government by simply subscribing to the *CBD* and responding to the project announcements. The new federal contracting environment employs methods used by all agencies that require a contractor to have a marketing program designated specifically for the government and not a sales program based only on low-bid opportunities.

Contracting methods used in the federal government today are as varied as the agencies themselves. DOD units use JOC and SOC as frequently as any other contracting method. JOCs and SOCs are IDIQ contracts that enable the agency to select one or multiple contractors to complete work using design-build methods on a particular military base or geographical region over an extended period of time, usually averaging 5 years. The United States Postal Service and other agencies also make use of similar contracting methods.

This type of contracting is used to compress overall project delivery schedules even more than a single project design-build contract can provide. The importance for contractors is that a JOC contract format can provide a steady source of work, often without further competition, for 5 to 10 years. However, not being successful can effectively lock out a contractor for doing work for this agency for this same period of time.

The contracting methods such as JOCs now used by federal agencies are constantly being changed to reflect the current needs of an agency for a specific program or requirement at a project or site. In addition, the contracting methods employed involve complicated proposal responses and include multiple projects that may not yet be defined in scope. Failing to market to and meet with the client early in the development stage of these projects might prevent a contractor from recognizing how to properly respond to the proposal request. Marketing early and being thoroughly informed of the client's expectations from the contracting format is crucial to responding properly and being selected as the successful contractor.

Similarly, privatization projects issued by the government must have marketing incorporated early in the process. Privatization projects are among the most complicated proposal requests to respond to, and often the government will leave it to proposing firms to identify the specific details on such components as financing, government financial guarantees, and length of contract terms. Marketers who wait for the proposal requests on privatization projects to begin determining how to respond will clearly not have enough time to respond competitively. Early involvement with the agency is necessary to gain sufficient knowledge about its desires and expectations on such a project and is mandatory if a contractor wants to be competitive during the proposal stage.

Although the government's intention is that each responding contractor be given equal information to respond via the proposal request and associated addendum, it is probable that no request can totally communicate the requirements of the agency and what it is specifically seeking from a supplier. Because the federal government prohibits any agency from discussing a proposal request once it has been issued, it is only before a project is released that marketing can be accomplished with any federal agency.

FAR regulations require that any information shared during the proposal stage with one contractor be shared in writing with all other interested contractors. This requirement effectively prevents any contracting officer from discussing the pending project once it has been advertised. Only by marketing and meeting with the prospective agency personnel early and at regular intervals can a federal marketing program be effective.

Nothing prevents a contracting officer from discussing the agency's expectations on a particular project before it has been released. This is the appropriate time for a marketer to learn about the requirements and prepare to respond to the published proposal request. This early knowledge might lead to appropriate teaming or subcontracting to fill a void in the contractor's capabilities that will be important to the client.

After a project has been released, it may be too late to secure the best teaming partner, and the contractor is left responding to the proposal request with a weak or inappropriate team. A foregone conclusion might be that the contractor should not waste valuable resources and costs to respond to the request if information was not gathered at the proper time—before the proposal was published and officially released.

Know Your Customer

Like any commercial marketing program, federal marketing strategy is based on gaining significant knowledge about a specific agency to enable the contractor to respond in accordance with the customers' requirements. Blindly preparing a sales proposal without sufficient knowledge of the specific client's needs is as futile in government contracting as it is in commercial marketing.

However, unlike commercial clients, the federal government is like an open book when it comes to fact finding and gaining knowledge about a specific agency or contracting program. As importantly, through the federal Freedom of Information Act (FOIA), contractors can learn valuable information about their competitor's successful proposal or current contract status, including how they are performing on a specific contract.

Are Business Relationships Possible?

By law the government must treat every service supplier equally, small or large business, new or repeat contractor. FAR Part 3 states, "Government business shall be conducted in a manner above reproach...with complete impartiality and with preferential treatment for none." Although this may appear to prohibit any preferential treatment to an individual company in the proposal stage, it is against human nature to treat everyone equally without any preference becoming apparent in the selection stage of the proposal process.

The federal proposal process usually includes a selection committee to review and make a decision on which company will be awarded the contract whenever the contract is based on items besides lowest bid. Because more and more government contracting is being awarded on issues other than low price, such as capabilities, the use of selection committees is becoming an important factor in government marketing.

The names of participants in the selection committee are usually not released, with the government preferring instead to maintain anonymity of the committee to prevent any bias during the selection process. However, effective early marketing before the proposal is advertised can lead to information about the project, including the expected members of the selection committee. As with all federal employees, marketers should never attempt to sway committee members, but perceptions certainly can be created early that lead to positive reviews of the contractor during the proposal process.

Often the contracting officer becomes a member of the selection committee or at minimum has direct input during the selection process. Marketing directly to the contracting officer before the proposal request is released is appropriate to discover as much information about the project and proposal process as possible. This is the time to inform the contracting officer about the capabilities of the contractor and its experience on any past or present projects with this agency or other agencies.

This early marketing can create the proper perception that the contractor is a capable and acceptable solution for the agency's needs that can be carried forward into the selection process. This early marketing can create a better impression than just about any written proposal response can. Humans are more likely to respond positively to personal meetings than they are the written word. This is a key concept in all types of marketing, the so called "warm and fuzzy feeling" in which people are more likely to buy from someone they know than someone they only read about.

This same type of relationship can be created with any of the selection committee members before the proposal is released. Any firm competing for work would prefer an individual reviewing the competitive proposals to be personally aware of the firm's capabilities and history instead of reading about it first. So although the government establishes regulations that treat everyone equally, it is human nature to acknowledge a contractor that has performed successful past contracts and is well known to a majority of the selection committee members. Try as it might, the regulations cannot create a level playing field when good marketers have done their homework.

Just as important as communicating knowledge about a contractor's capabilities to the contracting officer and potential selection committee members is the ability to learn details about the project requirements. The perception that a contractor is capable of completing a specific contract can be negated if the contractor's response to the proposal does not fit the agency's expectations. For example, if a selection committee is expecting a contractor to supply no more than five management personnel on a site for a cost-reimbursable project and the contractor submits a proposal that includes over ten management personnel, the committee is likely to reject the proposal as too costly regardless of how well liked the contractor is.

Successful preproposal visits and marketing not only create a positive perception about a contractor, but they also result in gaining the knowledge of specific contract information that enables the contractor to respond in the manner the agency expects. This is the appropriate time to present a justification of why the contractor believes ten management personnel are necessary and that five would be insufficient to satisfactorily perform the requirements of the proposal. Again, it is before the proposal is released that these important issues can be resolved to create a competitive advantage for the contractor. In this case, competitors might perceive wrongly that five is still what the agency expects when in fact the agency has actually released the information that the contractor's reasoning for ten is completely justified and required.

In this situation, had the contractor waited until the proposal was released and brought the requirement for ten in lieu of five management personnel to the agency's attention, the only possible result would be an addendum being issued to all contractors clarifying the requirement specifically to ensure that all competitors receive the same information. The contractor would then lose any competitive advantage in suggesting its better solution for the client. Further, if the contractor should decide to propose with ten instead of five managers without bringing the issue to the attention of the agency in the proposal preparation stage, it is likely that the selection committee would fault the contractor for this submission.

Repeat Clients: Is It Possible with Federal Work?

Because FAR requires that the government operate by judging each proposal as a separate project, with every firm to be treated equally, it is difficult but quite possible to expect the government and a specific agency to become a repeat customer. This situation depends greatly on positive and continuous marketing to the agency and not passively waiting for an opportunity to be advertised in the *CBD*.

Early marketing before projects are released not only creates the positive perceptions and fact finding discussed in the previous section, but it can also create a foundation for continued and repeat work with this agency and contracting officer. Again, people are more inclined to buy from someone they know and feel comfortable dealing with than attempting to try a new company or product. Maintaining a continuous business relationship with an agency's representatives creates a situation that will likely result in more proposals being won by the contractor even though the selection process is to be completely unbiased.

Certainly it is more difficult to create repeat clients with federal agencies than with private sector clients. However, the movement toward the use of new contracting methods, privatization, and outsourcing and the new FAR regulations that create the environment to permit agencies to use these methods will facilitate the federal government's use of preferred contractors.

The Government's Contractors Report Card Database

Many agencies have, in fact, begun to put in place review methods that give preference to contractors that have performed successful contracts for them in the past. The government maintains a central contractor and supplier reference history on work performed for the government. This reference resource is available to all government departments and is called Construction Contractors Appraisal Support System (CCASS). CCASS is managed by the Corps of Engineers, which maintains the database of information. Agencies can access the database to investigate the performance history of a contractor on other government contracts and use this information during the selection process.

CCASS will likely mean that contractors who perform well will become the repeat and preferred suppliers to the government. Contractors can ask through the FOIA about the current information regarding their companies contained in the database. In addition, by effectively marketing to satisfied government clients, the contractor can request that a contracting officer or other agency representative submit a positive review to CCASS at anytime.

By recognizing the new contracting methods being employed by the federal government and its move toward commercial practices, marketers should realize that the government can indeed become a repeat client by proactively marketing their firms to prospective federal clients.

Freedom of Information Act: A Major Marketing Resource

Although private clients are entitled to keep any information about their current and past suppliers confidential, the federal government, by law, is prevented from keeping any of their contracting information a secret. FOIA guidelines require that the government release all but the information judged proprietary to anyone requesting the information. All firms should take advantage of this law to increase their competitiveness by learning from and about their competitors through the government.

Besides the direct and early marketing discussed previously, a marketer should learn everything possible about contractors currently doing business for an agency. The contractor does not have to be responding to a proposal to request information under the act; it can be requested at any time by any company or U.S. citizen.

Contractors should also realize that any information they submit to the government might be released to their competitors under FOIA. Any information the contractor judges to be proprietary should be clearly marked in the proposal as such to prevent its release. FAR regulations requires this information to be marked "SOURCE SELECTION INFORMATION—SEE FAR 3.104." This prevents the release of this information unless the contracting officer determines it not to be proprietary. When a contractor decides to target a prospective agency's work, it can immediately research every company doing business for the agency using FOIA. Although FOIA will not cause the release of detailed proposal responses by competitors, much information can be obtained that can be extremely useful when competing against these same competitors on future work. Existing contract pricing is often requested under FOIA, and the government is obligated to release information on prices it is paying a supplier for goods or services because every taxpayer is entitled to know this information.

This type information is especially useful on term contracts such as JOCs or SOCs that are being recompeted for after a contractor's current contract expires. Marketers can learn what current pricing the agency is paying for the services, as a guideline in their own proposals. Marketers might request all other information pertaining to the current contract, including an organization chart or structure for the incumbent's management office, overhead rates, personnel rates, and other pertinent information that can assist the contractor in preparing to compete.

If the contract is a cost plus with an award fee, the marketer should specifically request all historical data on the fees awarded to the contractor to determine how a contractor is currently performing. In addition, the actual award fee letters from the government to the contractor should be requested to determine if the government is suggesting any methods that the current contractor can utilize to improve its operation and services to the agency.

This information can obviously be very useful when responding to the new proposal request by incorporating the pertinent information into the contractor's own proposal. For example, if a current contractor has been asked to improve in a particular area of service, the contractor should be sure to highlight this service in its own proposal response.

It should also be recognized that the information learned from a FOIA request can influence a contractor's decision not to propose on a particular project and assign personnel to a project that has better chances of success. Although a marketer should be able to gather information on the current contractor's performance and realize that the agency is very likely to reaward this contract to the incumbent, the FOIA information can be used to verify if this is the case. Many companies will have legal counsel participate in FOIA requests to ensure that all appropriate information is released.

Although the government exercises a significant amount of due diligence in determining what information it can release in a FOIA request, the marketer should be aggressive in requesting as much information as possible. This may include requesting that information withheld on the first request be released through a second FOIA. However, a marketer should realize that a contracting officer might resent repeated requests, and the efforts of the company in responding to the new proposal requests may be negatively affected.

FOIA letters are not necessarily complicated; they should reference the Freedom of Information Act and be very specific about the information being requested for release. If the contracting officer is asked to release all information on a current contract, the contractor will usually be asked what specific items it wants released. The marketer should request only information it judges might be useful in responding to proposal requests.

Debriefs: Another Means to Improve Competitiveness

When a contractor submits a proposal on a federal project, regardless if it is successful or not, it is entitled to a debriefing from the selection committee. The debriefing explains why the contractor was not chosen and presents specific information regarding the contractors proposal's weaknesses and strengths. The debriefing process is a very valuable tool, and the information exchanged should be used to better prepare the next government proposal response. Marketers should take advantage of the debriefing process on every proposal they submit to the federal government.

Debriefs will not usually provide information about the winning contractor's proposal; this information would have to be requested by the FOIA. Debriefs should, however, explain how the contractor's proposal ranked within the competitive range of others received.

Those attending a debrief should ask for detailed and specific answers regarding the company's response. The debrief committee should be able to explain in detail, by section of proposal response, how the proposal was judged. The marketer should not merely listen to the explanation but should ask questions whenever insufficient information is provided.

The debriefing, along with the FOIA information received, should form the foundation to improve a marketing effort with this agency and contracting officer on future work.

Getting to the Right Contact

As with corporate clients, marketers must target the appropriate contact within any federal agency. Federal agencies, like large corporate customers, have numerous levels of management and departments with multiple personnel directly responsible for individual projects or programs. Each agency has different organizational structures, and marketers must first determine the appropriate personal contact for the strategic marketing plan they are implementing.

In many cases, the initial information might come from one individual, but the proposal as discussed above is reviewed by multiple representatives. A marketer might start with the contracting officer who manages a current contract the contractor is targeting or the contracting officer within a department who is expected to manage projects on which the contractor will propose.

When marketing to an individual within the federal government, the marketer must be aware of and adhere to the legal guidelines pertaining to federal employees. As described previously, FAR prohibits federal employees from accepting gifts from suppliers, including lunches and other meals. Although federal employees are able to accept a cup of coffee when visiting a contractor's office, if a marketer takes those employees out of the office for a cup of coffee, the employees in most cases would be obligated to pay for their own coffee.

With the reduction in federal employees that has recently occurred, many marketers will find that contracting officers have little time to participate in general meetings that are not intended to accomplish anything besides improving the relationships between the parties. Marketers must be aware of an employee's time and not impose on it irresponsibly.

Because the government can "hold no secrets," federal agency representatives are usually very forthcoming with the information they can provide a marketer. All one has to do is ask.

Budget information is open to all taxpayers and contains specific details on each of the projects being funded by Congress as well as the funding requests by the agency to Congress. The budget requests reveal the projects the agency is planning, and the actual approved budgets detail which projects will move forward in the government's current fiscal year (October 1st to September 30th). Government Web sites can also provide this information, including the House, Senate, and Budget Office sites.

Often contacts via e-mail can accomplish as much as a personal visit after the marketer has established a relationship with the representative. Being respectful of a person's time is as important as meeting with the person. In fact, if a marketer is perceived as a nuisance, a company's chances of being successful with the agency, and in particular the agency representative, might be adversely affected.

A marketer should utilize all types of marketing procedures such as mailings, newsletters, phone calls, and e-mail messages rather than depending solely on personal visits with federal representatives. Personal contacts are important but should never interfere with someone's work.

Marketers should also be aware of the seminars, conferences, and association meetings that the agency representatives attend regularly. By attending these same meetings, the marketer has an opportunity to meet informally with the government representative away from the office. These meetings can move the relationship from business only to meeting on recreational terms. Again, people are more likely to purchase from friends and people they know well, and attending these seminars can greatly advance the marketer's goal of making this agency a customer or a repeat customer.

Establishing a close relationship with agency representatives enables the marketer to gather helpful details about the customer's needs, requirements, and work opportunities to facilitate better responses to announced proposals. An informed contractor is a better competitor.

CBD Searches

The *CBD,* while regarded more as a sales than a marketing resource, is a vital information source of current projects and business opportunities throughout the federal government. All supplies and services including designing and construction contracting are advertised in the *CBD.*

The only federal contracts that might be omitted in the *CBD* are subcontracts and services contracted by a prime contractor on a major contract or site for the federal government. For example, the Department of Energy (DOE) has a management contractor at each of its major nuclear sites, including Hanford, Washington; Oak Ridge, Tennessee; and Rocky Flats, Colorado. Each of these managers subcontracts a significant portion of its work but does not typically use the *CBD* for announcing this work. Often the work is announced locally to increase local contractor participation and is usually announced on the project's or manager's Web site. Marketers targeting the work at these sites should not only maintain contact with the appropriate representatives but should also regularly review information posted at the Web sites, using a Web page minder service if appropriate.

The *CBD* also is used by federal agencies for advertising preproposal and presolicitation announcements. These early announcements can be used by marketers to begin preparing to respond to a specific opportunity, including the formation of any teaming necessary for the proposal. As is the case with actual project announcements, once a preannouncement has been advertised, most contracting officers will no longer discuss the project personally. Even at this stage of the project, all information and changes must be shared in writing with all interested parties.

Once a project appears in the *CBD* it is too late for any type of marketing regarding this opportunity. The *CBD* can be a means to monitor marketers' interactions with the government by determining what projects they knew about before an announcement is made public and what projects they were not aware of until after the announcement.

The previous sections outlined the reasons why an opportunity appearing in the *CBD* may actually not be worth proposing if marketers have not done their homework on this opportunity before the advertisement is released.

Qualifications and Registration

Federal agencies are now moving to prequalify all service suppliers, including design and construction firms. Some agencies will use qualification formats on a project-by-project format. Other agencies are actually requiring companies desiring to do business with their division to be registered, qualified, or both before they can bid or propose on any of the agency's work.

The DOD has created a contractor registration database for all defense agencies, including the Air Force, Army, and Navy. The Defense Federal Acquisition Regulation Supplement requires contractor registration in the Central Contractor Registration (CCR) "prior to award of any contract, basic agreement, basic ordering agreement, or blanket purchase agreement." The DOD has created a Web site for all their agencies to investigate if a contractor is registered before awarding a contract.

Although this registration is only a formality to doing business today with DOD, it can be expected in the future that such registration will be used to qualify and monitor the performance of contractors and suppliers. Those firms not satisfactorily completing work might be prevented from proposing on or receiving further contracts with this federal agencies.

Today the CCASS is being used to review a contractor's performance on past contracts, and these structured databases will be used to disqualify poor performers from receiving future work with the federal government. This is another means of moving the government toward commercial practices. No corporate client would award work to a contractor that has not performed satisfactorily in the past, and the government is beginning to take steps to ensure that it too will no longer award work to unsatisfactory suppliers.

Secret and Top-Secret Work

The government also releases a considerable amount of work classified as secret or top secret. These opportunities are not advertised in the *CBD* or made public for obvious reasons. However, contractors that have or can receive a security clearance are entitled to bid or propose on this work. A firm desiring to receive a security clearance should investigate current regulations and seek appropriate guidance and consultant assistance as necessary.

Because these projects are not advertised, a contractor must depend upon its marketer's ability to determine what secret opportunities are available by developing relationships with agency representatives. Firms are invited to propose or bid on this work, and because no public announcement is made, such work can be determined only by asking to be advised directly about it by the contracting officer or other agency representative.

The agency representative will only discuss these opportunities with firms that already have a clearance. Therefore, the first step in marketing secured work is to obtain a security clearance. If a contractor is already cleared, it should make the agencies it is interested in completing work for aware of this fact.

Virtual Marketing through the Government's Extensive Web Sites

Regardless of where marketers are based geographically, they can be kept informed of all federal opportunities. Through the extensive network of government Web sites, it is possible to virtually market all federal government agencies.

Although the sites described previously in the section Federal Web Sites are for contracting opportunities only, each agency Web site contains a considerable amount of information that can be useful to a marketer. An agency's home page at its Web site goes beyond the listing of present opportunities that these previously identified sites are limited to.

The agency's Web site can provide information about the agency's goals and services available to its customers. This and related information can become very useful when responding to a proposal request for the agency. The sites also can provide information about the organizational structure of the agency and appropriate contacts within its organization for marketing construction and related services. This can be much more helpful than attempting to investigate the right contact by phone and cold calling.

The Web sites will generally include the names, titles, and e-mail addresses of appropriate contacts, which can eliminate the bureaucratic nightmare of investigating an agency manually. The Web sites can also be used to link to other government sites to research budget requests pending, projects sites that have contracting opportunities, and various divisions within the agency that also pertain to the marketer's goals in contacting potential clients.

Related marketing information on these sites also includes information about upcoming seminars the agency is hosting that might be appropriate for a marketer to attend. Most agencies will also list their long-term goals and planning as public information; this also becomes a valuable source of information for a marketer.

The following are some sites that are of interest to construction-related marketers:

- General Service Administration (www.gsa.gov)
- Environmental Protection Agency (www.epa.gov)
- Department of Energy (www.DOE.gov)
- Small Business Administration (www.sba.gov)
- U.S. AID (www.usaid.gov)
- U.S. Postal Service (www.usps.gov)
- Air Force (www.af.mil)
- Navy (www.navy.mil)
- Army (www.army.mil)

The extensive collection of federal Web sites is probably the most appropriate starting place for collecting information to establish a strategic federal marketing program. Once the program is in place, the Web sites provide continuous information to keep a marketer current on each agency's activities and programming.

Qualifications, Technical Proposals, and Oral Presentations

As the federal government has moved away from low-bid contract awards, the inclusion of prequalification and qualification statements, technical proposals, and often oral presentations in the government marketing process has become standard. Although most of these requirements begin only when a specific opportunity is identified, many agencies have begun to prequalify contractors and suppliers, as mentioned previously.

Although design firms have been involved in the technical and oral presentation process with the federal government, due to the selection process of the Miller act, contractors are often not prepared for the extensive proposal preparation process required to now respond to most larger government projects. As the government moves toward more commercial practices, expect to see this trend continue. This is especially true for contracting such as the JOCs and SOCs presented previously.

These proposals can be extremely costly, with some responses costing in the hundreds of thousands of dollars. The government has recognized that proposal costs have been excessive and has implemented programs to reduce the cost of proposal response. In many cases the government limits the number of firms expending these amounts to only those that have a reasonable chance of being successful through the process of shortlisting firms during an initial qualification stage.

In the two-step proposal process, firms are first invited to submit their qualifications and some limited technical responses. Then the government shortlists only a limited number of firms that have the best chance of being awarded the contract and invites these firms to submit cost and technical proposals. In the case of the two-step design-build proposal process, the contractors are required to submit conceptual designs only after being shortlisted.

Obviously it is imperative for the contractor to prepare an exceptional first-phase submittal if it wants to have the opportunity to be selected for phase 2. It is crucial that the contractor has completed its marketing before the release of the proposal, especially in situations where the contractor is relatively unknown to the agency. The two-phase proposal process is naturally biased toward contractors and suppliers that have performed exceptionally well for the agency previously. If the contractor has not previously had the opportunity to complete services for the agency, it is unlikely that it will pass the qualification phase unless it has clearly communicated its abilities to the agency's representatives during preproposal marketing.

As presented previously, clients naturally lean toward and buy from companies and people they are familiar with. The first-phase qualification process should never be the first time a contractor introduces its company to the agency.

A considerable amount of proposal dollars will be wasted if the agency has not already become familiar with the contractor's capabilities and performance history. With so many of the competitors having already completed work for the agency, it should not be surprising that the firms shortlisted will be those with a history of successful performance for the agency.

Marketers should make agency representatives aware of their capabilities and experience prior to ever submitting the first proposal or qualification statement. Having the customer be completely knowledgeable of the contractor's capabilities is as important as knowing the customer. A *CBD* announcement is not the time to begin marketing. By that time it is too late—literally because of federal regulations that prevent personal discussions pertaining to the project after an announcement has been made. Any attempts by a marketer to introduce a company to a government representative after an announcement is made will not be productive because the representative cannot meet with firms intending to propose on active solicitations. Even if the marketer says that he or she does not intend to discuss the announced project, the representative should decline to meet until the solicitation has been awarded.

Oral presentations have become a more frequent part of the selection process in federal contracting. Orals are used not only to give the supplier an opportunity to expand on its written presentation but also to allow the government to meet the personnel that it would be working with if the award is made to them.

Oral presentation for federal work has the same goal as it does with commercial clients: to create a warm and fuzzy feeling with the government representatives, indicating the contractor's personnel will be reasonable and enjoyable to work with. Although it is imperative that the presentations convey their personnel's capability to complete the proposed project successfully, they must also appear as people that the government representatives will want to be closely associated with over the length of the project.

Oral presentations are now common in federal procurements, and marketers should gather the resources necessary to successfully make these presentations. Often, superintendents and project managers who are technically capable are not experienced or practiced in presenting their capabilities orally.

Marketers are usually charged with the responsibility of preparing the team that will actually make the oral presentation. Marketers should be planning for this presentation as soon as a decision is made to propose on a project, even if it is a two-step process and the firm has yet to be shortlisted. Often there is too little time between being notified of making a shortlist and presenting orally for sufficient preparation and practice time for the participants.

Part of the proposal process is determining the best personnel to propose, keeping in mind their ability to present during oral presentations. Many firms will hire consultants to assist in coaching the presenting team; others will supply support and training to assist the presenters. However the organization decides to assist the team, it is imperative that the presenters be given ample time to prepare and practice their presentation. Often this is difficult because the team may be on current projects that limit its ability to participate in the proposal process or practices. The marketer or team leader should take whatever steps are necessary to have the participants spend a reasonable time in preparing, because the award of the proposal to the firm may be determined by the outcome of the oral presentation.

Proposal Preparation

Although it is not actually part of a marketing program to respond to proposals, nor is it the intention of this book to cover proposal preparation in detail, there are a few guidelines to note when responding to government procurements. Part of the proposal response depends upon the efforts of the marketing program, particularly in the area of references that are typically part of every proposal response.

In federal government proposal responses, it is especially critical to update and verify your references for acceptability before they are used, and this should typically be the responsibility of the firm's marketing department. The government will often require submittals of past experience, of the company and also of the proposed personnel. With federal solicitations this usually requires responding with most if not all previous federal contract experience.

Marketers have the responsibility to ensure that the appropriate government representative is listed as the reference contact and that this person will be willing to provide an appropriate reference for the company. If references do not respond to the request, it is just as damaging as if they gave a bad reference. Often the government solicitation request will stipulate that a nonresponsive reference will be considered a negative reference, or points will be deducted for a nonresponse. It becomes critical for the marketer, by maintaining contacts with past clients, to know which will respond promptly and those which are likely to neglect or fail to respond in a timely manner due to their own time pressures or other reasons.

The proposal preparation process also should involve the marketer's knowledge of the customer and their expectations and requirements in the solicitation. This input is extremely valuable in making sure that the client's expectations are presented in the response, especially for those items not precisely covered in the proposal requests but which the marketer has gained knowledge of through marketing the customer.

This information should be shared with the proposal team in the early response preparation stages and also in the red team review. The red team is a group of people who do not participate in the proposal preparation but thoroughly review the draft proposal response before it is sent to the client to ensure that all the requirements of the proposal response have been properly answered. Marketers, with their knowledge of the federal agency, can provide insight during the red team review to ensure that published and nonpublished requirements have been covered by the contractor's response.

Most agencies, in a continued effort to keep response costs to a minimum and keep all competitors on an equal footing, require that proposals not be overly elaborate. The solicitation will typically require that proposals limit or eliminate color and extensive graphics. The request for proposal (RFP) will explain in detail the page limitations, font size, and other word processing requirements to make all submittals equal in presentation.

The important issue is not having a proposal "look good" but "read good." Fancy and elaborate covers, extensive charts, and graphics often defeat the purpose of relaying the information to the prospective customer. If a proposal is not readable, it will usually not be taken seriously.

Spending time making the proposal legible and concise will bring more rewards than a colorful proposal that contains a poorly written response. Federal source selection committees include those who are very competent technically on the type of work that is being competed for. These representatives will carefully read the proposal to determine if the contractor has the ability to complete the required work successfully.

The written response must contain the information requested, and it must also be readable. Time is better spent doing grammar reviews and making corrections than creating additional graphics that are often not understood by the reviewers.

It is also extremely important that the proposal response addresses each and every question or issue raised and requested by the RFP. If reviewers cannot easily find the response to questions or requirements, the contractor will be negatively rated on this issue. Often proposal writers will submit the response they believe is intended, but this is not what the RFP actually requested. This is why there should be red team reviews of the proposal that double-checks that each request is responded to clearly and concise and that it appears in the proposal response where it should.

With the marketer's assistance, the proposal writing team should determine key issues that become the outline for stressing value rather than pricing. Because most proposals will be weighted on technical and price issues, the contractor must determine how to emphasize the added value their proposal brings over the competitors'. Marketers should provide valuable input in this planning, knowing the customer's requirements and important response issues. Marketers should also be able to provide information about expected competitors and structure a strategy to ensure the proposal addresses any weaknesses of the competition and overcomes any advantages another competitor might provide the customer.

This effort should effectively differentiate the contractor's proposal from all others submitted. It should never be the intent of a proposal response to just "answer the mail;" a proposal should emphasize the contractor's added value whenever appropriate. However, it is important to recognize that the client's proposal outline must be adhered to, and this differentiation must occur in the context of answers to the proposal requirements.

The proposal response should be answered in the exact order of the RFP outline. Do not make it difficult for the reviewers to find responses to their questions. Further, give only the information they request, no more no less. Trying to write around not having a specific capability or experience can create more confusion than just explaining succinctly that the contractor does not have the requested experience. It is better to offer an explanation of how this short-coming does not make the proposal uncompetitive.

Never use "boilerplate" responses, or standard written responses or general marketing explanations of a contractor's services. For example, a contractor will usually have a "canned" written presentation on quality, scheduling, and other contracting issues that are used on marketing literature or general qualification packages. The boilerplate rarely can provide a specific answer to a RFP. The reviewer will recognize boilerplate and judge the answer nonresponsive. Do not expect to use marketing literature as proposal responses; it only implies that the contractor does not have the knowledge to respond properly or that the contractor does not feel the question justifies a specific response. Either way, the contractor has little chance of being chosen when its competitors have spent the time to respond properly.

If a marketer and management concur that they do not have the internal resources to properly complete and manage a proposal response in-house, they should consider either hiring a consulting firm that specializes in proposal preparation or pass on submitting. Attempting to respond to a government proposal without supplying the necessary resources to compile a legitimate response will typically end up being a futile effort, with the response costs being better spent on other marketing or proposals that the contractor can better manage.

Marketing through teaming

The extent of government privatization, outsourcing, and new procurement methods often require that contractors form a team to provide a complete response to a proposal or government supplier opportunity. Throughout all government contracting there is an extensive amount of teaming that occurs to satisfy proposal requirements. Every contractor targeting federal opportunities can expect to become part of a teaming arrangement to be competitive.

Federal teaming occurs due to the following requirements:

- Architectural or engineering capability
- Small business requirement
- Size or magnitude of project
- Self-performing capability
- Specific work experience
- Government or agency experience
- Bonding or financial assistance
- Specific proposal requirement
- Financing requirements
- Maintenance and operations requirements
- Personnel resources
- Personnel experience
- Competitor knowledge of customer

Contractors have teamed to submit a proposal for all these reasons and others. Often teaming occurs between competitors who recognize that by joining forces they are a better competitor than if they submit individually.

Better teaming arrangements occur prior to the actual release of a proposal request. By marketing early and gaining sufficient knowledge of the agency's requirements for a specific request, a contractor can create the most competitive team. On the other hand, waiting until an opportunity has been advertised usually results in less-qualified teams, because the knowledgeable marketers have already put together the most qualified team for the proposal response.

Teaming has become common in federal procurement and can be a prime/subcontractor relationship between a contractor and designer or a joint venture between two competitors. A marketer must be able to recognize when an agency's request will be better responded to by forming a team rather than attempting to propose alone. Early identification of key elements of the requirement and related weakness in the contractor's abilities (e.g., personnel, financial, etc.) should alert management that a teaming arrangement is necessary and that steps should be taken to form an acceptable team.

Although having all the required resources internally may give a contractor a competitive advantage over competing teams, the government, unless it is specifically detailed in the proposal request, should have no objection to the formation of a team that better meets the RFP requirements. However, the teaming arrangement must be fully explained in detail, especially the management of the team.

With the recognized downsizing of government staffing, a team that the selection committee believes will place additional workload on the agency's employees will most likely be rejected. Any teaming arrangement should clearly establish to the reviewer that no additional oversight would be required of the team than there would be for a single-source supplier. Indications that this is not the case or an insufficient explanation that the team will be managed cohesively and report as a single entity will result in an unsuccessful proposal.

Successful past projects completed by the team or examples of the team members having worked together, although not necessarily as a team, should also be used to reinforce the proposal. In most cases, the proposal request will require teams to demonstrate past experience. This requirement should be recognized early in the preannouncement stage to ensure that the marketer has the time to review a possible teaming arrangement and select the most appropriate partners before someone else does.

16

Strategic Alliances

Introduction

Today's virtual companies speak of the electronic age's ability to foster and promote strategic alliances in all industries, including design and construction. No longer is it necessary to have extensive in-house resources and capabilities to provide complete services to a customer. Aggressive design and construction firms are providing services to clients on an as-needed basis through the use of strategic alliances using virtual age techniques.

This chapter introduces methodologies to help marketers use a host of resources so that they can proactively respond to all proposal and bid requests as a virtual organization using strategic alliances. In addition, many contractors, architects, and engineers are using carefully placed strategic alliances to manage the inherent risk of doing work in the construction industry and thereby enhancing their bottom lines.

All marketers can expect to become involved in marketing a strategic alliance in today's business environment. This chapter helps all marketers prepare to use alliances to their advantage as well as that of their customers.

An alliance can be defined as an advanced level of teaming between firms that lasts beyond a one-project joint venture. Strategic alliances are long-term business relationships between one or more firms but are usually not legal entities. Alliances can be structured between industry service companies, such as an architect and general contractor, and are created to provide a more complete level of service to clients. Alliances can also be between competitors, for example, two contractors that alone do not have the capability, financial or otherwise, to compete individually for a particular type of work. Strategic alliances can also be established between a service provider and its customer to foster a win-win business arrangement for both firms. For example, an architect aligns with a fast food chain to provide standard designs for its restaurants. The client is assured of continuity and conformity of their franchise designs and receives prompt and reliable service from the architectural firm in exchange. This architectural firm would also most likely provide value pricing to this client due to the repetitive nature of the designs.

Within the construction industry, alliances are also common between sub-contractors and general contractors. Alliances of this type provide a contractor with better pricing and estimating abilities through the subcontractor's input, and the subcontractor is assured of receiving a subcontract for the work if the contractor is successful. Subcontractors can also align with design firms, offering design input and assistance in exchange for recognition in their specifications. Finally, alliances can be structured between numerous partners, and one company can be involved in numerous separate alliances as appropriate.

However the alliance is structured, it will typically include a formal, written agreement outlining the responsibilities of each member and the basic measurement for success. Strategic alliances are typically reviewed by management on at least a yearly basis to monitor their progress and acceptability for continued operation.

Marketers can expect strategic alliances to become more common in the next few decades because virtual technology makes it possible to add services to a company's portfolio without having to add employees or other resources.

Alliance Benefits

Alliances are typically structured to provide benefits not only to the team members but also to the customers of the alliance. Some benefits of most alliances are

- Adds resources not available in-house.
- Better team yields better proposal responses.
- A standard practice in the new millennium.
- Expands individual company's marketing capabilities.
- Enhances bottom lines.

Resource support

Strategic alliances are most often arranged to provide resource support to the companies involved. The general purpose of an alliance is to supplement an organization's capabilities and resources without having to hire, purchase, or otherwise acquire additional resources to complete a certain type of work.

Alliances provide a means for the teaming companies to pool resources including personnel, financing, equipment, technical capabilities, and even customers. The latter is often necessary when a contractor has a client requiring services it cannot complete with its own resources. For example, a contractor's customer decides to implement a design-build contract for the first time and the contractor has no design capabilities. In such a situation a marketer should be able to suggest to management architectural firms with which to align so that it can continue servicing this customer.

In fact, virtual companies take this arrangement to the maximum extent, forming companies by bringing together all the resources needed to complete a particular project without possessing any in-house capability at all. For example, a virtual engineering company with no employees except the principal owner might receive a contract for the engineering design of a bridge. This virtual engineering company would then outsource the entire design by bringing together the best available engineers throughout the world through alliances, using cyberspace capabilities as presented in Chap. 2. Once the project is completed, the members disperse until another design is contracted.

Alliances are also used frequently on a local basis, particularly on design-build projects, when an architectural firm will team with a local contractor to complete work under one single contract with the client. Local alliances also include teaming between small business contractors who align with larger general contractors to complete a project that they might not otherwise be capable of completing on their own for financial or other reasons. In these alliances the large contractor could meet small business subcontracting goals required by the client, especially with federal and municipal governments.

Alliances should not be formed, however, without expecting a win-win situation for all the parties involved. If an alliance does not benefit all parties equally, with each member contributing an equal share of resources necessary to complete the targeted customers and work types, it will not be successful. A marketer is often the first person within a firm to suggest a particular alliance to capture a particular work type or customer. In this regard marketers should be aware of the function of an alliance and be prepared to support the alliance once it is initiated.

Better sales and proposal responses

Alliances are formed to make all the teaming companies more successful by adding resources and capabilities in a pooled fashion. This pooling should add considerable value to the sales, bidding, and proposal process. Marketers should become fully versed in the alliances' capabilities and be prepared to market these capabilities to the mutually agreed upon targeted customer base.

Because alliances are structured as long-term commitments, marketers for all companies involved should actually meet with their counterparts to discuss structuring a complete marketing program that operates independently of each company's own marketing program. This program should combine the capabilities, resources, experience, and personnel of all companies into a uniform marketing response. For example, if one company has 40 years experience in construction and the other has 61 years, the alliance marketing could state, "over 100 years of combined experience."

The marketers should use these combined benefits to their fullest extent when marketing to potential clients. It is inevitable that at some point a customer will question the alliance's ability to operate effectively as a single entity. In fact, competitors that have all the necessary resources in-house will certainly inform the client that your alliance is actually a group of companies, none of which alone has the capability of completing the project individually.

Alliance marketers should be prepared to discuss this situation with the potential customer and explain away any such concerns, highlighting the benefits the alliance brings to the customer. For example, overhead costs are lower because none of the firms has to individually maintain a full staff of personnel to bring these services to the client that the alliance can provide collectively. Marketers should be prepared to explain in detail to the client how management of projects will occur within the alliance. In most situations the customer will desire a single point of contact within the alliance and the agreement that all firms within the alliance will be jointly responsible for any work completed for the customer. This requires that the individual companies have a formal agreement that spells out how the alliance will be managed and that each firm will be responsible for the others' work under any alliance contracts.

Typically each alliance will have a new name, and contracts executed under the name will have to be signed by one or more of the individual companies doing business under the alliance name. The marketer should clearly explain this to potential clients long before a contract is completed to ensure that the customer is not taken by surprise when the contract is returned to the customer.

A standard practice in the new millennium

Alliances will become a common practice in the virtual age because they actually create virtual organizations. As corporations become used to dealing with and using virtual companies to their benefit, virtual alliances will be used by most construction and design firms. Clients will expect contractors and architects to supply the best talent to complete their work. If, for example, an architect overseas has the talent necessary to complete a particular design for a customer, this customer will expect the alliance to bring this architect's services to the project, virtually through electronic capabilities, rather than relocating the architect, which will save money for the client.

Strategic alliances will be forced to implement virtual capabilities to the fullest extent possible to increase the level of services offered to clients. In turn, alliances will become global rather than local or regional. Virtual alliances will offer clients their services anywhere in the world, bringing the best talent, resources, and abilities to their customers regardless of where they are located. Virtual capabilities are fully discussed in Chap. 2.

The virtual age will also change the nature of alliance structuring, making alliances more fluid and adaptable to a client's requirements at any given time. Strategic alliances will actually evolve to respond to a client's needs rather than to fill gaps in a contractor's capabilities, which is the reason, as explained previously, that most strategic alliances are formed today.

Expands individual company's marketing capabilities

Although alliances are usually structured to supplement a contractor's service capabilities, they typically have the added benefit of increasing a firm's marketing resources and capabilities. Once an alliance is formed, marketers from each of the individual companies are able to garner the support of their counterparts in the alliance. This support includes increased network contacts, expanded customer base, new markets, and often expanded geographical coverage.

Aggressive marketers will make full use of these resources, capitalizing on these capabilities not only for the alliance marketing but for their individual company marketing as well. For example, a contractor's marketer can network with a new architectural alliance marketer to determine if that marketer has any customer contacts at potential clients that the contractor has been targeting without success. The marketers may find that the alliance can provide an introduction to a potential client that would not have been possible otherwise.

Enhancing bottom lines

Additionally, all strategic alliances will have the ultimate goals of improving the team members' overall profitability. By increasing the size, number, and types of projects a contractor can compete for, an alliance will increase contract billings and result in an increase in the company's profits.

Finally, the contractor is increasing volume without the usual increase in personnel and resources to handle this workload because the teaming members contribute resources collectively; therefore, the margins for the contractor's profitability will also increase substantially.

The numerous benefits that result from strategic alliances should be fully understood by all marketers to ensure they recognize that an alliance can improve not only the company's performance but their own as well.

Putting Together an Alliance

As discussed previously, strategic alliances are typically formed to fill a specific need or void within an organization's capability. In addition, alliances are often formed as a natural outgrowth of an established working relationship with one or more companies. The latter is usually promoted by marketers that recognize the added value an alliance can bring in marketing to potential clients.

Whatever the reason for forming a strategic alliance, it should be based on creating an entity that is clearly more effective than any of the individual companies involved. The member companies should complement each other's capabilities rather than duplicate already available resources, talent, and experience. Alliances should not be established between competitors because they rarely can become successful. The firms involved will be reluctant to share information and resources to prevent one firm from gaining a competitive edge over the other.

Establishing a strategic alliance involves at least three steps to formalize the agreement and ensure a foundation for its long-term success:

1. Formal agreement
2. Marketing strategy
3. Communication links

Formal agreement

Although technically not a legal agreement, in most cases, a strategic alliance should begin with a formal agreement between the firms that clearly outlines the goals and responsibilities of the parties involved. The alliance is usually structured to operate for a set period of time, usually 1 year, after which time the agreement can be renewed should all parties agree. Most agreements include "escape" clauses that permit the dissolving of the alliance should one or more of the firms decide it is in their best interest to do so. Remember that an alliance is usually a business arrangement based on trust between parties and should never require an extensive legal document to outline responsibilities. Starting an alliance based on legalities defeats the intent of the parties to operate with openness and trust between the members.

The agreement should outline the responsibilities of each team member, specific personnel support, resources including funding for the costs of operating and marketing, and specific measurements to determine the success of the alliance. Each company should make clear its expectations for the alliances so everyone has a clear understanding of what it will take for the alliance to be judged successful.

Management of the alliance should be structured in the agreement, with one firm designated the final decision maker, if and when necessary. Typically this requires one firm to be the managing partner to ensure that no stalemates occur in operating decisions. For example, if an alliance is structured between two companies, each with an equal 50 percent share, one firm must be given managing partner authority. If one firm votes yes and the other no on a decision issue, each with an equal share, the managing partner will ultimately break the stalemate. In many alliances, the managing partner rotates among the team members based on a preestablished length of time; for example, every 6 months the other team member becomes the managing partner.

Marketing strategy

In addition to the operating agreement, most alliances have a marketing and sales understanding to ensure that the alliance is able to contract a sufficient amount of work to make it successful. The alliance marketing program should outline specific markets the team will target, ensuring that any conflicts over existing customer bases are prevented.

Marketers assigned to the alliance should be given specific budgets. The marketing program should also specifically outline the resources each member will contribute, including personnel (full or part time). Team members should know whether the personnel assigned to the alliance will be paid from the alliance budget or if the personnel salaries will continue to be absorbed directly by the team members' individual company.

Most often it is best to dedicate full-time resources to the alliance; if marketers are directed to contribute to the alliance as time allows, the alliance usually becomes an afterthought for the individual marketer. This is especially true if the marketer's performance continues to be judged solely on his or her results for the individual firm rather than for the alliance.

Communication links

Much of the success of strategic alliances is based on an ability to effectively communicate between team members. This is particularly true for alliances operating virtually. Team members should immediately establish communication links that include utilizing all electronic capabilities now available.

Much of the success of alliances is the ability to respond quickly to a potential customer's requests. Member firms of an alliance that are remotely located can only operate successfully if they establish a virtual communication link to make it appear to the customer that they are responding as one centrally located firm, not a group of disjointed companies.

Alliances must also be able to communicate uniformly with prospective customers and existing clients. This requires the alliance firms to establish communication links that bring together not only the member companies but also clients. For example, an alliance with one firm in Chicago and the other in London must be able to link and communicate in real time with a client in Brazil. If it does not have this capability, the client will recognize that it is dealing with two remotely located firms that are not operating cohesively enough to serve its needs.

Communication requirements are just as important to an alliance as they are to the individual team members. The initial alliance agreement should establish the parameters and commitments necessary to implement the communication links necessary for successful marketing and for subsequently managing alliance projects.

Marketing Strategic Alliances

After a strategic alliance has been established and operating agreements are in place, it becomes the marketers' responsibility to ensure the success of the alliance by effective marketing that leads to sales and resulting profitability. All marketers can expect their involvement with alliances to increase in the immediate future, and they must be prepared to market the alliance as aggressively as they would their own company.

Marketers within the individual companies should become fully familiar with the intended targets of the alliance and be capable of forwarding leads to the alliance marketers whenever appropriate. Aggressive marketers will use their company's alliance(s) and program(s) to their benefit even if they are not directly reporting to an alliance, for example, networking with the alliance marketers and marketers from the other member companies. Also, the marketer might ask to be introduced to the other team members' existing clients or receive a referral to contact customers directly.

Because strategic alliances are formed as an outgrowth of successful past relationships, marketers should make maximum use of the new resources that become available because of this teaming. Alliances usually push business relationships closer, and aggressive marketers should use this to their benefit to capitalize on these new information resources.

Highlights of establishing a marketing program for strategic alliances include:

- Expounding upon the team's expanded resources
- Concentrating on niche markets including design-build
- Adapting to virtual capabilities
- Better risk sharing
- Just-in-time talent capabilities

Better resources

All alliances should enable the marketer to present much better resources to prospective clients by combining all the member firms' abilities into a single cohesive presentation. By combining their historical data of past experience, the alliance builds a considerable past performance history. By combining resumés, the firms are able to show clients a vast "bench" of experienced personnel resources.

Marketers should compile a complete qualification statement response, as covered in Chap. 12, for the alliance by virtually combining the two or more companies into one. For example, one company has $14 million in equipment, the other $18 million; the alliance can then market an inventory of available equipment of over $30 million.

This combining of resources is usually done under the selected name of the strategic alliance. In many cases contractors prefer not to indicate that the alliance is comprised of two or more companies; they want the client to perceive that it is one individual company. This can be accomplished only if the alliance takes steps to operate as a joint venture, allowing contracts to be executed under the name of the alliance. Although there is nothing necessarily illegal about operating as such, it is preferable to advise prospective customers that this "new company" is operating as a joint venture between two or more existing reputable companies.

If the alliance has not been established as a legal entity, the client should be informed immediately of the situation and that this "association" combines the resources of the member companies to better service their clients. This situation is often true in design-build work, where the contractor has teamed with an architectural team but uses a subcontract for the design and does not establish a joint venture with the design firm. Because most architectural and engineering firms do not have the financial capacity to be bonded or assume the risk of the construction phase, the contractor must use a subcontractor for the design. However, in such situations, rather than only selling the contractor's capabilities, the contractor markets the alliance by including the resources and experience of the design team. Often this alliance will be referred to as an "association," emphasizing the combined capabilities without requiring the formation of a legal entity such as a joint venture. The marketer can then expound on the team's more effective resources and resulting capabilities and not be limited to using only the contractor's marketability.

Niche markets

Strategic alliances are formed to target a customer base or niche market that the individual companies could not sell without combining or pooling their experience or resources. At the same time it is the usually the intention of the parties to form the alliance to target specific markets and/or geographical areas. These markets should be clearly detailed in the alliance agreement to prevent confusion or overlap in the individual companies' marketing program or other alliances that the companies have formed.

Many alliances are formed between architects and general contractors or construction managers to market, sell, and complete design-build projects. Because contractors typically do not have in-house design capabilities and architects have no contracting experience, teaming occurs frequently for this type of contracting. Those teams that work well together and begin to form a substantial history of working together on successful work will often enter into an alliance to market their joint capabilities. Like other alliances, this design-build team would precisely detail its intended markets of clients and geography, preventing overlap of its traditional service customers that prefer to handle design and contracting separately. However, with the agreement of the team members, their marketers might share networking leads from the existing client base of each individual company to market these traditional contracts separately.

Alliance marketing is usually better defined than general marketing programs for a general contractor because of the concentration of niche markets for the alliance. For example, an alliance might target only food and beverage manufacturers for design-build work. Once the targets have been determined, the marketer can apply the techniques addressed in the previous chapters to implement a marketing program for the alliance.

Virtual capabilities

Today, alliances are formed between companies regardless of their locations relative to each other or their intended customer base. The communications capabilities now available include e-mail, video conferencing, and the ability to work together on documents in real time and permit alliances to service customers as if they were all in the same location at the same time. This enables a design-build alliance to complete a design in Tokyo with a contractor based in New York for a project in Mexico whose customer is located in London. Virtual capabilities now permit the level of service on this project to be equivalent to or better than if all the parties were in Mexico together.

Strategic alliance marketers should ensure that not only are they given the electronic communication resources necessary to market the alliance, but also that the alliance management team uses these and similar resources to service the client. Applications of virtual construction techniques and cyberspace marketing have been addressed in previous chapters.

Taking more risk

Alliances create a means for the member teams to complete work that they would not otherwise be able to manage with their own individual resources or capabilities. Strategic alliances not only permit companies to complete work that extends beyond their internal range of services, but also to undertake larger contracts that the companies could not financially manage on their own. This expanded capability increases the marketing opportunities for the alliance and the individual company marketers as well. For example, a contractor's marketer might uncover a large project that the company could not manage on its own but could continue marketing the lead for the alliance or turn the lead over to the alliance marketing team. If the lead does not fall within the alliance's marketing agreement, it still might and should be considered if the team can meet the qualifications necessary to complete the work satisfactorily for the client.

Marketers should also understand how the alliance's increased capacity to take on additional risk can be parlayed into a benefit for potential clients and use this information to their marketing advantage. For example, bonding capacity can increase, more technical personnel can become available, and more service types can be completed to increase the interest of a potential client and existing clients where appropriate.

Just-in-time talent

Many strategic alliances are today being formed on a "virtual" basis, with the team members called upon only when needed. This is particularly true in teaming for personnel resource capability and often for internal teaming between engineering and architectural firms. Such alliances bring in personnel on an as-needed basis to support design completion. The alliances market the extended resource capability where appropriate, usually on larger work or niche markets.

In construction, these virtual teams are created often between large and small businesses. Large contractors require the small business participation to meet subcontracting goals, and the small business uses the large business to supplement bonding and resources on an as-needed basis. In all these situations the teams can operate virtually and do not have to be located close geographically; they can use cyberspace communications until an actual project need arises.

Relationship Building between Subcontractors and General Contractors

Every morning in Africa, a gazelle wakes up.
It knows it must run faster than the lion or it
will be killed. Every morning a lion wakes up.
It knows it must outrun the slowest gazelle or
it will starve to death. (So) it doesn't matter
whether you are a lion or a gazelle, when the
sun comes up, you'd better be running.
GEORGE ALLEN

Introduction

The changes and paradigms affecting the construction industry as presented in Chaps. 1 and 2 require general contractors to reexamine their relationships with subcontractors. The virtual age requires a new attitude of project management that involves subcontractors meeting the industry's customers' expectations of compressed time schedules and better product value. Clients are now aligning their real estate needs with their general business goals, requiring that either the holdings themselves become profitable or the costs of these holdings be minimized.

As owners often turn to design-build as a means to compress their building schedules and transfer risk to the construction industry professionals, earlier and more frequent subcontractor involvement is necessary. Chapter 14 describes the necessary early involvement of subcontractors required in a design-build project.

The communication capabilities of the digitized economy also require cotractors to directly connect with their subcontractors to implement these techniques, including virtual scheduling addressed in Chap. 2. This requirement becomes even more important for contractors who self-perform little, if any, of the field construction activities. For most general contractors, subcontractors actually control the ability of the contractor to compete successfully on price when subcontractors complete the majority of the field construction.

These changes are reason enough for all contractors to implement programs that improve and strengthen their relationships with subcontractors. At the same time, it is as important for subcontractors to recognize that they depend almost exclusively on general contractors for their contracts. To this end, subcontractors must also implement programs necessary to build close business relationships with as many contractors as practicable.

Many of the paradigms facing the industry will also directly affect subcontractors, including the increasing use of design-build. This contracting method requires the contractor to select major subcontractors in the planning and design phases, well before any price competition is possible. Subcontractors should not only rely on notification from general contractors of a bid date for prospective work.

Subcontractors also must implement marketing programs to ensure they are aware of opportunities sufficiently in advance of contractor selection. As with general contractors, subcontractors should no longer depend solely on a sufficient market of low-bid opportunities to maintain a healthy backlog. Marketing as described in this book is for all levels of construction contractors and all can implement the programs suggested to improve their success.

Benefits of Early Subcontractor Involvement

Marketers should be fully informed of the benefits that early involvement of subcontractors can bring to the overall success of a project and be prepared to discuss these issues with prospective customers as appropriate. Marketers also need a thorough understanding of their management's direction in promoting subcontractors' involvement in their projects. As general contractors begin to implement management programs to facilitate this early participation, marketers should use them to their full advantage in marketing opportunities.

Because subcontractors are experts in their particular niche markets, they are often aware of new processes, materials, and equipment that can improve the success of any project. Industry customers, as previously mentioned, are more actively managing their real estate investments, and participation by subcontractors in the planning process can greatly improve the overall value of a project for clients. For example, a subcontractor might recognize that a proposed system has maintenance costs that are too high to satisfy the owner and can suggest a lower-cost maintenance system that provides better value for the customer even if the initial installation cost is higher.

Contractors can also use subcontractors to compress overall building schedules to meet their customers' requirements. Subcontractors' participation in the planning stages can provide input into better procedures and systems that shorten the building schedule and can discuss how their work will affect or be affected by other subcontractors so that possible delays can be pinpointed before they affect the project's progress.

Virtual construction techniques including digitized communications and virtual scheduling presented in Chap. 2 also require early and close relationships between contractors and subcontractors. Contractors cannot fully implement the hardware and software improvements possible without being connected electronically to all their major subcontractors. Subcontractors' marketing programs should include the steps necessary to ensure they recognize how a contractor's management can affect their capability to market and negotiate work with each contractor they intend to target as a customer.

Relationships between general contractors and subcontractors should be considered a two-way street. The general contractor enjoys improved success and competitiveness on projects if there is early subcontractor involvement and better relationships with the subcontractors who control the actual costs of field construction. Better relationships benefit subcontractors by securing more negotiated versus low-bid work, faster payments by contractors because of electronic connectivity, and overall improved profitability by all parties due to compressed schedules that result in lower overhead costs. These relationships leverage resources unique to each, to achieve results that are profitable for the whole.

Comarketing Programs between Contractors and Subcontractors

Subcontractors present prime opportunities for general contractor marketers to expand their resources and networks to increase their productivity. Subcontractors are not only a good source of internal industry information, but they also can provide leads on potential projects and customers. An article in *NREI* described this strategic alliance in five steps:[1]

1. *Alignment.*   The first thing that has to be done is to align expectation among the various partners who come from different industries and different cultures.
2. *Control.*   It's important to establish quality control because reputations are on the line. If the partnership falters, then both companies are tainted, and if it goes bad, business is lost.
3. *Goals and objectives.*   There needs to be clear objectives for both organizations. If one organization doesn't really understand what it is supposed to achieve, the relationship is going to fall short, and the client will suffer.
4. *Performance objectives.*   Both organizations have to be able to achieve a reasonable level of quality and profitability. A situation with one organization performing 90% of the work and the other achieving 80% of the profit will not stand for very long.
5. *New business.*   Getting people to work together takes one set of skills, but getting people motivated to develop new business takes even more work.

Team building and strategic alliances are powerful when they work well. For a contractor and its subcontractors to transform these individual companies into a team can take considerable effort by all parties. The success of this venture can spell better cost controls, minimize scope discrepancies, provide timely decision making, lower risks, and result in profits for the team due to the trust, camaraderie, common goals, sharing of resources, and leadership that has developed.

Establishing a network of associated subcontractor marketers provides the following benefits:

- Knowledge of competitors
- Learning of new trends, services, and products
- Information on local architectural and engineering firms capabilities
- Leads for additional potential customers and projects
- Increased technical capability for proposal responses

Competitors

Subcontractors can provide information about competitors to construction marketers that is helpful in determining the types of customers and projects competitors are targeting successfully. Subcontractors are often aware of strategic alliances being formed by other general contractors, new services competitors are providing customers, and even the financial condition of a competitor.

Marketers should apply this information to better position their own companies in maintaining a competitive stance. No subcontractor should be asked to violate a code of ethics in releasing information about competitors, but marketers should not hesitate to ask questions to ensure they are fully aware of the current trends in the local marketplace.

Trends, services, and products

Subcontractors have special market intelligence regarding their specific niche within the construction community. They are interacting daily with various contractors, designers, and owners. They are reading a diverse influx of information from channels a contractor may not have access to. This different perspective on the supply and demand of their sub field can shed new light on trends developing. Many subcontractors can provide guidance regarding market-pricing strategies. Knowledge about shortages, surpluses, and imbalances within sub markets can provide considerable advantages to contractors who work closely with their subcontractors, making them a part of their team and supporting their personal goals and needs.

Subcontractors regularly monitor conditions within their field of expertise to find the latest services and products that are being developed. One or several of these items may make it possible for a contractor to land an order with a new project, by providing a leading-edge view of where the market is directed, or with a distinct value-engineering advantage.

An example of this is the concept of sharing information about new design and construction methods currently not practiced within a market area. Many times, selective markets have developed a new productivity solution for a problem they have uncovered and have kept it to themselves. However, subcontractors on that job may have been exposed to this new concept and can redirect the plans of a contractor with this critical information.

Information about design teams

Marketers are becoming more directly involved with architectural and engineering organizations for design-build projects, alliances, and mutual marketing programs. Subcontractors can provide information about design teams they are working with successfully, information a marketer can use. Several subcontractors have established themselves as the industry leader in their respective field. Before any pencil is laid to the paper, designers will call these companies first to collaborate their ideas and visions. Such a business partner for the contractor can be invaluable because the partner is in on the ground floor of many opportunities and is preferred by the designers because of its experience and successful track record.

For example, a subcontractor's marketer might be aware of a design team that specializes in applying subcontractors' input to improve the overall value of a project for their clients. This information might lead to a design-build proposal being submitted with this design team or even a strategic alliance being structured with the team. This caliber of subcontractor within the marketing mix of the team is just the type of strategic partner that can ensure new leads and instant respect and recognition by the decision makers for the contractor. Their combined strength can provide a new level of service to the designers and owner of the projects the team chooses to pursue.

As importantly, subcontractors can alert marketers to design teams that have poor reputations in a particular market type or with certain customers, preventing a contractor from teaming with them and submitting a proposal that has no chance of success. However, marketers should confirm any information received to be sure it is not biased based only upon the subcontractor's poor performance with this client.

Ten specific reasons for team building and teaming up (and the inherent benefits) are

1. Creates a climate of collaboration and cooperation through open and honest communications

2. Allows you to bid on larger contracts and take on more challenging and interesting work

3. Increases your employee base without hiring unskilled workers

4. Provides better decision-making by the implementation of similar sets of project goals and objectives

5. Allows you to expand geographically in new territories with existing clients while minimizing risks

6. Enhances your capabilities, strength, knowledge, and market intelligence

7. Allows you to share expenses to expand a relatively tight budget and limited cash flow

8. Provides increased productivity with the total team with a clear structure of responsibilities and project duties

9. Allows for the encouragement of continuous improvement between subcontractors and contractors because of management commitment and support

10. Minimizes the potential for disputes and project conflicts

Providing marketing assistance

Aggressive subcontractors have established marketing programs in place that should be capitalized on by all construction marketers. Site and civil works subcontractors and testing laboratories can be especially useful in networking because these organization are often the first to be aware that a client is considering a potential project.

For example, a site contractor might be hired to perform soils testing on a site that an owner is contemplating purchasing for a planned project. The site subcontractor, assuming the client has not requested confidentiality in the contract, can pass this lead along to general contractor marketers that it has established networking relationships with.

Subcontractors often perform maintenance contracts and other direct services to local owners, during which a subcontractor might become knowledgeable about a customer's planned expansion or new project. Again, this information can be shared when appropriate with other marketers. An elevator installation subcontractor, for example, might have a maintenance contract with a developer that has asked the subcontractor for information about new technology in elevators for an upcoming project. A lead such as this is often shared with the subcontractor's marketing network.

Construction marketers can also ask to be introduced to a subcontractor's customers in an effort to begin building a business relationship with these clients for the general contractor. Subcontractors are certainly good sources of potential marketing and sales leads that all marketers should aggressively use to the fullest extent possible. Although marketing as high up the food-chain as possible is most often acknowledged as a marketing strategy (discussed in Chap. 13), marketing down the food chain also can provide benefits to marketers. In fact, the most successful marketers use all the resources available to them, expanding their network to the largest extent possible, and include subcontractor marketers whenever possible.

Increasing technical competency with subcontractors

As customers, such as the federal government, increasingly turn toward selections based upon technical capability rather than only low price, subcontractor assistance during proposal preparation is becoming increasingly important. To show direct experience in completing similar work, the contractor, through teaming with a subcontractor that has the expertise required, can offer job performance assurance to the customer. Contractors can also grow their business without the costs, risks, and management hassles with full-time permanent employees. Subcontractors in turn can concentrate their efforts on doing what they do best, providing quality services without needing to worry about their overall marketing.

For example, proposals for restoration of a historic structure might require the general contractor to establish a team of technically competent subcontractors experienced with the restoration techniques required on the project, such as masonry restoration. The technical expertise voids are better filled with a competent subcontractor than by attempting to downplay the significance of this work item or advising the client that the work will be subcontracted as appropriate. Better proposal responses would allow the client to verify that the contractor has the team capable of completing the work already in place.

Contractors are also expanding their market share by targeting niche markets with expanded scope or financial resources that subcontractors typically have managed in the past. For example, contractors are now providing customers with complete maintenance services for the building they build—build it, then maintain it. These service contracts usually require contractors to hire subcontractors for specific portions of the work (e.g., elevator maintenance). Subcontractors not only provide these services, but they are a source of leads for customers that are contemplating outsourcing their maintenance and service requirements.

Design-build contracting highlights the necessity to maintain effective two-way relationships between subcontractors and general contractors. Most design-build proposals require the contractor to submit pricing information well before design documents are complete, so firm fixed-price quotes must be received from subcontractors. The contractor must then select a subcontractor to bring onto the team without price comparison, but based only on past relationship and technical competency. This early selection also requires the subcontractor to have a marketing program in effect to take advantage of these situations by making contractors aware of their abilities before a specific proposal.

Design-build proposals also require technical input from major subcontractors before plans are completed and proposals submitted. In many situations, the subcontractor is called upon to complete an initial design for their scope of work to be included in the proposal package, For example, a mechanical subcontractor would prepare schematic drawings of the proposed mechanical design for the contractor to submit to the client.

Establishing a comarketing program between contractors and subcontractors can provide these mutual benefits. By making each other more successful, the team achieves a win-win situation for all involved.

Reaching Out to Contractors and Subcontractors

Most subcontractor marketers must recognize that general contractor marketers will be emphasizing higher food-chain networks than subcontractors. This should not prevent subcontractors from attempting to forge relationships with contractor's marketing teams. Any subcontractor marketer who can prove his or her productivity to a contractor's marketer will certainly be readily accepted into a productive marketing network. It is, however, most often left to the subcontractor to take the initial steps required to begin such a relationship.

Marketing-driven contractors will usually have an established program to build relationships with subcontractors. Unfortunately for the subcontractor marketers, it is usually contractor personnel such as superintendents, office engineers, and estimators that are assigned to building these relationships, not the contractor's marketing personnel. Although this promotes effective business relationships, it usually does not result in a win-win-marketing program for the subcontractor. These contractor representatives are able to provide details only on existing and planned bid opportunities, not always on future prospects that can expand a subcontractor's marketer search for future work.

This situation requires that a subcontractor marketer implement the ideas and suggestions presented in Chap. 13, marketing to owners, in their program to achieve a relationship directly with the contractor's marketing team. It is most important with such a program to immediately inform the contractor's marketer of the expanded resources that the subcontractor can bring to benefit the contractor's marketing success. An example is marketing leads generated from the subcontractor's maintenance contracts as described previously. After a contractor's marketer recognizes that the subcontractor can be an important new source of information, the contractor will take the steps necessary to advance this relationship to a win-win situation.

Selection of a subcontractor by a contractor is not just a process of picking the subcontractor with the lowest prices. In developing a marketing relationship and team there are several key factors, beyond basic dollars, that contractors should consider. Some of the fundamental quality standards for contractors to consider, according to industry professionals, are attitude and professionalism, business management, capacity and resources, experience, fairness, honesty, reasonableness, reputation, safety record, technical knowledge, trustworthiness, and workforce.[2] These same traits are what the contractor also hopes the owner is basing its decisions on. Capability, experience, and trustworthiness are the core of a successful, team-building experience for the next project and for a long-term relationship.

As difficult as it is for a contractor to market directly to a client, a subcontractor will experience similar difficulties in attempting to form a long-term marketing relationship with a contractor. In addition, the subcontractor must recognize that a contractor might not be awarded all the work of a client, and the contractor should not be expected to award all its work to the subcontractor. If a subcontractor loses a contractor's project, it must still continue marketing this contractor, establishing a long-term relationship.

Although this book has concentrated on general contracting marketing, all the principles, suggestions, and ideas can effectively be used by subcontractors, not only for marketing to contractors but directly to owners. Aggressive subcontractor marketers will use all these resources to ensure the profitability of their companies.

Bid teaming

A new strategy is beginning to be seen around the country, called bid teaming. This methodology has been enhanced in large part due to the growth of design-build and its natural teaming requirements. This premise evolved from Dr. W. Edwards Deming, who suggested ending the practice of awarding business on the basis of price tag alone.

Design-build and other negotiation-oriented procurement formats have formulated the concept that practicing companies of project partnering can, through cooperation rather than confrontational competition, reduce costs and increase profits. The whole group, owner, end users, designers, contractors, subcontractors, and suppliers, make a sincere effort to assemble, structure, and motivate the project stakeholders to work together as a team and coordinate efforts toward common project goals.

The project team, along with the owner, focuses on bringing the entire project in within budget and on time, in lieu of cutting the individual subcontractor segment costs. The net results for all participants can be higher percentages of profitability for each stakeholder and success for the entire project. Add to that the potential for repeat work from the owner and you have a team-building approach that can shape the way we view project relationships.

Creating a market-driven sales culture

Creating a market-driven sales culture in the construction business by developing the skills of your superintendents, project managers, estimators, foremen, etc., can be summed up in one word: relationship.

Contractors have the distinct misperception that low dollar is the deciding feature in the subcontractors who get the job. This is far from the truth. Contractors today must perform. They need the best, most professional subcontractors they can find. Owners demand quality, speed, and a team that works well together to get each job completed without legal intervention. Repeat business is the mantra for new business development. Subcontractors and contractors alike must take it upon themselves to educate the construction community about their capabilities. No one knows your business better than you do, and there is an ongoing need to regularly reacquaint your clients with your capabilities and unique talents.

The proverbial philosophy of the 80/20 rule is true: 80 percent of your business is derived from 20 percent of your customers. An inexpensive first step is to prepare a letter that thanks them for past work and alerts them to any special achievements that your company has accomplished. Customers you have sold to and satisfied once are much more likely to buy from you again. This initial letter should include a basic overview of your team players with their direct phone and fax numbers and e-mail addresses. After this initial step, you must follow it up with a call. Did they receive your letter, do they have any questions, is there any more you can do for them now? Your dialogue can be something like: "We just wanted to give you our current phone numbers, new employees, and some of the new services/products we have added." This action can be duplicated at various positions within the organization: superintendent to superintendent, project manager to project manager, estimator to estimator. Getting this activity to a more personal one-on-one level can add a new dimension to your success.

To successfully follow up this initial step, strive to make yourself available when clients need you. As we discussed in Chap. 6, preconstruction can be a valuable stage at which to lend a hand to the target client firm. Doing this on multiple levels within their organization and yours can help you learn the client's actual needs, concerns, and office politics. Who will make the final buying decision, what scope of work does the contractor actually need? What options would be important? What are its current subcontractor relationships? What is its ability to self-perform this work? What edge can you give it to facilitate its overall success to obtain the work?

In Chap. 4 we discussed the wealth of information that is available on-line to give you a head start over your competition. Both Construction Market Data and F. W. Dodge have preconstruction reports that provide a forewarning of new forthcoming work. Call a client and ask the status of the project. Is there any assistance you can provide? Sometimes, just by being there, you can hear about other projects and leads that may open a new door of opportunity. Drop by the office or job site with doughnuts at 7:00 a.m. Or swing by the office after work with a six-pack of beer. The worst that will happen is you'll be told everyone is too busy to talk right now. Don't push it. Just by being friendly, respecting the client's need for time management, and asking when a good time in the future would be, you've earned points for being an individual who understands the demands of a construction career.

Guide for subcontractors

The steps described above can also be used to create a market-driven construction business for a subcontractor. Ten other sure-fire ways to do so are

1. Make your proposals neat, professional, and thorough. Getting your scope sheet in to the contractor as early as possible will enhance your reputation.

2. If you need a crane to load or unload a truck, let the contractor know before the start of the job, so it can be scheduled. Nothing is more frustrating than having a subcontractor announce at the last minute that a crane is needed.

3. Address all of your correspondence to a specific individual. This shows you know who the individual is and you want to establish an ongoing relationship. Customers receive many pieces of mail. But by making it personalized,

you can gain their immediate attention. Another approach is to communicate your message on statement invoices, estimates, proposals, or other routine business correspondence by, for example, attaching a brief note on your safety record or concern for client service, or on time-delivery methods.

4. Recognize your clients for achievements they make in the community, for any special effort they made on your behalf, for a new project, or the completion of one. This can be done with a simple postcard, stationery with your letterhead, a fax, or e-mail. Everyone likes to be recognized for achievements. The idea that someone actually noticed your achievements is significant. Relationship marketing is not easy. It takes time and considerable effort. The benefits, however, can be substantial. Companies can spend less time on poorly directed marketing programs and now focus specifically on building sales and customer loyalty through enhancing the customer database with behavior-driven information and using a personalized, focused customer relationship.

5. Always give a little more than you need to. You will be remembered as a valuable member of the team by constantly doing more than what you were contracted to do. This goodwill and overachievement will come back to you tenfold. Provide something new, different, and valuable each time you meet with your clients. It doesn't have to be a lot, just something new like new information on a recent bid, a new product in the market, or a bit of information from a recent newspaper or magazine article. It is up to you to build the relationship.

6. State your ideas for value engineering clearly and regularly to the contractor. Going to the contractor after a problem occurs and stating that you knew something wouldn't work and that you have an option is frustrating, irritating, and not in the best interest of anyone.

7. Provide an easy to understand schedule of how long you believe your work will actually take. Be honest. Many contractors fail because of accepting an aggressive schedule they knew they could not maintain. This will ruin a relationship faster than anything. By being honest and backpedaling aggressively, you will be more important to the long-term success than pushing with everything you have. Simply saying, "That is an aggressive schedule and we believe we could not provide the quality necessary to the overall project" will immediately interest the contractor. Contractors usually hear confirmations. A response that backpedals your ability will make the contractor want to find a way to get your firm, one that is honest and concerned about quality, as a member of the team.

8. Document changes. Regularly correspond to the contractor with changes. Changes come from field changes, design modifications, and owner demands. Quick, responsive correspondence about changes and their impact to the contractor in time and money will keep everyone on the same team, working toward a common goal and knowledgeable about forthcoming pitfalls.

9. Do what you say you will do. This will get you the immediate respect of the contractor; the contractor will know you can be counted on to meet the commitments you make. This can apply to meetings, proposals, executions, and support.

10. Perform, perform, perform. All is for naught it you cannot perform with the quality and effectiveness necessary to bring your segment of the project in within budget, safely, and on time. Quality of action, word, and deed are not forgotten art forms on jobs today. Subcontractors who can consistently perform well the job they have been contracted to do will be popular with all contractors.

Notes

1. Steve Bergsman, "Strategic Alliances: Tying the Knot for a Stronger Future," *National Real Estate Investment,* November 1996, p. 87.
2. Denise Norberg, "What to Consider When Choosing a Subcontractor," *Engineering News-Record,* June 1996, pp. 18–20.

Bibliography

Briggs, Pam. 1989. "Marketing Coordination." *Professional Services Management Journal.* Vol. 16, February, pp. 13–17.

Britt, Stuart Henderson, and Norman F. Guess, eds. 1983. *Marketing Manager's Handbook.* Chicago: Dartnell Publishers.

Carnegie, Dale. 1981. *How to Win Friends and Influence People.* New York: Simon and Schuster.

Cohen, William A. 1987. *Developing a Winning Marketing Plan.* New York: John Wiley & Sons.

Cohen, William A. 1986. *Winning on the Marketing Front: The Corporate Manager's Game Plan.* New York: John Wiley & Sons.

Coxe, Weld. 1986. "The Marketing Plan." *A/E Marketing Journal.* Vol. 13, August, p. 3.

Crandall, Rick, 1996. *Marketing Your Services—For People Who Hate to Sell.* Chicago: Contemporary Books.

Doherty, Paul. 1997. *The Internet Guide for Architects, Engineers & Contractors.* New York: R. S. Means Company. An excellent overview of everything you ever wanted to know about the Internet but were afraid to ask.

Friedman, Warren. 1984. *Construction Marketing and Strategic Planning.* New York, McGraw-Hill.

Garfunkel, Stanley J. 1980. *Developing a Marketing Plan: A Practical Guide.* New York: Random House.

Hayes, Rick Stephan. 1985. *Marketing for Your Growing Business.* New York: John Wiley & Sons.

Hellebust, Karsten G., and Joseph C. Krallinger. 1989. *Strategic Planning Workbook.* New York: John Wiley & Sons.

Hickman, Craig R., and Michael A. Silva. 1984. *Creating Excellence: Managing Corporate Culture, Strategy, and Change in the New Age.* New York: Signet.

Jones, Gerre. 1983. *How to Market Professional Design Services.* New York: McGraw-Hill. One of the original leaders who led the marketing profession into the front lines.

Kerruish, Karen. 1991. *The Marketing Plan.* SMPS Core Series.

Lea, Bruce. 1982. "The Marketing Plan." *SMPS Marketing Information Reports.*

Lehmann, Donald R., and Russel S. Winer. 1988. *Analysis for Marketing Planning.* Homewood, Ill.: BPI/Irwin.

Luther, William M. 1982. *The Marketing Plan: How to Prepare and Implement It.* New York: AMACOM.

Mackay, Harvey. 1997. *Dig Your Well before You're Thirsty.* New York: Doubleday.

Naisbitt, John. 1982. *Megatrends: Ten New Directions Transforming Our Lives.* New York: Warner.

Parkinson, Hank. 1987. "Consensus: The Key to a Successful Marketing Plan." *Construction Specifier.* Vol. 40, January, pp. 26–30.

Parmerlee, David. 1995. *Preparing the Marketing Plan.* Lincolnwood, Ill.: American Marketing Association, NTC Publishing Group. Great source of blank forms for preparing and monitoring your marketing plan.

Peters, Thomas J., and Robert H. Waterman, Jr. 1982. *In Search of Excellence: Lessons from America's Best Run Companies.* New York: Warner.

Spaulding, Margaret, and William D'Elia. 1989. *Advanced Marketing Techniques for Architecture and Engineering Firms.* New York: McGraw-Hill.

Steiner, George A. 1979. *Strategic Planning: What Every Manager Must Know: A Step-by-Step Guide.* New York: The Free Press.

Waterman, Robert H. 1987. *The Renewal Factor: How the Best Get and Keep the Competitive Edge.* New York: Bantam.

Glossary

Included here are selected terms related to marketing and project delivery systems, and their definitions. The definitions are consistent with the way the terms are used in this book, and many vary from standard definitions used elsewhere. Our special thanks to the Associated General Contractors of America for allowing us to utilize several definitions from its new publication, *Project Delivery Systems for Building Construction*, by Robert W. Dorsey.

Agreement, form of agreement A document setting forth the provisions, responsibilities, and obligations of parties to a contract. Standard forms of agreement for building construction are available from the American Institute of Architects and the Associated General Contractors of America and are designed to allow the insertion of data relevant to particular projects.

Alliance A long-term relationship between parties for services on several projects. Owner-contractor alliances are sometimes called partnering or preferred suppliers.

American Institute of Architects A national association that promotes principles, standards, and activities that are important to the practice of architecture, including ethics, education, legislation, and professional advice. The AIA also publishes many documents that guide design and construction processes.

American Subcontractors Association A national trade association made up of companies that perform traditional trade work.

Architect A professional person who is duly licensed by a state (by examination or reciprocity) to perform services in that state involving the design of buildings.

Associated Building and Contractors   A national trade association made up primarily of general contracting companies but also including related occupations and professions.

Associated General Contractors of America   A national trade association made up primarily of general contracting companies but also including related occupations and professions. There are four divisions: Building, Highway, Heavy & Industrial, and Municipal-Utilities.

Association   An organization established to serve the interests of similar parties. Examples are trade associations, manufacturers' associations, and vendors' associations.

Award   The act by one party of granting a contractual opportunity to another party typically as a response to a proposal, as in an owner awarding a contract to a low bidder or a general contractor awarding a subcontract.

Bid   A proposal submitted in various forms, oral or written, to perform remunerative work or to buy an object. Related definitions in construction are:

> **Bid, competitive**   Proposals are compared to each other on some prescribed basis, such as a set of contract documents, and the "lowest and best" bid is usually accepted.

> **Bid shopping, bid peddling**   The practice by a few general contractors of continuing to shop for lower bids from subcontractors after receiving initial bids from those and other subcontractors, sometimes by exposing (peddling) the bids. Subcontractors also sometimes "shop around" to determine competitors' bids and then offer second or third proposals. These practices are considered unethical but not illegal.

Bonus   A payment over and above a basic amount for superior performance; may be part of an incentive clause, in which case it must be contractual.

Brochure   An illustrated document, part marketing, part informational, used to describe products, capabilities, or companies. Sometimes equipment brochures are part of the documents turned over to the owner by the contractor at closeout.

CAD(D)   Computer-aided drafting (and design); a general term for a wide array of operations and techniques.

Cash flow   The income stream from projects; a measure of liquidity.

Catalog   A compendium of information, usually illustrated, about a product or range of products used by designers and contractors to help select and properly install building components. Catalog "cuts" (illustrations) containing descriptive information about products are frequently part of submittals for approval during construction.

Change order   An amendment to a contract based on a change initiated by the owner, designer, contractor, or building official and documented by written amendment signed by the owner and contractor after price and schedule adjustment are agreed on.

Closeout   A process of completing a construction project and turning it over to the owner. It is usually a multiweek sequence of approvals, partial occupancies, a punch list, documentation, and celebrations.

Competitive advantages   Characteristics of your company's products or services that make them better than or different from the competition.

Construction   A term for a multitude of activities that integrate to become a built product. The term is normally applied to the industry responsible for the construction of the vast array of buildings, public works, and monuments of modern society. Branches of construction include:

Building construction   The segment of construction involved in commercial, institutional, and some industrial buildings, but excluding houses.

Heavy and highway construction   Related to roads, bridges, dams, airports, etc.; sometimes called civil construction.

Industrial construction   Factories, refineries, power plants, etc.

Public works construction   Related to service facilities such as water treatment and sewage disposal plants.

Residential construction   Primarily home building but may also include multifamily and group housing.

Construction management   A project delivery system based on an agreement whereby the construction entity provides leadership to the construction process through a series of services to the owner, including design review, overall scheduling, cost control, value engineering, constructability, preparation of bid packages, and construction coordination. In agency CM the construction entity is typically retained at the same time as the design team and provides continuous services to the owner without taking on financial risks for the execution of the actual construction. In at-risk CM the construction entity, after providing the agency services during the preconstruction period, takes on the financial obligation to carry out construction under a specified cost agreement. The construction manager in at-risk CM frequently provides a guaranteed maximum price. At-risk CD is sometimes called CD/GC because the construction entity becomes essentially a general contractor through the at-risk agreement.

Construction Writers Association   A nonprofit, nonpartisan, international organization for professional journalists, writers, editors, and publicists serving the information needs of the construction industry.

Constructor   The term adopted by practitioners who execute construction to define the persons who are responsible for all or part of the building process. With some contractors, constructor is used to designate the party directly responsible for the execution of the project.

Contractor   A person or company that accepts responsibility to perform the obligations of a contract; a term usually applied to one who engages in contract execution as regular employment. More-specific terms include:

Construction contractor   A person or company that performs construction as a business.

General Contractor   An entity that takes responsibility for whole projects through agreement with owners.

Prime contractor   One that has a contract directly with an owner.

Specialty contractor   An entity that focuses on trade work entailing particular skills, such as control systems, ornamental works, and finishes; sometime the term is used interchangeably with trade contractor.

Subcontractor   One that has a contract with a prime contractor.

Sub-subcontractor   One that has a contract with a subcontractor, sometimes called a second-tier subcontractor.

Trade contractor An entity that performs one or more traditional branches of work, such as masonry, roofing, or carpentry; usually works as a subcontractor to a general contractor but may take on prime contracts directly with an owner, as is frequently the case in agency construction management.

Crisis management plan A plan developed in advance for handling media and public inquiries during a crisis.

Database Accumulated information in an organized filing system.

Demographic characteristics Characteristics of people that can be measured or quantified.

Design A process of composing ideas and requirements into an understandable scheme or plan for a product. Building design involves architects, engineers, consultants, and sometime contractors working together to develop drawings and written descriptions (specifications) for a building.

Design-build A project delivery system based on an agreement whereby the design service and construction service are formed into a single entity and that entity is obligated to the owner for the combined services. The design services may be provided by in-house designers employed by the construction company or by retained consultative designers. Design-build-lease, turnkey, and bridging are types of design-build.

Developer An entrepreneur who invests in land and buildings and who sometimes manages the construction involved in those investments.

Domain name A Web site address, also referred to as a URL (Uniform Resource Locator).

Engineer A professional person who is duly licensed by a state (by examination or reciprocity) to perform services in that state involving the design of buildings, transportation systems, environmental facilities, etc.

Estimating Forecasting the costs of labor, material, equipment, and related items prior to the execution of a project, usually based on units of historical data in the contractor's files, published indexes, and information supplied by subcontractors and suppliers.

Extranet A password-protected Web site used by project teams to manage the flow of documents during a construction project.

Fast-track Accelerated scheduling that involves commencing construction prior to the completion of contract documents and then using means such as bid packages and efficient coordination to compress the overall schedule.

Flash-track A considerably accelerated fast-track schedule.

Intranet A password-protected Web site used by companies to communicate internally with employees.

Job A vernacular term for a construction project; frequently used in such other terms as job site, job costs, and job-related activity.

Joint venture A contractual collaboration of two or more parties to undertake a project, examples being an architect-architect, an architect-contractor, and a contractor-contractor.

Key selling point   The single most important selling point or competitive advantage that sets your company apart from your competitors.

Liability   A bond term denoting any legally enforceable obligation.

Marketing   Provides a direct step to customer satisfaction and provides customers with the goods, products, or services they want, resulting in a handsome profit for the organization.

Marketing communication   Communication efforts designed to support marketing goals.

Marketing mix   A mix of activities that meshes with customers' wants and needs, integrated with other external, environmental factors to create a successful marketing program.

Owner   The party to the contract that has legal possession of the property or is duly selected to represent the property owner, and that typically provides the financing for the construction.

Partnering   A formal structure to establish a working relationship among all the stakeholders through a mutually developed strategy of commitment and communication.

Preconstruction services   A range of activities by a contractor prior to execution of construction, including value engineering, constructability, cost and schedule studies, procurement of long-lead-time items, and staffing requirements.

Profit   The amount of money remaining after all expenses on a project have been paid, including both job and office overhead; the amount on which company income taxes must be paid.

Project delivery system   A comprehensive process wherein designers, constructors, and various consultants provide services for design and construction to deliver a built project to the owner.

Psychographic characteristics   Characteristics of people that group them by homogeneous segments based on their psychological makeup and lifestyles.

Schedule   An organized array of information to illustrate resource allocation, interrelationships of activities, costs, and performances.

Society for Marketing Professional Services   An association of professional service firm leaders that sharpens skills, pools resources, and works together to create business opportunities. Its mission is to be the premier resource for education, information, and resources in marketing professional services for the built and natural environment.

SWOT analysis   An analysis reviewing systematically all major internal and external environmental conditions that could affect an organization.

Target audience   The group of people most likely to buy your products or services.

Virtual construction   When an organization compresses its product- or service-development time by shrinking the interval between need identification for a new product or service and the date at which that product or service becomes available. A virtual construction company is required to completely revise its structure to control and instantaneously access an increasingly sophisticated flow of electronic information.

Washington Building Congress   A nonprofit association made up of professionals from a variety of disciplines, all with an active interest or involvement in the Washington, D.C., area's real estate, design, and construction community. The organization was established in 1937 to represent the collective interests of its members by providing education and networking opportunities and by promoting the advancement of the building industry.

Web site   A collection of information in the area of the Internet known as the World Wide Web that can be used to sell or promote.

Index

ABOUT THE AUTHORS

MICHAEL T. KUBAL is President of the Enviromental Division, J.A. Jones Construction Co., in Washington, D.C. He has extensive experience as a project manager of multimillion-dollar building projects and as a supervisor of renovations and historic restorations. Mr. Kubal is the author of McGraw-Hill's *Waterproofing the Building Envelope* and *Engineering Quality in Construction*.

KEVIN MILLER is President and founder of the Frost Miller Group, a marketing communications firm based in Washington, D.C., that represents a broad range of businesses nationwide. He was previously Director of Marketing Communication for one of the nation's largest construction firms.

RONALD D. WORTH is Executive Vice-President of the Washington Building Congress, the largest building trade association in the Washington, D.C., region. In a career that has spanned two decades, he has, among other positions, worked coast-to-coast as Director of Marketing for a national construction services corporation, during which time he interacted with leading contractors, subcontractors, architects, engineers, and owners. He is a regular industry lecturer and writer, and is on the boards of several national trade associations.

DISK WARRANTY

This software is protected by both United States copyright law and international copyright treaty provision. You must treat this software just like a book, except that you may copy it into a computer in order to be used and you may make archival copies of the software for the sole purpose of backing up our software and protecting your investment from loss.

By saying "just like a book," McGraw-Hill means, for example, that this software may be used by any number of people and may be freely moved from one computer location to another, so long as there is no possibility of its being used at one location or on one computer while it also is being used at another. Just as a book cannot be read by two different people in two different places at the same time, neither can the software be used by two different people in two different places at the same time (unless, of course, McGraw-Hill's copyright is being violated).

LIMITED WARRANTY

McGraw-Hill takes great care to provide you with top-quality software, thoroughly checked to prevent virus infections. McGraw-Hill warrants the physical diskette(s) contained herein to be free of defects in materials and workmanship for a period of sixty days from the purchase date. If McGraw-Hill receives written notification within the warranty period of defects in materials or workmanship, and such notification is determined by McGraw-Hill to be correct, McGraw-Hill will replace the defective diskette(s). Send requests to:

> McGraw-Hill, Inc.
> Customer Services
> P.O. Box 545
> Blacklick, OH 43004-0545

The entire and exclusive liability and remedy for breach of this Limited Warranty shall be limited to replacement of defective diskette(s) and shall not include or extend to any claim for or right to cover any other damages, including but not limited to, loss of profit, data, or use of the software, or special, incidental, or consequential damages or other similar claims, even if McGraw-Hill has been specifically advised of the possibility of such damages. In no event will McGraw-Hill's liability for any damages to you or any other person ever exceed the lower of suggested list price or actual price paid for the license to use the software, regardless of any form of the claim.

McGRAW-HILL, INC. SPECIFICALLY DISCLAIMS ALL OTHER WARRANTIES, EXPRESS OR IMPLIED, INCLUDING, BUT NOT LIMITED TO, ANY IMPLIED WARRANTY OF MERCHANTABILITY OR FITNESS FOR A PARTICULAR PURPOSE.

Specifically, McGraw-Hill makes no representation or warranty that the software is fit for any particular purpose and any implied warranty of merchantability is limited to the sixty-day duration of the Limited Warranty covering the physical diskette(s) only (and not the software) and is otherwise expressly and specifically disclaimed.

This limited warranty gives you specific legal rights; you may have others which may vary from state to state. Some states do not allow the exclusion of incidental or consequential damages, or the limitation on how long an implied warranty lasts, so some of the above may not apply to you.